Geophysical Monograph Series

Including

IUGG Volumes

Maurice Ewing Volumes
Mineral Physics Volumes

GEOPHYSICAL MONOGRAPH SERIES

Geophysical Monograph Volumes

1. Antarctica in the International Geophysical Year A. P. Crary, L. M. Gould, E. O. Hulburt, Hugh Odishaw, and Waldo E. Smith (Eds.)
2. Geophysics and the IGY Hugh Odishaw and Stanley Ruttenberg (Eds.)
3. Atmospheric Chemistry of Chlorine and Sulfur Compounds James P. Lodge, Jr. (Ed.)
4. Contemporary Geodesy Charles A. Whitten and Kenneth H. Drummond (Eds.)
5. Physics of Precipitation Helmut Weickmann (Ed.)
6. The Crust of the Pacific Basin Gordon A. Macdonald and Hisashi Kuno (Eds.)
7. Antarctic Research: The Matthew Fontaine Maury Memorial Symposium H. Wexler, M. J. Rubin, and J. E. Caskey, Jr. (Eds.)
8. Terrestrial Heat Flow William H. K. Lee (Ed.)
9. Gravity Anomalies: Unsurveyed Areas Hyman Orlin (Ed.)
10. The Earth Beneath the Continents: A Volume of Geophysical Studies in Honor of Merle A. Tuve John S. Steinhart and T. Jefferson Smith (Eds.)
11. Isotope Techniques in the Hydrologic Cycle Glenn E. Stout (Ed.)
12. The Crust and Upper Mantle of the Pacific Area Leon Knopoff, Charles L. Drake, and Pembroke J. Hart (Eds.)
13. The Earth's Crust and Upper Mantle Pembroke J. Hart (Ed.)
14. The Structure and Physical Properties of the Earth's Crust John G. Heacock (Ed.)
15. The Use of Artificial Satellites for Geodesy Soren W. Henricksen, Armando Mancini, and Bernard H. Chovitz (Eds.)
16. Flow and Fracture of Rocks H. C. Heard, I. Y. Borg, N. L. Carter, and C. B. Raleigh (Eds.)
17. Man-Made Lakes: Their Problems and Environmental Effects William C. Ackermann, Gilbert F. White, and E. B. Worthington (Eds.)
18. The Upper Atmosphere in Motion: A Selection of Papers With Annotation C. O. Hines and Colleagues
19. The Geophysics of the Pacific Ocean Basin and Its Margin: A Volume in Honor of George P. Woollard George H. Sutton, Murli H. Manghnani, and Ralph Moberly (Eds.)
20. The Earth's Crust: Its Nature and Physical Properties John C. Heacock (Ed.)
21. Quantitative Modeling of Magnetospheric Processes W. P. Olson (Ed.)
22. Derivation, Meaning, and Use of Geomagnetic Indices P. N. Mayaud
23. The Tectonic and Geologic Evolution of Southeast Asian Seas and Islands Dennis E. Hayes (Ed.)
24. Mechanical Behavior of Crustal Rocks: The Handin Volume N. L. Carter, M. Friedman, J. M. Logan, and D. W. Stearns (Eds.)
25. Physics of Auroral Arc Formation S.-I. Akasofu and J. R. Kan (Eds.)
26. Heterogeneous Atmospheric Chemistry David R. Schryer (Ed.)
27. The Tectonic and Geologic Evolution of Southeast Asian Seas and Islands: Part 2 Dennis E. Hayes (Ed.)
28. Magnetospheric Currents Thomas A. Potemra (Ed.)
29. Climate Processes and Climate Sensitivity (Maurice Ewing Volume 5) James E. Hansen and Taro Takahashi (Eds.)
30. Magnetic Reconnection in Space and Laboratory Plasmas Edward W. Hones, Jr. (Ed.)
31. Point Defects in Minerals (Mineral Physics Volume 1) Robert N. Schock (Ed.)
32. The Carbon Cycle and Atmospheric CO_2: Natural Variations Archean to Present E. T. Sundquist and W. S. Broecker (Eds.)
33. Greenland Ice Core: Geophysics, Geochemistry, and the Environment C. C. Langway, Jr., H. Oeschger, and W. Dansgaard (Eds.)
34. Collisionless Shocks in the Heliosphere: A Tutorial Review Robert G. Stone and Bruce T. Tsurutani (Eds.)
35. Collisionless Shocks in the Heliosphere: Reviews of Current Research Bruce T. Tsurutani and Robert G. Stone (Eds.)
36. Mineral and Rock Deformation: Laboratory Studies —The Paterson Volume B. E. Hobbs and H. C. Heard (Eds.)
37. Earthquake Source Mechanics (Maurice Ewing Volume 6) Shamita Das, John Boatwright, and Christopher H. Scholz (Eds.)
38. Ion Acceleration in the Magnetosphere and Ionosphere Tom Chang (Ed.)
39. High Pressure Research in Mineral Physics (Mineral Physics Volume 2) Murli H. Manghnani and Yasuhiko Syono (Eds.)
40. Gondwana Six: Structure, Tectonics, and Geophysics Gary D. McKenzie (Ed.)
41. Gondwana Six: Stratigraphy, Sedimentology, and Paleontology Garry D. McKenzie (Ed.)
42. Flow and Transport Through Unsaturated Fractured Rock Daniel D. Evans and Thomas J. Nicholson (Eds.)
43. Seamounts, Islands, and Atolls Barbara H. Keating, Patricia Fryer, Rodey Batiza, and George W. Boehlert (Eds.)

44 **Modeling Magnetospheric Plasma** *T. E. Moore and J. H. Waite, Jr. (Eds.)*

45 **Perovskite: A Structure of Great Interest to Geophysics and Materials Science** *Alexandra Navrotsky and Donald J. Weidner (Eds.)*

46 **Structure and Dynamics of Earth's Deep Interior (IUGG Volume 1)** *D. E. Smylie and Raymond Hide (Eds.)*

47 **Hydrological Regimes and Their Subsurface Thermal Effects (IUGG Volume 2)** *Alan E. Beck, Grant Garven, and Lajos Stegena (Eds.)*

48 **Origin and Evolution of Sedimentary Basins and Their Energy and Mineral Resources (IUGG Volume 3)** *Raymond A. Price (Ed.)*

49 **Slow Deformation and Transmission of Stress in the Earth (IUGG Volume 4)** *Steven C. Cohen and Petr Vaníček (Eds.)*

50 **Deep Structure and Past Kinematics of Accreted Terranes (IUGG Volume 5)** *John W. Hillhouse (Ed.)*

51 **Properties and Processes of Earth's Lower Crust (IUGG Volume 6)** *Robert F. Mereu, Stephan Mueller, and David M. Fountain (Eds.)*

52 **Understanding Climate Change (IUGG Volume 7)** *Andre L. Berger, Robert E. Dickinson, and J. Kidson (Eds.)*

53 **Plasma Waves and Instabilities at Comets and in Magnetospheres** *Bruce T. Tsurutani and Hiroshi Oya (Eds.)*

54 **Solar System Plasma Physics** *J. H. Waite, Jr., J. L. Burch, and R. L. Moore (Eds.)*

55 **Aspects of Climate Variability in the Pacific and Western Americas** *David H. Peterson (Ed.)*

56 **The Brittle-Ductile Transition in Rocks** *A. G. Duba, W. B. Durham, J. W. Handin, and H. F. Wang (Eds.)*

57 **Evolution of Mid Ocean Ridges (IUGG Volume 8)** *John M. Sinton (Ed.)*

58 **Physics of Magnetic Flux Ropes** *C. T. Russell, E. R. Priest, and L. C. Lee (Eds.)*

59 **Variations in Earth Rotation (IUGG Volume 9)** *Dennis D. McCarthy and Williams E. Carter (Eds.)*

60 **Quo Vadimus Geophysics for the Next Generation (IUGG Volume 10)** *George D. Garland and John R. Apel (Eds.)*

61 **Cometary Plasma Processes** *Alan D. Johnstone (Ed.)*

62 **Modeling Magnetospheric Plasma Processes** *Gordon R. Wilson (Ed.)*

63 **Marine Particles: Analysis and Characterization** *David C. Hurd and Derek W. Spencer (Eds.)*

64 **Magnetospheric Substorms** *Joseph R. Kan, Thomas A. Potemra, Susumu Kokubun, and Takesi Iijima (Eds.)*

65 **Explosion Source Phenomenology** *Steven R. Taylor, Howard J. Patton, and Paul G. Richards (Eds.)*

66 **Venus and Mars: Atmospheres, Ionospheres, and Solar Wind Interactions** *Janet G. Luhmann, Mariella Tatrallyay, and Robert O. Pepin (Eds.)*

67 **High-Pressure Research: Application to Earth and Planetary Sciences (Mineral Physics Volume 3)** *Yasuhiko Syono and Murli H. Manghnani (Eds.)*

68 **Microwave Remote Sensing of Sea Ice** *Frank Carsey, Roger Barry, Josefino Comiso, D. Andrew Rothrock, Robert Shuchman, W. Terry Tucker, Wilford Weeks, and Dale Winebrenner*

69 **Sea Level Changes: Determination and Effects (IUGG Volume 11)** *P. L. Woodworth, D. T. Pugh, J. G. DeRonde, R. G. Warrick, and J. Hannah*

70 **Synthesis of Results from Scientific Drilling in the Indian Ocean** *Robert A. Duncan, David K. Rea, Robert B. Kidd, Ulrich von Rad, and Jeffrey K. Weissel (Eds.)*

71 **Mantle Flow and Melt Generation at Mid-Ocean Ridges** *Jason Phipps Morgan, Donna K. Blackman, and John M. Sinton (Eds.)*

72 **Dynamics of Earth's Deep Interior and Earth Rotation (IUGG Volume 12)** *Jean-Louis Le Mouël, D.E. Smylie, and Thomas Herring (Eds.)*

73 **Environmental Effects on Spacecraft Positioning and Trajectories (IUGG Volume 13)** *A. Vallance Jones (Ed.)*

74 **Evolution of the Earth and Planets (IUGG Volume 14)** *E. Takahashi, Raymond Jeanloz, and David Rubie (Eds.)*

75 **Interactions Between Global Climate Subsystems: The Legacy of Hann (IUGG Volume 15)** *G. A. McBean and M. Hantel (Eds.)*

76 **Relating Geophysical Structures and Processes: The Jeffreys Volume (IUGG Volume 16)** *K. Aki and R. Dmowska (Eds.)*

77 **The Mesozoic Pacific: Geology, Tectonics, and Volcanism—A Volume in Memory of Sy Schlanger** *Malcolm S. Pringle, William W. Sager, William V. Sliter, and Seth Stein (Eds.)*

78 **Climate Change in Continental Isotopic Records** *P. K. Swart, K. C. Lohmann, J. McKenzie, and S. Savin (Eds.)*

79 **The Tornado: Its Structure, Dynamics, Prediction, and Hazards** *C. Church, D. Burgess, C. Doswell, R. Davies-Jones (Eds.)*

80 **Auroral Plasma Dynamics** *R. L. Lysak (Ed.)*

81 **Solar Wind Sources of Magnetospheric Ultra-Low Frequency Waves** *M. J. Engebretson, K. Takahashi, and M. Scholer (Eds.)*

82 Gravimetry and Space Techniques Applied to Geodynamics and Ocean Dynamics (IUGG Volume 17) *Bob E. Schutz, Allen Anderson, Claude Froidevaux, and Michael Parke (Eds.)*

83 Nonlinear Dynamics and Predictability of Geophysical Phenomena (IUGG Volume 18) *William I. Newman, Andrei Gabrielov, and Donald L. Turcotte (Eds.)*

84 Solar System Plasmas in Space and Time *J. Burch, J. H. Waite, Jr. (Eds.)*

85 The Polar Oceans and Their Role in Shaping the Global Environment *O. M. Johannessen, R. D. Muench, and J. E. Overland (Eds.)*

86 Space Plasmas: Coupling Between Small and Medium Scale Processes *Maha Ashour-Abdalla, Tom Chang, and Paul Dusenbery (Eds.)*

87 The Upper Mesosphere and Lower Thermosphere: A Review of Experiment and Theory *R. M. Johnson and T. L. Killeen (Eds.)*

88 Active Margins and Marginal Basins of the Western Pacific *Brian Taylor and James Natland (Eds.)*

Maurice Ewing Volumes

1 Island Arcs, Deep Sea Trenches, and Back-Arc Basins *Manik Talwani and Walter C. Pitman III (Eds.)*

2 Deep Drilling Results in the Atlantic Ocean: Ocean Crust *Manik Talwani, Christopher G. Harrison, and Dennis E. Hayes (Eds.)*

3 Deep Drilling Results in the Atlantic Ocean: Continental Margins and Paleoenvironment *Manik Talwani, William Hay, and William B. F. Ryan (Eds.)*

4 Earthquake Prediction—An International Review *David W. Simpson and Paul G. Richards (Eds.)*

5 Climate Processes and Climate Sensitivity *James E. Hansen and Taro Takahashi (Eds.)*

6 Earthquake Source Mechanics *Shamita Das, John Boatwright, and Christopher H. Scholz (Eds.)*

IUGG Volumes

1 Structure and Dynamics of Earth's Deep Interior *D. E. Smylie and Raymond Hide (Eds.)*

2 Hydrological Regimes and Their Subsurface Thermal Effects *Alan E. Beck, Grant Garven, and Lajos Stegena (Eds.)*

3 Origin and Evolution of Sedimentary Basins and Their Energy and Mineral Resources *Raymond A. Price (Ed.)*

4 Slow Deformation and Transmission of Stress in the Earth *Steven C. Cohen and Petr Vaníček (Eds.)*

5 Deep Structure and Past Kinematics of Accreted Terranes *John W. Hillhouse (Ed.)*

6 Properties and Processes of Earth's Lower Crust *Robert F. Mereu, Stephan Mueller, and David M. Fountain (Eds.)*

7 Understanding Climate Change *Andre L. Berger, Robert E. Dickinson, and J. Kidson (Eds.)*

8 Evolution of Mid Ocean Ridges *John M. Sinton (Ed.)*

9 Variations in Earth Rotation *Dennis D. McCarthy and William E. Carter (Eds.)*

10 Quo Vadimus Geophysics for the Next Generation *George D. Garland and John R. Apel (Eds.)*

11 Sea Level Changes: Determinations and Effects *Philip L. Woodworth, David T. Pugh, John G. DeRonde, Richard G. Warrick, and John Hannah (Eds.)*

12 Dynamics of Earth's Deep Interior and Earth Rotation *Jean-Louis Le Mouël, D.E. Smylie, and Thomas Herring (Eds.)*

13 Environmental Effects on Spacecraft Positioning and Trajectories *A. Vallance Jones (Ed.)*

14 Evolution of the Earth and Planets *E. Takahashi, Raymond Jeanloz, and David Rubie (Eds.)*

15 Interactions Between Global Climate Subsystems: The Legacy of Hann *G. A. McBean and M. Hantel (Eds.)*

16 Relating Geophysical Structures and Processes: The Jeffreys Volume *K. Aki and R. Dmowska (Eds.)*

17 Gravimetry and Space Techniques Applied to Geodynamics and Ocean Dynamics *Bob E. Schutz, Allen Anderson, Claude Froidevaux, and Michael Parke (Eds.)*

18 Nonlinear Dynamics and Predictability of Geophysical Phenomena *William I. Newman, Andrei Gabrielov, and Donald L. Turcotte (Eds.)*

Mineral Physics Volumes

1 Point Defects in Minerals *Robert N. Schock (Ed.)*

2 High Pressure Research in Mineral Physics *Murli H. Manghnani and Yasuhiko Syona (Eds.)*

3 High Pressure Research: Application to Earth and Planetary Sciences *Yasuhiko Syono and Murli H. Manghnani (Eds.)*

Geophysical Monograph 89

Natural and Anthropogenic Influences in Fluvial Geomorphology
The Wolman Volume

John E. Costa
Andrew J. Miller
Kenneth W. Potter
Peter R. Wilcock
Editors

American Geophysical Union

Published under the aegis of the AGU Books Board.

Cover photograph by R. Newbury

Library of Congress Cataloging-in-Publication Data

Natural and anthropogenic influences in fluvial geomorphology / John
  E. Costa . . . [et al.], editors.
      p. cm. — (Geophysical monograph, ISSN 0065-8448 ; 89)
    Includes bibliographical references (p.).
    ISBN 0-87590-046-1
    1. River channels—Congresses. 2. Watersheds—Congresses.
  I. Costa, John E.   II. Series.
  GB561.N38   1995
  551.4' 42—dc20   95-19002
                                                               CIP

ISSN 0065-8448

ISBN 0-87590-046-1

This book is printed on acid-free paper.

Copyright 1995 by the American Geophysical Union, 2000 Florida Avenue, NW, Washington, DC 20009, USA

Figures, tables, and short excerpts may be reprinted in scientific books and journals if the source is properly cited.

Authorization to photocopy items for internal or personal use, or the internal or personal use of specific clients, is granted by the American Geophysical Union for libraries and other users registered with the Copyright Clearance Center (CCC) Transactional Reporting Service, provided that the base fee of $1.00 per copy plus $0.10 per page is paid directly to CCC, 222 Rosewood Dr., Danvers, MA 01923. 0065-8448/95/$01.+.10.
This consent does not extend to other kinds of copying, such as copying for creating new collective works or for resale. The reproduction of multiple copies and the use of full articles or the use of extracts, including figures and tables, for commercial purposes requires permission from AGU.

Printed in the United States of America.

CONTENTS

Preface
John E. Costa, Andrew J. Miller, Kenneth W. Potter, and Peter R. Wilcock　　ix

Channel Initiation, Geomorphic Effectiveness, and Variable Geomorphic Processes

Conservation Principles and the Initiation of Channelized Surface Flows
Terence R. Smith and George E. Merchant　　1

Microtopography of Hillslopes and Initiation of Channels by Horton Overland Flow
Thomas Dunne, Kelin X Whipple, and Brian F. Aubry　　27

Geomorphically Effective Floods
John E. Costa and Jim E. O'Connor　　45

Valley Morphology and Boundary Conditions Influencing Spatial Patterns of Flood Flow
Andrew J. Miller　　57

Morphology and Processes of Valley Floors in Mountain Streams, Western Cascades, Oregon
G. E. Grant and F. J. Swanson　　83

Relative Scales of Time and Effectiveness of Watershed Processes in a Tropical Montane Rain Forest of Puerto Rico
F. N. Scatena　　103

Contribution of Heavy Rainfall to Rainfall Erosivity, Runoff, and Sediment Transport in the Wet Tropics of Australia
Bofu Yu　　113

Magnitude, Frequency, and Special Factors Affecting Channel Form and Processes in the Seasonal Tropics
Avijit Gupta　　125

Regulated Rivers, Stream Management, and Land-Use Impacts

Rivers and the Art of Stream Restoration
Robert Newbury　　137

Effective Discharge and the Design of Channel Maintenance Flows for Gravel-Bed Rivers
E. D. Andrews and James M. Nankervis　　151

Managing Bedload Sediment in Regulated Rivers: Examples from California, U.S.A.
G. Mathias Kondolf　　165

Regulated Streamflow, Fine-Grained Deposits, and Effective Discharge in Canyons with Abundant Debris Fans
John C. Schmidt and David M. Rubin　　177

Hydrologic Factors in Regeneration of Fremont Cottonwood Along the Fremont River, Utah
Benjamin L. Everitt　　197

Fluvial Processes on an Urbanizing Alluvial Fan: Eilat, Israel
A. P. Schick　　209

Spatial Controls on Patterns of Land-Use Induced Stream Disturbance at the Drainage-Basin Scale—An Example from Gravel-Bed Streams of the Ozark Plateaus, Missouri
Robert B. Jacobson　　219

M. Gordon (Reds) Wolman, Jones Falls, Baltimore County, Maryland, June, 1972, following Tropical Storm Agnes.

PREFACE

This book concentrates on the development and evolution of fluvial systems by natural and human processes, highlighting new insights and knowledge about the concept of magnitude and frequency, effects of land-use, and effects of human regulation of streamflow. The papers describe new research developments and approaches to understanding the origins of channels, water and sediment movement on slopes and in streams, the magnitude and frequency of physical processes that affect water and sediment movement, and the implications of these processes for policy decisions such as stream restoration. This is an appropriate time to pause and evaluate what is known about the hydrologic interactions of natural processes, changing land-use, and forced physical constraints on hillslope and fluvial systems. The recent emphasis on ecosystems by the federal government, and the inseparable connection to hydrology, make these prime subjects to ponder in this volume. Included are papers that directly address some of the most difficult and pressing issues related to the role of geomorphology in strategic science.

Contributions are grouped into two sections, one containing papers that investigate near-natural fluvial systems, and their response to geomorphic processes of different magnitude and frequency, and a second group of papers that describe fluvial systems under the strong imprint of human interference. The section on channel initiation, geomorphic effectiveness, and variable geomorphic processes includes nine papers that document the theoretical and field evidence for the origins of channels, the conditions necessary for significant landscape disruption in upland basins, a model that replicates patterns of erosion and deposition of a high-magnitude, low-frequency flood in the Appalachians, the connection between hillslopes and channels in a mountain setting, and three papers on tropical watershed processes and flood-channel morphology that discuss effective fluvial processes in this relatively unknown and under-studied environment. The section on regulated rivers, stream management, and land-use impacts contains seven papers that discuss the diverse response of fluvial systems to human-induced flow and form modifications. Topics include stream restoration, the design of channel-maintenance flows for gravel-bed rivers, consequences and management of coarse- and fine-grained sediment and riparian vegetation in highly regulated rivers, and short- and long-term effects of changing land use on fluvial processes and morphology.

This book is a tribute to one dimension of a multi-dimensional and sagacious man, M. Gordon Wolman, known to friends, colleagues, and students as "Reds." The book focuses on a scientific area that interested Reds, and to which he devoted much of his teaching and research career—the study of watershed and river processes and morphology, and modifications and responses of fluvial systems to human actions. Most of the papers published herein were presented by colleagues and former students as part of a special symposium in honor of the career of M. Gordon Wolman at the Spring 1995 American Geophysical Union meeting in Baltimore.

The papers address issues and problems of natural and human-influenced fluvial systems for which Reds helped formulate the current scientific and policy framework. The diversity of material reflects the landscape of Reds' curiosity, and conveys the breadth of his influence. It would be difficult to overstate the scientific, technical, and philosophical contributions of Reds to the profession of Geography and Geology. After study with Kirk Bryan at Harvard, Reds worked for the U.S. Geological Survey and was a major player in the renaissance of river studies in the 1950s and 1960s. Publications based on work during this period have had a profound impact not only on geomorphology and hydrology, but on sedimentology, stratigraphy, pedology, and environmental management. The small watershed that he studied and used to formulate his ideas and theories of the workings of rivers (Brandywine Creek, Pennsylvania) became known throughout the world as the gauge against which the processes of all other streams and rivers were measured. Some of these classic USGS Professional Papers by Reds and colleagues, long out-of-print, are still in high demand by students and professionals. Many believe that the book coauthored by Reds with Luna Leopold and John Miller, *Fluvial Processes in Geomorphology*, is among the most insightful and broad scholarly contributions in geoscience; it became a benchmark in physical geography and geomorphology for over two decades and may be among the most cited references on fluvial processes.

Another series of papers, separated by nearly 20 years, stands out as a significant contribution to geomorphology. The opportunity to study the geologic impacts of a major hurricane on rivers in New England in 1955 had an obvious impact on his thinking. A few years later, Reds and John Miller wrote one of the most cited and quoted papers in geomorphology and hydrology, "Magnitude and frequency of forces in geomorphic processes" (*Journal of Geology*, 1960). The ideas expressed in this paper have become part of the fundamental theory in geomorphology. In this paper Reds argues that the frequency distribution of streamflow and sediment transport in most streams suggests 90% of the work (i.e., 90% of the load) is removed from watersheds by ordinary discharges that recur at least once every 5 or 10 years. The geomorphic work accomplished by these smaller, more frequent flows overshadows the catastrophic, but infrequent, work accomplished by large, rare floods. The implications of this paper are fundamental to understanding and predicting the impacts of many scientific issues facing our country today, such as global climate change, impacts of shifting land use, and the quality of the nation's streams and rivers. Two decades later Reds, in collaboration with Ran Gerson, wrote a sequel to his earlier paper in which he refined the concept of "effective force" in geomorphology and earth-surface processes. They established that the "effectiveness" of a geologic process is determined by the magnitude and frequency of the process, the resistance of earth materials to the imposed force, and the constructive or restorative processes that occur during the intervals between events. The geomorphology community continues to enjoy the benefits of Reds' sharp mind and his research leadership in fluvial processes by virtue of his long and prolific stream of studies and students.

Our appreciation is extended to Richard Vogel for help and advice in completing the book. Stephen Burges, President of the Hydrology Section, provided enthusiastic endorsement. This book would not have been possible without the support of a group of geomorphologists who, as part of the review process, agreed to receive, review, and return papers within a very short time period. The following individuals offer this effort as part of their contribution to honoring Reds: Victor Baker, Michael Church, David Furbish, William Graf, Alan Howard, Edward Keller, R. Craig Kochel, James Knox, Hugh Mills, Waite Osterkamp, Milan Pavich, Karen Prestegaard, Fred Swanson, Christopher Waythomas, and Ellen Wohl.

John E. Costa
U.S. Geological Survey
Vancouver, Washington

Andrew Miller
University of Maryland
Catonsville, Maryland

Kenneth W. Potter
University of Wisconsin
Madison, Wisconsin

Peter R. Wilcock
U.S. Geological Survey
Vancouver, Washington

Conservation Principles and the Initiation of Channelized Surface Flows

Terence R. Smith and George E. Merchant

Department of Computer Science and Department of Geography, University of California, Santa Barbara, California

A family of models governing the evolution of surface drainage systems is introduced. The models involve an equation for water flow based on the continuity equation and special cases of the Navier-Stokes equations, and an erosion equation that is based on the conservation of sediment and sediment transport laws. These models embody the concept of a free water surface and include the equations of Smith and Bretherton as special cases. Numerical solutions, based on a 2-step MacCormack predictor-corrector finite difference scheme, were computed for a variety of special cases of the family of models. The special cases have the form

$$-\alpha = \nabla \cdot \frac{\nabla(z+h)}{|\nabla(z+h)|} nh^{5/3} |\nabla(z+h)|^{1/2} \quad \alpha \geq 0$$

$$\frac{\partial z}{\partial t} = \nabla \cdot \frac{\nabla(h+z)}{|\nabla(h+z)|} kq_w^\gamma |\nabla(z+h)|^\delta \quad \gamma \geq 0, \delta \geq 0$$

in which α is the rainfall rate, h is water depth, n is a Manning-type coefficient, z is surface elevation, $q_w = nh^{5/3}|\nabla(z+h)|^{1/2}$ is the discharge of water per unit width down the gradient of the water surface, and γ and Δ are exponents of water discharge and water surface slope terms, respectively, that characterize the sediment-transport law. Results obtained from an analysis of these numerical solutions indicate the emergence of an organized flow pattern resulting from the erosion of an initial, randomly perturbed surface. While the initial water flow pattern is quasi-random and with little apparent order, the emergent pattern involves rill-like features with a quasi-uniform spacing. This pattern is characterized by a dominant wavelength of rill spacing that is functionally related to parameters of the sediment transport law as well as to rainfall and slope parameters. The relationship for the associated "dominant" frequency M_w of rill-like features is well represented by the empirical equation

$$M_w = 2.19 \left[\frac{\gamma^{0.63} \delta^{0.06} \theta^{0.46} D^{0.16}}{I^{0.07}} \right]$$

in which θ is the slope of the unperturbed surface in degrees; I is a measure of "erosion time"; and D is the distance upslope. These results differ significantly from those obtained in the analysis of simpler models. In particular, they indicate that rill-like patterns can emerge for cases of the sediment transport law in which $\gamma = 1.0$. The results appear to have significant implications for the evolution of patterned surface flow phenomena in real landscapes and for associated geomorphic phenomena.

1. INTRODUCTION

Understanding the process by which overland flows organize themselves into well-structured networks of channels is at once the most fundamental problem of geomorphic science and the most elusive. One may categorize past approaches to solving this problem into those based on established physical principles, those based on statistical arguments, and those based on arguments invoking variational principles. All three approaches have provided useful insights. The statistical approaches, however, invariably assume the existence of channelized flow patterns and, at best, provide partial explanations, while the variational approaches are typically based on assumptions that cannot currently be justified. The analysis that we present in this paper is based on a contention that conservation principles still provide the most satisfactory approach to understanding the initiation of channelized flow patterns. In particular, a reasonable strategy for constructing models of channelized flow initiation is to choose equations embodying conservation and constitutive laws for the flows of water and sediment. While it is desirable to base such models on the most fundamental sets of equations available, it is clearly advantageous to choose the simplest cases of such equations that, on the one hand, adequately model the phenomena of interest and, on the other hand, are matched in terms of their degree of accuracy. Once a basic understanding of channel initiation has been established by this means, it is likely that the statistical and variational approaches will lead to further insights.

The current paper makes two contributions to the present literature. The first contribution is a family of models representing the flow of water and sediment over erodible surfaces. These models embody a representation of a free water surface, and include, as special cases, previously analyzed models [*Smith and Bretherton*, 1972; *Loewenherz*, 1991]. The second and main contribution of the paper is a set of numerical solutions for specific members of the class of models. These solutions describe how flows over randomized surfaces initially become organized into rill-like features. They also indicate the manner in which the spacing of emergent, rill-like features is dependent on the nature of the sediment transport law and parameters characterizing the environment. While these results are preliminary and are derived for only a few special cases of the general model, they indicate the potential value of a systematic exploration of the family of models over a large number of its members. It is argued that this family, and its extensions, provide an experimental facility by means of which a large number of interesting geomorphic issues may be addressed.

The paper is structured as follows. We first review the basis for our previous family of models and discuss their shortcomings. We then introduce and justify the new family of models. Finally, we present an analysis of numerical solutions for particular cases of the models; an analysis of the implications of these results; and a discussion concerning further applications of the family of models.

2. A FAMILY OF MODELS FOR THE EVOLUTION OF SURFACE RUNOFF

The flow of water over a surface is most generally modeled in terms of an equation for the conservation of mass and the Navier-Stokes equations. While the conservation of mass for sediment is a fundamental equation that governs the relation between the transport of sediment over a surface and the erosion of the surface, there are no sets of equations governing sediment transport that match the Navier-Stokes equations in terms of their general applicability. There are, however, many empirically-based constitutive relations that provide adequate approximations to sediment transport. Hence in applying the modeling strategy described above, it is appropriate to construct simplified versions of the Navier-Stokes equations that both model adequately those aspects of the flow of water important for channel initiation and that match appropriately the accuracy of the sediment transport equations.

2.1 *Previous Models and Their Shortcomings*

Based on the preceding strategy, various classes of models of the evolution of channelized flow have been suggested and investigated [*Smith and Bretherton*, 1972; *Kirkby*, 1986; *Loewenherz*, 1991; *Willgoose et al*, 1991, *Kramer, and Marder* 1992; *Loewenherz-Lawrence*, 1994]. While these analyses have contributed tantalizing insights into the process of rill initiation, they have failed to provide an acceptable explanation for the emergence of channelized flow. In order to motivate the introduction of a new family of models, we briefly discuss a previously investigated class of models that provide a useful basis for constructing improved models. These models are now well understood [*Smith and Bretherton*, 1972; *Loewenherz*, 1991] and involve:
1. a partial differential equation governing the flow of water based on:
   (a) a simple statement of the conservation of mass;
   (b) the assumption that the direction of water flow is down the gradient of the land surface;
2. a partial differential equation governing the transport of sediment based on:
   (a) a simple statement of the conservation of mass;
   (b) constitutive laws of the form $q_s = F(q_w, |\nabla z|)$ in which q_w is the discharge of water per unit width of the

surface and $|\nabla z|$ is the magnitude of the gradient of the surface z;

(c) the assumption that the direction of sediment flow is the same as the direction of water flow and down the gradient of the land surface;

3. equations dependent on two spatial dimensions and time;
4. an assumption that at the time scale at which significant erosion occurs, the flow of water over the surface is steady.

A simple set of equations satisfying these constraints is:

$$-\alpha = \nabla \cdot \frac{\nabla z}{|\nabla z|} q_w \qquad (1)$$

$$\frac{\partial z}{\partial t} = \nabla \cdot \frac{\nabla z}{|\nabla z|} F(q_w, |\nabla z|) \qquad (2)$$

where α is the rainfall rate; z is the elevation of the eroding surface; and q_w is the discharge of water per unit width over the eroding surface.

Analyses of these equations [*Smith and Bretherton*, 1972; *Loewenherz*, 1991] led to the establishment of conditions on $q_s = F(q_w, |\nabla z|)$ under which steadily eroding but unchanneled surfaces are stable or unstable. The main difficulty with the model is that in the unstable regions of a surface, the perturbations that grow into channel-like forms at the fastest rate are the smallest-length perturbations. Hence, there is no perturbation of finite, non-zero wavelength that can grow to dominate the pattern of channels. The essential insight into this troublesome form of instability is that, since water flows down the land surface gradient, the areas of greatest convergence of water and hence of greatest instability occur in the disturbances with the steepest sides. These are the disturbances of smallest wavelength. In the unstable regions of the surface, such convergence entails a cycle of greater erosive power, greater downcutting, steeper sides, and further increases in the rate of convergence.

2.2 Generalizing Previous Models

Various means for modifying and generalizing equations (1) - (2) so that instabilities in unchanneled solutions grow into well-structured patterns of channels have been suggested [*Smith and Bretherton*, 1972; *Loewenherz*, 1991]. Such means include:

1. the introduction of empirically-determined length scales, below which the kinematic model (1) - (2) loses its validity, together with appropriate modifications of the model;
2. the application of smoothing operators to various terms in the equations;
3. the introduction of more sophisticated models of the flow of water and, in particular, the introduction of a free water surface;
4. the introduction of stabilizing mechanisms involving other processes of erosion, transportation, and deposition.

While empirically-determined length scales that control the initial spacing of rills may be introduced in terms of a parameter representing conditions under which small-scale hydrodynamic factors become important [*Loewenherz*, 1991], it is, in practice, very difficult to justify and to manipulate specific limits. Concerning the introduction of smoothing operators, it is well-known [*Smith*, 1976; *Loewenherz*, 1991] that integral transformations of the dependent variables in equations such as (1) - (2) filter out high frequency effects and reduce instability at small wavelengths of disturbance. While there is some empirical basis for introducing smoothing operators, such as when hydraulic radius defined in terms of an integral is used to model the "depth" of flow, it is generally both difficult and unsatisfying to introduce specific smoothing operators.

Given the simplicity of equation (1), which merely represents the conservation of some abstract quantity flowing down the gradient of the land surface, we are clearly justified in introducing more realistic models of the flow of water. In relation to the problem of channel initiation, it appears important that we embed in the model some representation of a free water surface and the nature of water flow. This point has also been made by other researchers in the recent literature in which models representing both free-water surfaces and land surfaces are presented [*Kramer and Marder*, 1992; *Loewenherz*, 1994]. It is intuitively clear, for example, that the introduction of a free water surface entails flows whose direction is not necessarily down the gradient of the land surface, and it is no longer necessarily the case that the greatest convergence of water will occur in disturbances of smallest wavelength. Hence a major problem with the simple model (1) - (2) is potentially removable.

With the introduction of a free water surface, furthermore, one may immediately model situations in which there are two natural and distinct directions for the motion of sediment. The first direction is that of the motion of water, and involves sediment moving under the influence of the tractive forces of flowing water. The second direction is down the gradient of the land surface, and involves sediment moving largely under the influence of gravitational forces. This is a useful observation, since it is well-known that down-gradient motions of sediment under the influence of gravity are stabilizing influences whose effects increase as the wavelength of disturbance decreases. Finally, we note that the introduction of simple models of flow involving a free water surface may entail the natural introduction of smoothing operators. For example, the empirical concept of hydraulic radius, which is often used as a representation of the concept of "effective depth", essentially has the form of an integral across some width of flow.

Given such extensions to (1) - (2), an important question concerns the discovery of simple and natural sets of mechanisms that are sufficient to ensure an orderly emergence of channelized flow patterns.

2.3 A New Family of Models

Based on the previous consideration, we now introduce a family of models that provides a significant generalization of the family (1) - (2). In particular, this family involves the introduction of a free water surface.

Apart from the obvious continuity conditions, the equation characterizing the flow of water is based on two assumptions. The first assumption is that the direction of water flow is down the gradient of the water surface

$$-\frac{\nabla(z + h)}{|\nabla(z + h)|}$$

in which z is the surface elevation, h is the depth of water, and $z + h$ is the elevation of the water surface. The second assumption is that, in analogy with Chezy and Manning type equations, the magnitude of the flow may be characterized as some function of the depth of flow and the slope of the water surface

$$q_w = q_w(h, |\nabla(z + h)|) \qquad (3)$$

The equation characterizing erosion and the transport of sediment is based on the assumption that, in the general case, some sediment is entrained and transported by the flow of water according to a law of the form

$$q_{sw} = F(q_w, |\nabla(z + h)|) \qquad (4)$$

in the direction of water motion. It is also reasonable to assume that another component of sediment flux moves down the gradient of the land surface $-\nabla z/|\nabla z|$ according to a law of the form

$$q_{sg} = G(q_w, |\nabla z|, |\nabla(z + h)|) \qquad (5)$$

In general the direction of water motion is not the direction of the surface gradient. This, for example, is the case in channelized flows in which the direction of maximum surface slope may have a component towards the centerline of the channel. Hence bedload materials that are set in motion by the tractive force of water will tend to move, under the influence of gravity, down the slope of the surface rather than up the slope when they settle on the channel bed [*Smith*, 1974]. Furthermore, situations in which the motion of sediment is very largely driven by the force of gravity, as in the case of soil creep, can be described as special cases of (5) in which $q_w = 0$. Hence the family of models is given by the equations

$$-\alpha = \nabla \cdot \frac{\nabla(z + h)}{|\nabla(z + h)|} q_w(h, |\nabla(z + h)|) \qquad (6)$$

$$\frac{\partial z}{\partial t} = \nabla \cdot \frac{\nabla(z + h)}{|\nabla(z + h)|} F(q_w, |\nabla(z + h)|) +$$

$$\nabla \cdot \frac{\nabla z}{|\nabla z|} G(q_w, |\nabla z|, |\nabla(z + h)|) \qquad (7)$$

in which q_w is given by (3). We have, therefore, a system of two partial differential equations in the unknowns (h) (water depth) and (z) (surface elevation). It is clear that the previous family of models (1) - (2) may be viewed as special cases of the new family.

2.4 Special Cases of the New Family of Models and a Justification

For current purposes, and in line with our strategy of seeking the simplest set of equations that leads to the initiation of channelized flow, we restrict attention to certain special cases of (6) - (7) in the remainder of the paper. In particular, we place three restrictions on the general forms (3), (4), and (5):

1. In equation (6), we employ an equation of Manning type

$$q_w = nh^{5/3}|\nabla(z + h)|^{1/2} \qquad (8)$$

in which n is a Manning type coefficient. We note below that it may be of value to generalize the values of the exponents in (8).

2. In equation (4), we employ a power law of the form

$$q_{sw} = kq_w^\gamma |\nabla(z + h)|^\delta, \quad \gamma \geq 0, \quad \delta \geq 0 \qquad (9)$$

3. In equation (7), we restrict attention to a situation in which we ignore the gravitational component of sediment transport i.e.

$$q_{sg} = G(q_w, |\nabla z|, |\nabla(z + h)|) = 0$$

Hence the subfamily of models on which we focus our attention takes the form

$$-\alpha = \nabla \cdot \frac{\nabla(z+h)}{|\nabla(z+h)|} n h^{5/3} |\nabla(z+h)|^{1/2} \quad (10)$$

$$\frac{\partial z}{\partial t} = \nabla \cdot \frac{\nabla(z+h)}{|\nabla(z+h)|} k q_w^\gamma |\nabla(z+h)|^\delta \quad (11)$$

We now provide a brief justification for each of these restricted forms.

Concerning the equation for water flow over the surface, we may first argue, on the basis of hydraulic considerations, that surface flows may be locally approximated as one-dimensional flows whose movement is down the gradient of the water surface and whose magnitude in that direction is given by an equation of Manning type. Such an approximation is reasonable under conditions in which any convergence and divergence in the flows is relatively small. A somewhat more convincing argument may be based on a rational modification of an equation that is both similar to (10) and derivable from the continuity equation and the Navier-Stokes equations [*Todini and Venutelli*, 1991].

Starting with the equation of continuity and the Navier-Stokes equations, Todini and Venutelli assume almost-horizontal flows, and drop all terms relating to the vertical (i.e. z) components of velocity and acceleration. They then formally integrate all four equations over the depth of flow to obtain depth-averaged quantities. From the z-component of the Navier-Stokes equations they thus obtain an equation indicating that hydrostatic conditions exist throughout the fluid column. This condition, together with the assumption that the mass forces in the x- and y-directions are zero, is used to simplify the depth averaged momentum equations. The remaining three equations are further simplified by dropping the convective and local accelerations and obtaining equations for momentum in which the water slope coincides with the friction slope. Finally, substitution of the two momentum equations into the continuity equation leads to an expression that, in our notation, takes the form

$$-\alpha = \frac{\partial}{\partial x}\left[n_x h^{5/3} \left|\frac{\partial H}{\partial x}\right|^{1/2} \frac{\frac{\partial H}{\partial x}}{\left|\frac{\partial H}{\partial x}\right|}\right] + \frac{\partial}{\partial y}\left[n_y h^{5/3} \left|\frac{\partial H}{\partial y}\right|^{1/2} \frac{\frac{\partial H}{\partial y}}{\left|\frac{\partial H}{\partial y}\right|}\right] \quad (12)$$

in which $H = z + h$ is the elevation of the water surface.

This equation is intended to provide an approximate description for flows occurring on relatively low slopes. It is similar to equation (10), and differs only in the slope terms and for cases in which $n_x \neq n_y$. In (10) it is always the case that the magnitude of the flow is proportional to the square root of the surface gradient while its direction is down the gradient of the water surface. In the case of (12), however, one notes that

1. if the slope of the water surface is zero in either the x-direction or the y-direction, then the magnitude of the flow is proportional to the square root of the slope of the water surface in the y- or x-direction respectively, while the flow is down the gradient of the slope in these directions. In these two cases, which depend on the orientation of the coordinate system, there is agreement with (10).

2. If the water surface has non-zero components of slope in both the x- and the y-directions, then the magnitude of the flow is no longer proportional to the square root of the slope of the water surface and the direction of flow is no longer down the gradient of the slope.

The occurrence of such representation-dependent anisotropy in a general flow law is clearly disconcerting. Hence we introduce isotropy into the flow law (12) by extending the flow-law that holds for situations in which there are only x- or y-components of slope in (12) to situations in which the gradient of the water surface has an arbitrary orientation. Hence the terms

$$[n_x h^{5/3} \left|\frac{\partial H}{\partial x}\right|^{1/2} \frac{\frac{\partial H}{\partial x}}{\left|\frac{\partial H}{\partial x}\right|}, n_y h^{5/3} \left|\frac{\partial H}{\partial y}\right|^{1/2} \frac{\frac{\partial H}{\partial y}}{\left|\frac{\partial H}{\partial y}\right|}]$$

in (12) become

$$[n_x h^{5/3} |\nabla H|^{1/2} \frac{\frac{\partial H}{\partial x}}{|\nabla H|}, n_y h^{5/3} |\nabla H|^{1/2} \frac{\frac{\partial H}{\partial y}}{|\nabla H|}]$$

in which $|\partial H/\partial x|$ and $|\partial H/\partial y|$ have been replaced by $|\nabla H|$, and in which $n_x = n_y = n$. We note that, with this modification, we have provided a derivation of (8). This derivation makes it clear that equation (8) refers to a flow situation that is represented in terms of depth-averaged, horizontal velocities. This assumption is implicit in the remainder of our analysis.

The justification for neglecting the sediment transport term q_{sg} follows from our strategy of seeking the simplest forms of (10) - (11) that lead to the formation of channelized flow. Clearly, if the stability conditions inherent in the simple model (10) - (11) are insufficient to support the development of well-defined channel forms, then it would be both reasonable and worthwhile to examine extensions of the models in which q_{sg} is non-zero. Our justification for choosing the class of simple power laws (9) for flow-induced sediment transport is that such forms are in accordance with a large body of empirical evidence relating to sediment transport, both in channels and over slopes [*Julien and Simons,* 1985].

3. MODELING THE INITIATION OF CHANNELIZED SURFACE FLOWS

A key question concerning the family of models (10) - (11) is whether they are capable of providing insight into the processes by which initially planar or disorganized flows over a surface become organized into channel-like patterns. In this and the following section, we present preliminary results that indicate that this class of models indeed contains "reasonable" instances that provide useful insights into these processes. Although the goal of our current analysis is to indicate the potential value of the class of models, and not to provide definitive explanations for specific geomorphic phenomena, these results appear to justify an extensive and detailed investigation of the models (10) - (11).

Equations (10) - (11) are non-linear, partial differential equations, and as such are virtually impossible to solve in explicit form. We have, therefore, applied numerical techniques in solving special cases of the equations. The issue that we address is whether relatively small, random perturbations of a planar surface grow spontaneously into channel-like features that are characterized by some form of regularity.

Important decisions relating to such numerical modeling involve the choice of numerical scheme, the choice of modeling parameters, and the choice of boundary and initial conditions. Before presenting the results of our numerical methods, we therefore discuss our choice of schemes, numerical parameter values, and boundary and initial conditions.

3.1 *Numerical Solutions of the Equations*

Numerical solutions to (10) - (11) were obtained with the use of an explicit, two-step MacCormack predictor-corrector finite difference scheme that has previously been adapted for 2-dimensional overland flow [*Zhang and Cundy,* 1989]. This scheme is second-order accurate in time and space, and previous research has shown it to possess good stability and convergence properties. Shock waves, for example, are damped by this method, although artificial viscosity must generally be added after each time step to smooth the water surface [*Fennema and Chaudhry,* 1990].

A sequence of predictor-corrector operations is employed to obtain a solution at each time step. In the predictor step, the method leads to an approximate solution at a half time step $t_{n+1/2}$ for the computational point (i, j) using a forward difference for spatial derivatives. This approximate solution is then modified using a backward difference to get an improved solution for the corrector step at time $t_{n+1/2}$. The reverse procedure is then applied at time step t_{n+1} in order to obtain a symmetrical solution.

The procedure is comprised of the difference equations (13) - (20) shown below. This scheme was applied to equation (10) and to equation (11), with the variables in the scheme having the following interpretation in terms of our application: P is either water depth or land elevation; U is the flow of either water or sediment in the x-direction; V is the flow of either water or sediment in the y-direction; and E is the rate of rainfall. At time t_n, predictor (backward difference) is given by:

$$\overline{P}_{i,j}^{n+1/2} = P_{i,j}^n - \frac{\Delta t}{2\Delta x}(U_{i,j}^n - U_{i-1,j}^n) - \frac{\Delta t}{2\Delta y}(V_{i,j}^n - V_{i,j-1}^n) + \frac{\Delta t E_{i,j}^n}{2} \qquad (13)$$

At time t_n, corrector (forward difference) is given by:

$$P_{i,j}^{n+1/2} = \frac{1}{2}[P_{i,j}^n + \overline{P}_{i,j}^{n+1/2} - \frac{\Delta t}{2\Delta x}(\overline{U}_{i+1,j}^{n+1/2} - \overline{U}_{i,j}^{n+1/2}) - \frac{\Delta t}{2\Delta y}(\overline{V}_{i,j+1}^{n+1/2} - \overline{V}_{i,j}^{n+1/2}) + \frac{\Delta t E_{i,j}^{n+1/2}}{2}] \qquad (14)$$

At time t_{n+1}, predictor (forward difference) is given by:

$$\overline{P}_{i,j}^{n+1} = P_{i,j}^{n+1/2} - \frac{\Delta t}{2\Delta x}(U_{i+1,j}^{n+1/2} - U_{i,j}^{n+1/2}) - \frac{\Delta t}{2\Delta y}(V_{i,j+1}^{n+1/2} - V_{i,j}^{n+1/2}) + \frac{\Delta t E_{i,j}^{n+1/2}}{2} \qquad (15)$$

At time t_{n+1}, corrector (backward difference) is given by:

$$P_{i,j}^{n+1} = \frac{1}{2}[P_{i,j}^{n+1/2} + \overline{P}_{i,j}^{n+1} - \frac{\Delta t}{2\Delta x}(\overline{U}_{i,j}^{n+1} - \overline{U}_{i-1,j}^{n+1}) - \frac{\Delta t}{2\Delta y}(\overline{V}_{i,j}^{n+1} - \overline{V}_{i,j-1}^{n+1}) + \frac{\Delta t E_{i,j}^{n+1}}{2}] \qquad (16)$$

Finally, the artificial viscosity, which is required in order to damp out numerical instabilities in the computation of the water surface, is given by the following scheme, in which P is the variable that one is attempting to smooth (i.e. water depth in our case):

$$P_{i,j}^{n+1} = P_{i,j}^{n+1} + D_x P_{i,j}^{n+1} + D_y P_{i,j}^{n+1} \qquad (17)$$

$$D_x P_{i,j} = [\epsilon_{i+1/2,j}(P_{i+1,j} - P_{i,j}) - \epsilon_{i-1/2,j}(P_{i,j} - P_{i-1,j})] \qquad (18)$$

$$V_{x_{i,j}} = \frac{|P_{i+1,j} - 2P_{i,j} + P_{i-1,j}|}{|P_{i+1,j}| + |2P_{i,j}| + |P_{i-1,j}|} \qquad (19)$$

$$\epsilon_{x_{i-1/2,j}} = k\frac{\Delta t}{\Delta x} \max(V_{x_{i-1,j}}, V_{x_{i,j}}) \qquad (20)$$

in which k is a parameter that regulates the amount of dissipation and D_x, D_y are operators. We note that there is a similar set of equations for y.

In applying the scheme in an integration of (10) - (11), we were able to take advantage of a "decoupling" between the two equations that arises because erosion is assumed to take place at a time scale at which the flow of water is steady. It particular, the nature of equations (10) - (11) allows us to compute an equilibrium flow of water for a fixed elevation surface and some time t and, given this equilibrium flow, to compute the amount of erosion that would occur at each surface point in the next (small) interval of time Δt. If the interval Δt is chosen to be sufficiently small, we may assume that the flow pattern has not changed enough to change the pattern of erosion. Hence our computational scheme involved an iterative cycle of finding the equilibrium flow of water over a fixed surface z; eroding the surface by a very small amount; and recomputing the equilibrium flow.

The parameter values chosen for the computational schema for the water flow equation were $\Delta t = 0.0025$ s, $\Delta x = \Delta y = 1$ m, and $k = 5.0$ for the artificial viscosity. These values ensured the stability of the computations. We employed an initial flow consisting of a very thin (0.001 mm) film of water over the surface to avoid singularities in the computations. The equilibrium of the water surface was defined in terms of the maximum rate of change in water depth being less than some very small value.

The value of Δt for the sediment transport equation was chosen to be sufficiently small to ensure that the maximal amount of erosion per iteration did not exceed a value of 0.5 mm at any point on the grid. In particular, the erosion equation was halted whenever the maximal amount of erosion at any point over the grid exceeded 0.5 mm, and a new equilibration of the water flow equation was initiated. As we note from the results below, such a value would almost certainly be first achieved near the lower boundary of the surface, while the amount of erosion over the greater part of the surface at any iteration would be significantly less that 0.5 mm.

3.2 Modeling Parameters and the Initial and Boundary Conditions

We obtained numerical solutions for equations (10) - (11) over a 100 m x 100 m section of a planar surface. In obtaining the solutions, we employed a grid spacing of 1 m in both the x- and y-directions to give a total of 10,000 grid points. In Figure 1, we illustrate an example of such a surface in which the initially planar surface has been perturbed locally with small variations in elevation. We computed numerical solutions over a range of parameter values, including discharge exponent (γ) values ranging from 1.0 to 3.0; slope exponent (δ) values ranging from 1.0 to 5.0; rainfall rates (α) ranging from 1 cm/hour to 5 cm/hour; and surface slopes (θ) ranging from 3 deg to 15 deg. In relation to the water and sediment transport laws, we used a value of n = 33.3 for Manning's n (i.e. for a "smooth" flow situation in SI units).

The boundary conditions employed in obtaining the numerical solutions involve (see Figure 1)
1. a symmetrical ridge at the upper boundary, with no flow of water or sediment over the ridge;
2. a periodic boundary at the edges of the ridge (i.e. the ridge "wraps around" and the flows of water and sediment from one side are continued over to the other side);
3. an absorbing boundary at the lower end of the ridge, with the water depth held at a constant depth of 20 mm at the boundary.

In particular, we may view the lower absorbing boundary as a body of water having the capacity to carry away all water and sediment reaching it.

With respect to initial conditions, the planar surface was perturbed at each point with randomly generated disturbances, except for the points on the lower absorbing boundary and the upper symmetric boundary. We employed pseudo-random numbers in representing relatively small disturbances in the range 0.0 cm to 10.0 cm. These values were added to the elevation of the initial planar surface to produce an initial randomized surface. The initial flow over this randomized surface was computed for a given rainfall rate starting with an initial thin film of water having a depth of 0.001 mm at all points on the surface.

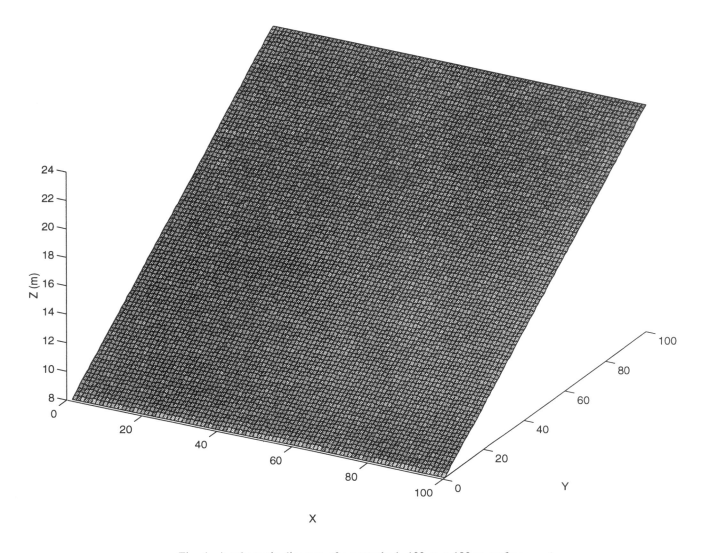

Fig. 1. A schematic diagram of a perturbed, 100 m x 100 m surface.

4. RESULTS OF NUMERICAL SIMULATIONS: THE INITIATION OF RILLS

We present the results of our simulations in two steps. We first examine a numerical solution to (10) - (11) for a given set of parameter values in order to illustrate characteristics of many of the numerical solutions. This examination also provides an illustration of the emergence of organized flow patterns from the randomized flow patterns. In particular, we examine the case $\gamma = 2.0$, $\delta = 4.0$, $\alpha = 3$ and $\theta = 9$ degrees. We attach no particular empirical significance to these parameter values, which we denote by [2.0, 4.0, 3.0, 0.16]. We then examine statistically the results of several runs in order to answer the following question: do the patterns of flow that emerge depend in a systematic manner on the parameters that characterize the various situations ? In particular, we examine the dependence of the patterns on γ and δ (the sediment transport law coefficients); α (rainfall); and θ (the land surface slope).

From among the many characteristics concerning the numerical solutions that we could describe, we have chosen to focus primarily on the elevation of the surface and the depth of the flow over the surface at both the beginning and the end of the run. The depth of flow in particular is a reasonably diagnostic variable with respect to the emergence of patterned flows, while the elevation of the land surface provides useful ancillary information.

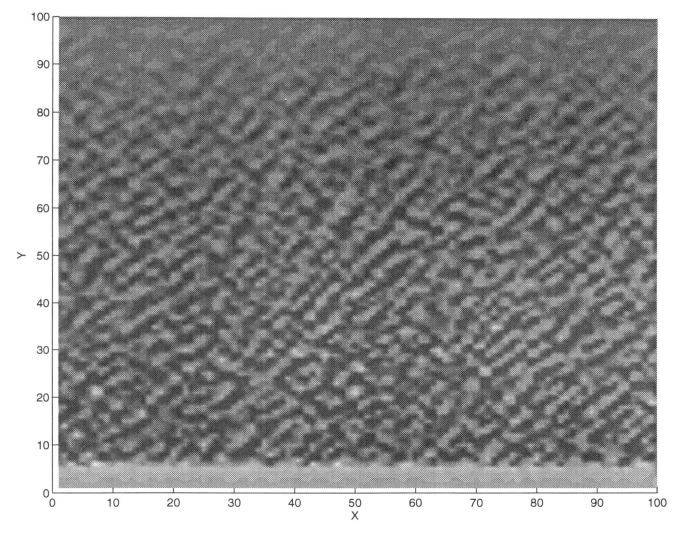

Fig. 2a. Initial depth of water flow: perturbed surface.

4.1 *The Emergence of Pattern*

In Figure 2a we show a representation of the depth of runoff over the initial, randomized surface for the case of interest. This may be compared with Figure 2b, in which we show a representation of the depth of flow after an initial run of the model. In Figures 2c and 2d we show, for the sake of comparison, representations of the depth of surface flow evolving from different initial surfaces and under different sets of parameter values. In particular, the two sets of parameter values are $\gamma = 3.0$, $\delta = 5.0$, $\alpha = 5.0$, and $\theta = 0.27$ deg ([3.0, 5.0, 5.0, 0.27]) for Figure 2c and $\gamma = 1.0$, $\delta = 1.0$, $\alpha = 5.0$, and $\theta = 15$ deg ([1.0, 1.0, 5.0, 15]) for Figure 2d. We do not show the initial, randomized flow patterns for these two cases, since they appear similar to those in Figure 2a. It is clear from these figures that the initial, randomized patterns of flow have developed a noticeable degree of structure and coherence.

We first characterize the general nature of the initial and final elevations and depths corresponding to the case [2.0, 4.0, 5.0, 0.16]. In Table 1, we show the values of the average elevation of the initial land surface at cross-sections taken every 10 m from y = 10 m to y = 100 m as well as the standard deviation, the maximum, and the minimum values for this variable. In this table, we also show similar information for the same variables for the eroded land surface at the same set of cross-sections, while in Table 2 we show analogous information for the depth of

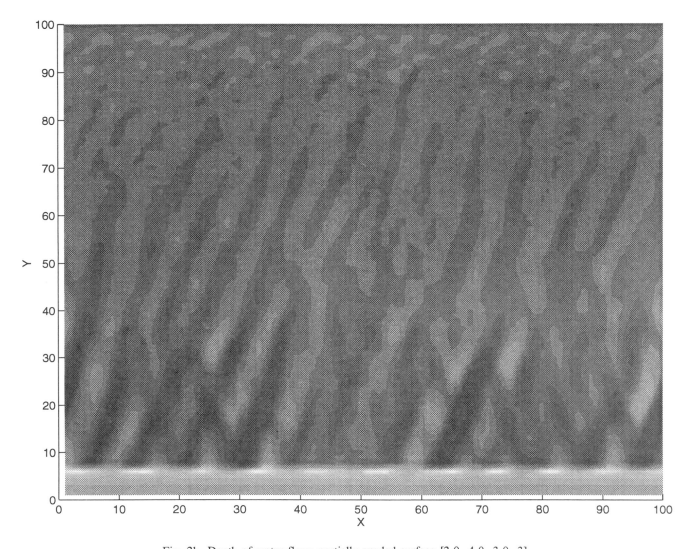

Fig. 2b. Depth of water flow: partially eroded surface [2.0, 4.0, 3.0, 3].

flow. In Figures 3a and 3b we show plots of various cross-sections through the initial and final surfaces, while in Figures 4a and 4b we show plots through the same set of cross-sections for the depth of water.

We note several points of interest from these tables and figures. As a result of erosion

1. the average surface elevation decreased on each of the cross-sections by amounts ranging from 23.56 cm at y = 20 m to 0.0054 cm at y = 100 m, with the average erosion for much of the surface being in the range 1.0 - 5.0 cm. The erosion decreases significantly in the higher cross-sections.
2. The minimum cross-sectional elevation for the final surface is a little greater (of the order of 1 cm) than for the initial surface at cross-sections above 50 m.
3. The difference between the maximum and minimum cross-sectional elevations is greater at all the cross-sections on the initial surface than it is on the final surface except for the upper boundary of the surface.

Since values for erosion are of the same order of magnitude as the randomized disturbances, and are typically less than 1 % of the relative relief (16 m) for the land surface, it is fair to say that we are observing the early stages in the development of patterned flows. The cross-sections for which the minimum elevation in the final surface exceeds the minimum elevation in the initial surface are all cross-sections at or above the 50 m cross-section. Such a change could only occur with the deposition of materials. Hence the deeper initial disturbances on the upper section of the surface have experienced infilling.

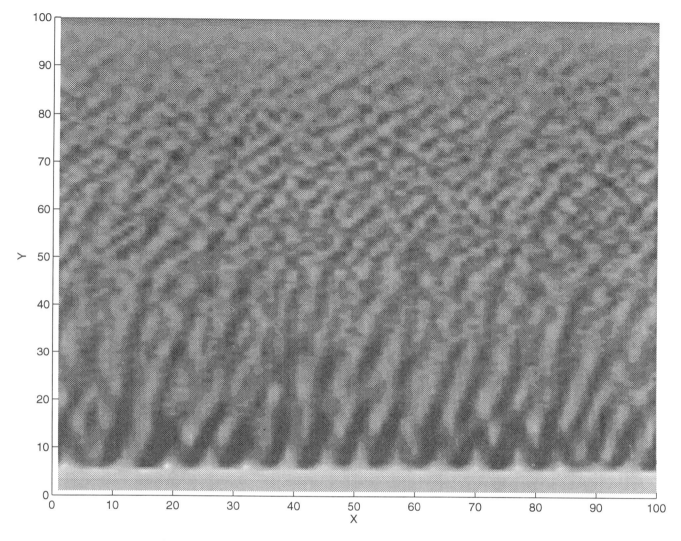

Fig. 2c. Depth of water flow: partially eroded surface [3.0, 5.0, 5.0, 15]

Considering also the magnitudes of the minimum-maximum differences, we may infer a general reduction in the relative relief at cross-sections that is in part attributable to the lowering of local elevation maxima and to some infilling of the local elevation minima.

Concerning the information on water depth we note that
1. the differences between the average initial and final water depths at all cross-sections are positive.
2. The maximum flow depths are greater in the final surface at cross-sections at 10 m - 30 m, while the differences between the maximum and minimum flow depths are also greater in the final surface at those cross-sections.

We note that the greater average depth of flow on the initial surface (corresponding to a lower average velocity of the initial flows) and the increase in the maximum-minimum differences in the lower cross-sections are consistent with some form of channelization.

In order to assess the degree of pattern and regularity implicit in Figure 2b and in the cross-sections shown in Figures 3b and 4b, we computed the spectral density function at each of the cross-sections shown in Figures 3a, 3b, 4a, and 4b. In Figure 5 we show the spectral density functions at these cross-sections both for the initial (uneroded) surface and for the eroded surface. In particular, the spectra for the initial surface are shown as hatched lines. In Figure 6 we show analogous information for the initial and final water depths.

As one would expect, the spectral density plots for the cross-sections of the initial, randomized surface show that

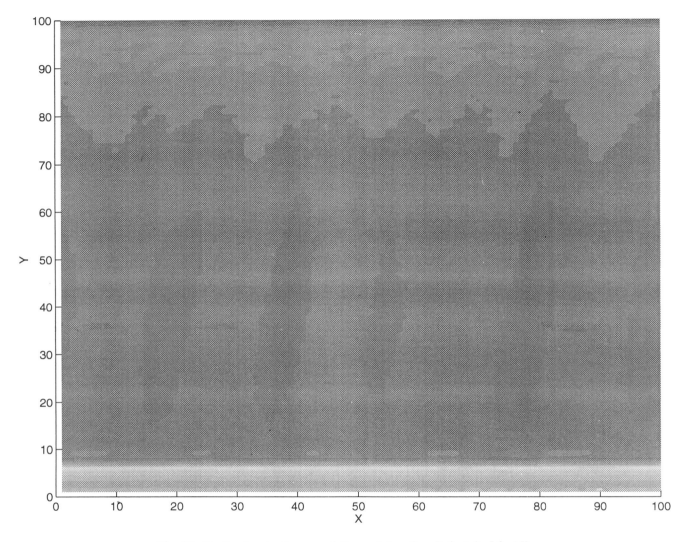

Fig. 2d. Depth of water flow: partially eroded surface [1.0, 1.0, 5.0, 15].

power is initially distributed over a large range of frequencies. The effect of erosion is to cause a significant relative decrease in this power over most of these frequencies. In order to show these patterns to greater effect, we have plotted in Figure 7 values of the spectral power that were computed by averaging the power spectral densities over each of the cross-sections from 10 m to 40 m above the base of the surface, taken every 1 m. These averages are shown for both for the initial elevation, the final elevation, the initial depth, and the final depth. On comparing the initial/final averages, it becomes clear that a dominant frequency of about 11 cycles per 100 m in both the eroded surface and the water depth has emerged from the broad range of initial frequencies. Above the 50 m cross-section, while a significant reduction in power over a large range of frequencies is still apparent, the dominant frequency is no longer clustered so clearly around 11 cycles per 100 m.

Finally, it is of value to examine measures of the coherency that the flow pattern attains as a result of erosion. A simple measure of such coherency is the correlation between neighboring cross-sections for both the elevation and the depth surfaces. In Table 3, we show computed values of correlation coefficients for a subset of pairs of cross-sections. In this table, we exhibit coefficients for pairs of cross-sections both 1 m apart and 10 m apart.

Table 1. Surface Elevation Characteristics

Loc (m)	Mean Elevation and Std (m)	Min Elevation	Max Elevation	Δ (Max, Min)	Δ (Initial, Eroded)
		Initial Surface			
10	9.490 (0.028)	9.440	9.536	0.090	-
20	11.086 (0.027)	11.043	11.139	0.096	-
30	12.690 (0.029)	12.640	12.739	0.099	-
40	14.293 (0.030)	14.242	14.340	0.098	-
50	15.890 (0.028)	15.840	15.939	0.099	-
60	17.498 (0.029)	17.440	17.540	0.100	-
70	19.086 (0.030)	19.041	19.138	0.097	-
80	20.685 (0.027)	20.641	20.739	0.098	-
90	22.291 (0.029)	22.242	22.340	0.098	-
100	23.840 (0.000)	23.840	23.840	0.000	-
		Eroded Surface			
10	9.255 (0.010)	9.234	9.275	0.041	0.236
20	10.993 (0.013)	10.966	11.029	0.063	0.093
30	12.646 (0.011)	12.620	12.669	0.049	0.044
40	14.258 (0.007)	14.234	14.270	0.036	0.035
50	15.864 (0.008)	15.845	15.883	0.038	0.026
60	17.468 (0.006)	17.452	17.483	0.031	0.024
70	19.071 (0.007)	19.056	19.088	0.032	0.015
80	20.676 (0.010)	20.654	20.697	0.043	0.009
90	22.285 (0.019)	22.249	22.325	0.076	0.006
100	23.840 (0.000)	23.839	23.840	0.001	0.001

The lack of coherency in the initial surface is in striking contrast to the coherency in the final surface, particularly in those cross-sections below 60 m. There is an equally marked increase in the coherency of the depth patterns between the initial and final surfaces.

From the changes in the flow patterns seen in Figures 2b, 2c, and 2d, the changes in the overall surface and depth characteristics, the changes in the power spectra for surface and depth characteristics, and from the increasing coherency of the flow, it is clear that the initially randomized flow patterns have developed a significant degree of structure. In particular, this structure is most marked in the lower half of the surface where a channelized flow pattern is a dominant aspect of the structure. We are justified, therefore, in asserting that patterns of rill-like features emerged during the course of erosion. For the particular case that we have examined these rills are characterized by a dominant wavelength of separation of about 9 m.

4.2 The Dependence of the Dominate Rill Frequency on Parameters

A question of immediate interest is whether similar or different dominant wavelengths are characteristic of other flow conditions. The significance of this issue is clearly indicated by a comparison of Figures 2c and 2d, in which there appear to be rill-like features at both a greater and a lesser spacing than that shown in Figure 2b. Hence we

Table 2. Depth of Flow Characteristics: Initial

Loc (m)	Mean Depth and Std (mm)	Min Depth	Max Depth	Δ (Max, Min)	Δ (Initial, Eroded)
		Initial Depth			
10	2.787 (0.189)	2.427	3.303	0.876	-
20	2.639 (0.168)	2.298	3.030	0.732	-
30	2.427 (0.168)	2.074	2.815	0.741	-
40	2.223 (0.128)	1.944	2.555	0.611	-
50	2.001 (0.133)	1.719	2.340	0.621	-
60	1.744 (0.102)	1.516	1.991	0.475	-
70	1.483 (0.079)	1.271	1.653	0.382	-
80	1.165 (0.064)	1.036	1.300	0.264	-
90	0.784 (0.037)	0.713	0.862	0.149	-
100	0.276 (0.006)	0.262	0.290	0.028	-
		Eroded Depth			
10	2.742 (0.238)	2.276	3.372	1.096	0.045
20	2.575 (0.292)	1.969	3.232	1.263	0.064
30	2.412 (0.198)	2.027	2.850	0.832	0.015
40	2.207 (0.121)	1.990	2.491	0.501	0.015
50	1.981 (0.084)	1.789	2.141	0.352	0.021
60	1.734 (0.056)	1.611	1.870	0.259	0.011
70	1.460 (0.049)	1.348	1.571	0.223	0.023
80	1.150 (0.040)	1.065	1.241	0.176	0.015
90	0.773 (0.030)	0.711	0.835	0.124	0.011
100	0.274 (0.067)	0.259	0.290	0.031	0.002

have generated and examined numerical solutions for a range of situations in which we varied (1) the transport law parameters, namely γ (discharge exponent) and δ (slope exponent); (2) rainfall rate α; and the land surface slope θ. The goal is to establish functional relationships between characteristics of the rills, such as their spacing, and these parameters. We note that this is also a goal of various analytic approaches to the problem [*Loewenherz-Lawrence*, 1994], although such analyses are typically tractable only when the equations have been linearized. This restriction to linearized equations in turn implies that the results of the analysis are typically valid only for the earliest stages of rill initiation.

In order to examine systematically the dependence of these patterns on the four parameters of interest, we employed the mean value and the variance of the emergent frequencies of water depth at given cross-sections as two "dependent" measures of the pattern of channelization.

Apart from the effects of the four independent variables on the emergent frequencies, there are two other effects that must be represented in the analysis in order to avoid biased results. First, it was noted that the mean rill frequency typically varied as a function of the distance upslope for many of the cases examined. In the case [1.0, 5.0, 1.0, 0.27], for example, the mean frequency at each of the cross-sections 10 m, 20 m, 30 m, 40 m, 50 m, 60 m, 70 m, 80 m, and 90 m after approximately 16,000 iterations was 7.7, 7.9, 8.1, 8.6, 9.4, 10.0, 10.5, 10.8, and 11.4 cycles per 100 m, respectively. Hence for each of the runs, we analyzed the mean frequencies observed at these nine cross-sections.

Second, it was noted that there were systematic changes in time in the mean value of the frequency at given cross-sections in a given run. Hence the analysis involved a measure of the "length" of time for which the surface in a given case eroded. The measure of time employed was the number of erosive iterations undergone by a given case and the basic measure of an iteration was the number of pairs of water-equilibrating steps and eroding steps in the numerical computation, with each transition determined by the criterion of the occurrence of 0.5 mm of erosion at some point in the surface. The number of iterations for the various runs ranged from 450 iterations to 16,300 iterations.

We investigated the dependence of our two dependent measures on these six parameters of interest with the use of multiple regression analysis. The values for five of the six independent variables employed in the analysis are shown in Table 4. Each row in the table represents one numerical experiment for a given set of parameter values. In relation to the parameters of the sediment transport law, the particular set of parameter values examined was chosen to span the range of most of the theoretical and empircial estimates of their values [*Julien and Simons*, 1985]. In particular, they span the values that have been used in a variety of applications, such as a model of channel geometry that was based on conservation principles and similar water and sediment transport laws [*Smith*, 1974].

Each of these runs was initiated on a surface that was perturbed by an independently generated set of surface disturbances. The mean and variance of the frequencies at the cross-sections for the initial, perturbed depth of flow were similar across all runs, and typical values for the mean and the variance of the number of cycles per 100 m were 20.0 and 70.0 respectively. We also show in Table 4 the averaged values of the mean frequencies that were observed to emerge after some number of iterations in the various experiments. In particular, the mean values of the frequencies that emerged at each of the cross-sections were

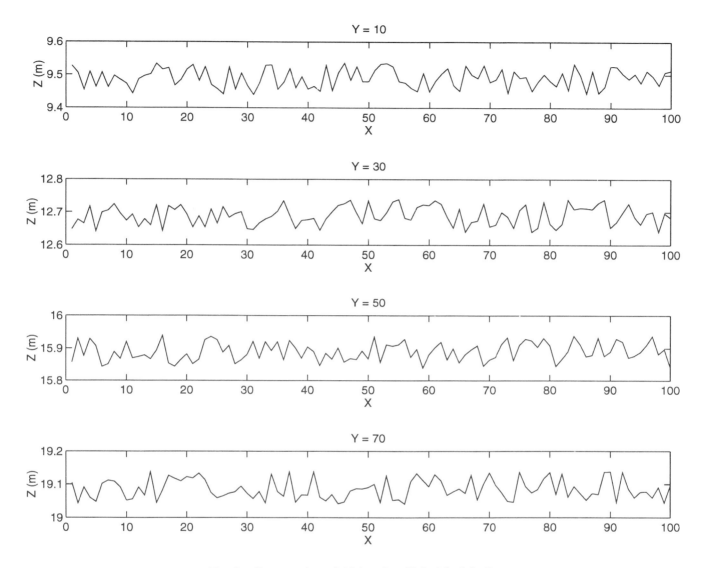

Fig. 3a. Cross-sections: initial surface [2.0, 4.0, 3.0, 3].

averaged over all cross-sections between 10 m and 90 m above the base of the slope for each run separately. These values range from 5.0 to 21.0 cycles per 100 m. Similarly, we show in Figure 4 the averaged variances at each of the cross-sections, ranging from 1.5 to 79.8. (We note that these averages over cross-sections were not the values used in the regression analysis).

We examined both raw variable values and logarithmic transformations of these dependent and independent variable values in the regression analysis. The regression equations employing the logarithmic transformations gave a better fit to the data than did the equations using untransformed values. We show in Table 5 the results of the regression analysis that employed the logarithmically transformed values of 162 observations (9 cross-sections for each of 18 runs). It is clear from the F-statistics that the regressions for both dependent variables are statistically highly significant. In particular, the analysis explains over 84 % of the variance in the values of the mean frequency and over 65 % of the variation in the variance of the frequency. Furthermore, five of the six in-dependent variables have statistically significant coefficients in both regressions. The following points summarize the results of the analysis:

1. the exponent of discharge per unit width is highly significant in both regressions, and is positively related both to the mean frequency of rill-like features and to the variance of the frequency. Hence larger values of γ

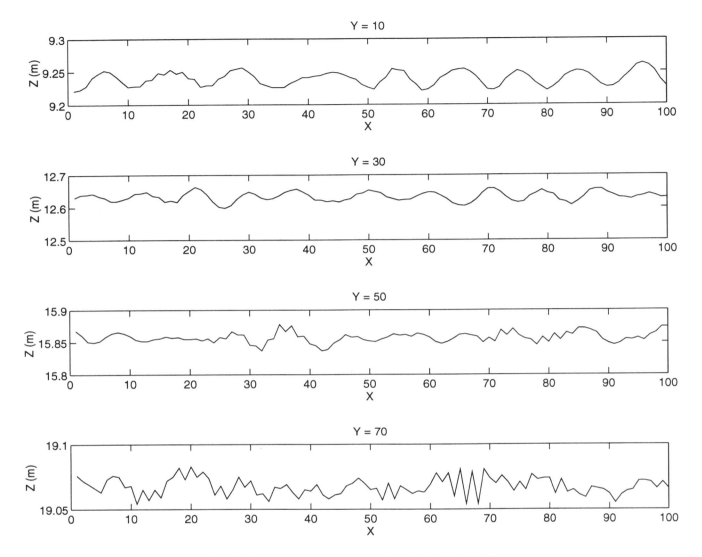

Fig. 3b. Cross-sections: eroded surface [2.0, 2.0, 3.0, 3].

result in a greater number of rill-like features and also a greater variability in this number.
2. The exponent of slope in the sediment transport law is significantly and positively related to the mean frequency of rills and negatively related to the variance of this number. Both effects, however, are relatively small in magnitude.
3. The rainfall rate is not related in a statistically meaningful manner to the mean frequency of rills but is negatively related to the variance in the number.
4. The effect of variations in surface slope is similar to the effect of variations in γ, and is highly significant. Hence increases in slope lead to an increased frequency of rill-like features and an increased variation in their numbers.
5. The measure of "eroding time" (number of computational iterations) is significantly related to the mean frequency of rill-like features, but is not significantly related to the variance of this measure. In particular, the mean value tends to decrease with the number of iterations.
6. The distance upslope is strongly related to both the mean frequency of rill-like features and to the variance in this number. Hence as one observes successive cross-sections upslope, the mean frequency and the variance in this frequency increases.

Hence we are justified in writing the following relationship between the mean rill frequency (M_w) and the five

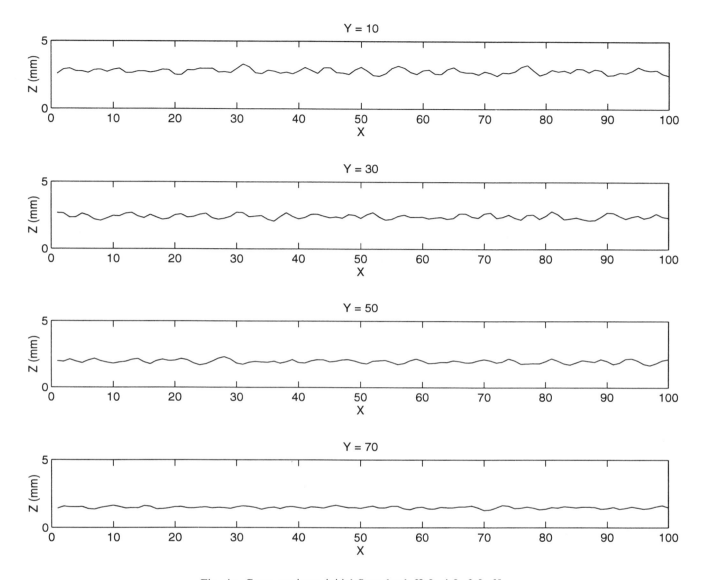

Fig. 4a. Cross-sections: initial flow depth [2.0, 4.0, 3.0, 3].

significant parameters:

$$M_w = 2.19 \left[\frac{\gamma^{0.63} \delta^{0.06} \theta^{0.46} D^{0.16}}{I^{0.07}}\right] \quad (21)$$

in which I is the number of iterations and D is the distance upslope. We are also justified in writing the following relationship between the variance in the rill frequency (V_w) and the five significant parameters:

$$V_w = 1.56 \left[\frac{\gamma^{1.37} \theta^{0.48} D^{0.59}}{\delta^{0.22} \alpha^{0.18}}\right] \quad (22)$$

Apart from demonstrating the ability of such a simple model to generate "channelized" surface flows, these empirical results give strong support to the hypothesis that the emergence of rill-like patterns and the characteristics of such patterns are dependent on the nature of the sediment transport law.

While the strong dependence of the mean rilling frequency on γ is particularly noteworthy, it is to be expected on the basis of earlier analyses that explained rilling instabilities in terms of the increased transporting capacity of converging flows of water for values of γ greater than unity [*Smith and Bretherton*, 1972]. In particular, as the exponent governing this non-linear effect increases, one would expect its ability to destabilize flows to become pronounced in effect. A point that requires

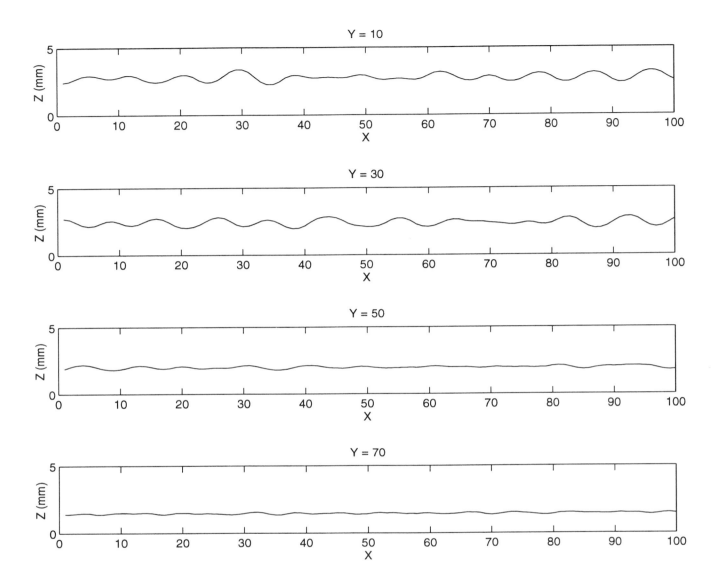

Fig. 4b. Cross-sections: eroded flow depth [2.0, 4.0, 3.0, 3].

explanation in the light of the current results, however, is a mechanism to explain the apparent emergence of rilling patterns for cases in which $\gamma = 1.0$. In analyses of equations (1)-(2), for example, [*Smith and Bretherton*, 1972; *Loewenherz*, 1991], this case was found to be neutrally stable and the reasons for its neutral stability are well-understood: the transporting capacity of water per unit of mass is unaffected by the convergence of flows of water and by the subsequent increases in discharge per unit width. The relation (21) indicates that even sub-linear transport laws are capable of leading to channelized flow patterns, although such predictions are outside its range of derivation. Apart from the indication that q_w-linear transport laws can lead to rilling, we must consider alternate possibilities that range from problems with our numerical scheme to difficulties with our data analysis. We note, for example, that the regression analysis employed in this case is a relatively blunt instrument and that it is possible that our analysis failed to disclose interesting non-linearities in the form of (21). Further investigation of this issue is clearly warranted.

The dependence of the means and variances of rill frequency on time (the number of computational iterations) and on the distance upslope may be given a rational explanation. First, Figure 4 shows that most of the emergent mean frequencies in the set of runs have values less

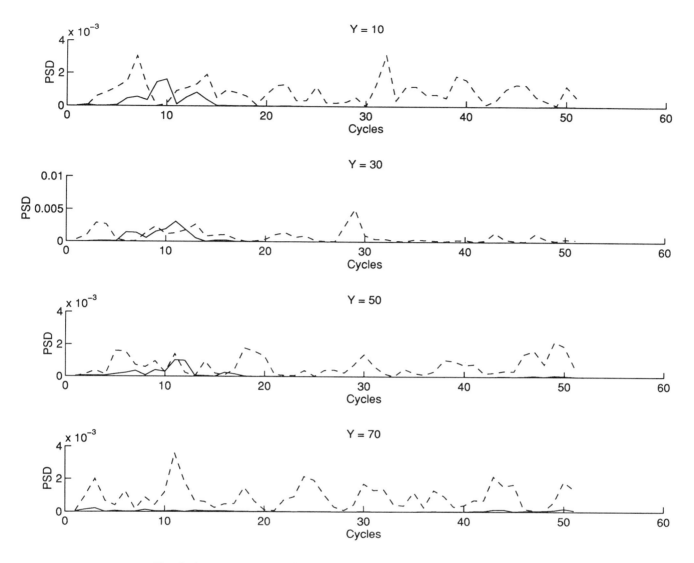

Fig. 5. Spectral power density plots: initial/eroded surface [2.0, 4.0, 3.0, 3].

than the random mean value of about 20 cycles per 100 m. Hence as erosion occurs over time, the mean value clearly must decline in these cases. Given the different "erosion times" of the various runs, one would indeed expect to observe a negative relation between the number of iterations and the value of the mean frequency. Similarly, the observed relation between the upslope distance and the mean and variance of the frequency may be explained in terms of the reduced ability of the flows at higher cross-sections to erode away the initial perturbed surface into a more structured form. Since the average discharge per unit width decreases monotonically as one moves up the surface, because of the lesser accumulation of rainfall, it should take longer for erosion to establish the stable, and generally lower, rill frequencies. Hence there will be an expected increase in the mean frequency as one moves upslope, at least in the early stages of rill emergence. An issue that is currently under examination, therefore, is to determine whether the emergent rill frequencies gradually "migrate" up the surface.

The lack of any significant relationship between the rate of rainfall and the mean frequency of rill-like features is also a little surprising at first sight, in view of the well-known empirical relationship between rainfall and drainage density. One may, however, view an increase in rainfall as essentially equivalent to some translation of the cross-section of observation to one that is further down the slope under conditions of lower rainfall rates. In particular,

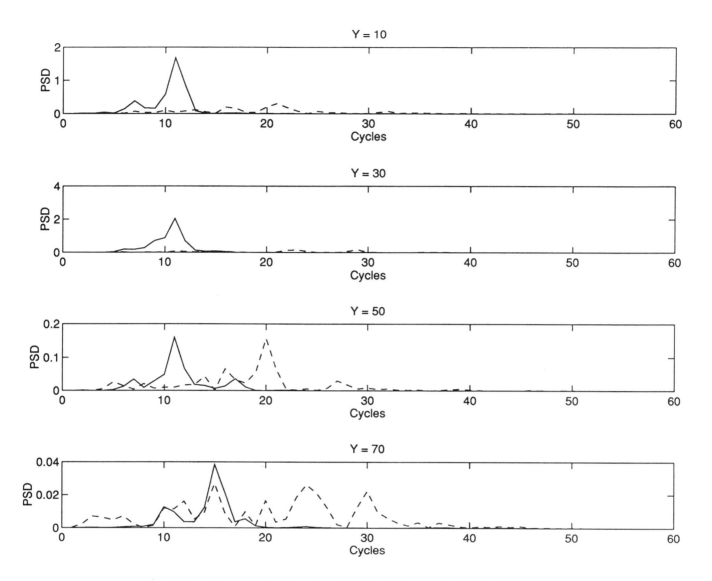

Fig. 6. Spectral power density plots: initial/eroded water depth [2.0, 4.0, 3.0, 3].

increased rainfall may be viewed as causing an increase in the discharge at a given cross-section. One may hence argue that changes to the rainfall rate do not affect the fundamental parameters determining rill frequency, such as those relating to the sediment transport law or the slope of the surface. Since, as argued above, it is a strong possibility that we observe a decrease in rill frequency as we go downslope because that is where rills first develop, then it is not surprising that the rainfall rate fails to influence the spacing of rills. Again, however, this is an issue that requires further investigation.

While the preceding results are both interesting and pleasing, it is clear that they must be regarded as preliminary indicators of the nature of the relationships between emergent rill frequencies and the nature of the sediment transport laws and environmental parameters. Further analysis and experimentation is required in order to both map out and understand the nature of such relationships.

5. SIGNIFICANCE OF THE RESULTS

We discuss the significance of the results, first in terms of the appropriateness of the models (10)-(11), then in terms of the validity of the numerical solutions, and finally in terms of the applicability of the results to geomorphic

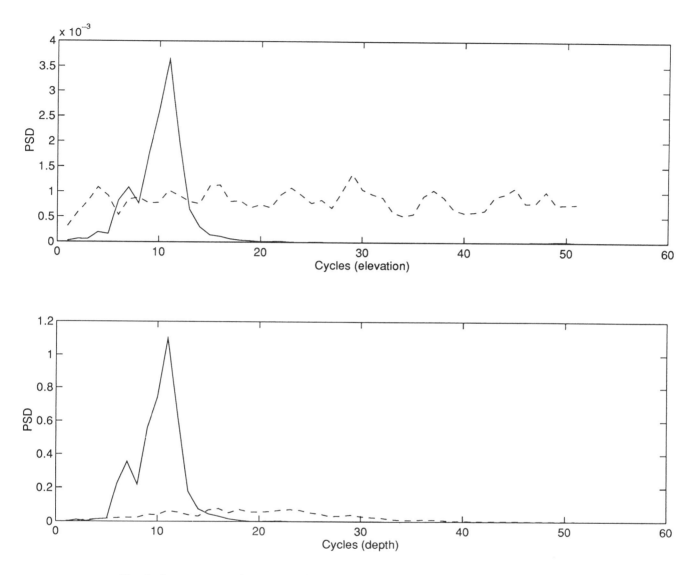

Fig. 7. Average spectral power density plots: initial/eroded water depth [2.0, 4.0, 3.0, 3].

phenomena.

5.1 *Appropriateness of the Models*

Clearly the model (10) - (11) is a great simplification of reality. The two equations intuitively appear to be well-matched in terms of their degree of applicability. It is to be noted, however, that equation (10) involves a low-slope approximation to the Navier-Stokes equations which is based on depth averaged flows and a two-dimensional approximation to three-dimensional flows. The manner in which the performance of the equation declines with increasing slope values is currently unclear.

It would be of value to compare the flow predictions of this equation with field results. The real issue, however, is whether the equations provide insight into real-world phenomena. The preceding numerical solutions seem to indicate an affirmative initial answer to the question. An adequate answer, however, will require a combination of further numerical studies as well as experimental and field studies.

5.2 *Validity of the Numerical Solutions*

An important concern with numerical solutions to non-linear partial differential equations is whether solutions

Table 3. Correlations between cross-section

Cross Section	Initial Surface	Eroded Surface	Initial Depth	Eroded Depth
At one meter separation				
10/11m	-0.08	0.94	0.55	0.99
20/21m	-0.10	0.96	0.49	0.99
30/31m	0.13	0.93	0.50	0.99
40/41m	-0.10	0.82	0.40	0.98
50/51m	0.03	0.64	0.50	0.97
60/61m	-0.03	0.21	0.51	0.96
70/71m	0.08	0.17	0.48	0.93
80/81m	-0.11	0.05	0.49	0.84
90/91m	0.04	-0.04	0.58	0.70
At ten meters separation				
10/29m	-0.08	-0.43	0.19	0.15
20/30m	-0.08	0.21	-0.01	0.34
30/40m	-0.06	0.34	0.18	0.32
40/50m	0.11	0.13	0.07	0.37
50/60m	0.04	0.25	-0.18	0.34
60/70m	-0.06	-0.01	0.04	0.16
70/80m	-0.06	0.02	0.09	0.10
80/90m	-0.09	0.05	-0.05	0.02
90/100m	-0.00	-0.08	0.21	-0.21

to the approximating difference schema are also approximate solutions to the differential equations. This is particularly the case when one is examining the growth of instabilities. The time and space steps employed in finding numerical solutions to (10) were chosen conservatively after tests over ranges of such values. The difference scheme was tested successfully against well-established solutions for 1-dimensional cases of (10). The criteria for terminating iterations of both (10) and (11) were also carefully investigated, and conservative values were adopted for these criteria.

Other concerns relate to the influence of the boundary conditions, and whether they have a significant impact on the resulting patterns of flow. While unwanted effects from the top boundary conditions are not an issue, the periodic boundary conditions at the edges of the surface may lead to effects that do not exist on indefinitely long ridge-like surfaces. Further investigation is required to resolve this issue and, in particular, the width of a ridge that is necessary so that unwanted effects are negligible over much of the surface. The precise effects of the absorbing boundary at the base of the surface also require investigation. It would appear safe to assume, however, that on sufficiently long surfaces, the upslope propagation of any boundary-dependent effects would diminish rapidly in effect.

In order to assure the validity of the results reported above, it would be of great value both to have analytical results concerning the stability characteristics of equations (10) - (11) and to compare them with independently obtained numerical solutions over a range of surfaces and boundary conditions.

5.3 The Results and Their Geomorphic Significance

The main results of the numerical analysis of equations (10) - (11) that have a bearing on geomorphological issues may be summarized as follows.

1. The introduction of a free water surface, simple Manning-type flow resistance equations, and the assumption that water flows down the water surface gradient into models of the form (1)-(2) appears sufficient to model the emergence of stable patterns of channelized flow. In particular, it does not appear necessary in modeling such patterns to introduce other stabilizing mechanisms, such as gravity-induced sediment transport down the gradient of the land surface.
2. The cases of equations (10) - (11) that were examined indicate that well-structured, rill-like flow patterns emerge on an initially randomized planar and erodible surface after a relatively small amount of erosion has occurred. The patterns appear to emerge first in the lower parts of the surface where the flow of water is deeper and more erosive.
3. The emergent patterns appear to be the result of both deposition and erosion of surface materials.
4. The dominant initial patterns of rill-spacing on the surfaces are well-characterized by a relationship of the form

$$M_w = 2.19 \left[\frac{\gamma^{0.63} \delta^{0.06} \theta^{0.46} D^{0.16}}{I^{0.07}} \right]$$

The main geomorphic significance of the results is that simple conservation principles and sediment and water transport laws appear sufficient to explain the emergence of channelized drainage. In particular it is not necessary to resort to variational principles or statistical arguments. Another result of geomorphic significance is the manner in which the emergent drainage patterns are related to the underlying sediment transport law. It is well-established in

Table 4. Parameter values employed in numerical experiments

Experiment	γ-value (discharge)	δ-value (slope)	rainfall rate	slope	number of iterations	average mean of frequency	average variance of frequency
1	1.0	1.0	1.0	0.27	82	7.58	5.27
2	1.0	1.0	5.0	0.05	42	5.98	7.29
3	1.0	5.0	1.0	0.05	16	6.24	2.59
4	1.0	5.0	1.0	0.27	326	8.97	13.80
5	1.0	5.0	5.0	0.05	10	6.53	3.79
6	1.0	5.0	5.0	0.27	27	17.67	37.28
7	2.0	2.0	3.0	0.16	15	11.82	12.30
8	2.0	3.0	2.0	0.16	25	11.27	11.69
9	2.0	3.0	3.0	0.16	23	11.32	7.02
10	2.0	3.0	3.0	0.21	30	11.24	8.00
11	2.0	3.0	4.0	0.16	22	10.68	8.38
12	2.0	4.0	3.0	0.16	32	10.25	5.83
13	3.0	1.0	1.0	0.05	16	16.06	74.32
14	3.0	1.0	1.0	0.27	38	19.72	74.78
15	3.0	1.0	5.0	0.05	34	9.94	14.24
16	3.0	1.0	5.0	0.27	9	21.02	71.5
17	3.0	5.0	1.0	0.27	13	20.33	69.66
18	3.0	5.0	5.0	0.27	39	16.92	23.20

the geomorphic literature that drainage density is dependent on a variety of environmental parameters, such as rainfall, lithology, and slope. The current results are not only in accordance with these observations, as in the case of rainfall and slope, but also indicate the manner in which the nature of the sediment transport law is a fundamental determinant of drainage densities through equation (21).

Since it is likely that the entrainment and transport of different materials are appropriately described by different forms of equation (4), it follows that they will be associated with differing patterns of surface drainage. Furthermore, it seems plausible in the light of the current results that variations in the parameters of the flow resistance law (8) should have similar effects. It is well-known [*Dingman,* 1984; *Julian and Simons,* 1985] that different flow regimes are approximately described by different exponents in (8). Hence it appears likely that different regimes will have different patterns of channelization. This is a matter that can easily be investigated with generalizations of (10) - (11) in which the Manning exponents are permitted to vary away from the usual values of 2/3 and 1/2.

An immediate and interesting corollary of these observations is that emergent patterns in real landscapes may be heterogeneous in terms of the fundamental patterns of channelization in response to both the heterogeneity of flow regimes and sediment transport regimes. In other words, a given landscape may embody different dominant channel frequencies at different locations, depending on the nature of the water and sediment transport conditions in effect at those locations.

We note that a similar effect is to be expected from the strong relationship between the dominate frequency of rill-like features and the slope of the surface. In particular, (21) indicates that the spacing between such features becomes greater as the slope of the land surface becomes smaller. It is well-established that, in river systems of any extent, typical gradients of channels and side-slopes tend to decrease as one moves from the upper fringes of a watershed to the lower areas associated with higher-order channels. This is clearly another source of heterogeneity that affects the spacing of channel-like features in geomorphic systems. In particular, if equation (21) is applicable at all scales of channel-like phenomena in a drainage basin, one would expect to find a greater spacing between channels in areas of lower surface gradients and a lesser spacing between channels on areas of higher surface gradient. This dependence of channel frequency on slope may even be a mechanism of some importance in explaining the observed geometry of channel networks in drainage basins. More generally, the effects on channel spacing that arise from spatial heterogeneity in dominate surface slope and in appropriate characterizations of both

Table 5. Results of Regression Analysis

Mean of frequency at cross-sections (cycles/100m)

Coefficients	Value	Std. Error	t value	Prob(t)
Intercept	2.19	0.15	15.10	0.0000
Discharge exponent	0.63	0.04	16.92	0.0000
Slope exponent	0.06	0.02	2.55	0.0118
Rainfall rate	-0.00	0.02	-0.19	0.8531
Surface slope	0.25	0.02	10.90	0.0000
Number iterations	-0.07	0.02	-3.14	0.0020
Distance upslope	0.16	0.02	8.01	0.0000

Residual standard error: 0.1746 on 155 degrees of freedom
Multiple R-Squared: 0.84
F-statistic: 131.7 on (6,155) df; p-value = 0

Variance of frequency at cross-sections (cycles/100m)

Coefficients	Value	Std. Error	t value	Prob(t)
Intercept	1.56	0.57	2.75	0.0067
Discharge exponent	1.37	0.15	9.43	0.0000
Slope exponent	-0.22	0.09	-2.48	0.0141
Rainfall rate	-0.18	0.08	-2.16	0.0324
Surface slope	0.48	0.09	5.28	0.0000
Number iterations	-0.11	0.08	-1.32	0.1894
Distance upslope	0.59	0.08	7.45	0.0000

Residual standard error: 0.68 on 155 df
Multiple R-Squared: 0.65
F-statistic: 48.44 on (6,155) df: p-value = 0

water motion and sediment transport as a result of (21) are an exciting problem for investigation.

5.4. *Future Research and the Applications of the Family of Models*

Apart from the technical concerns about the solutions to (10) - (11) that were discussed briefly above, there are many other issues concerning equations (10) - (11) and their solutions that require further investigation. For example, solutions need to be obtained for larger surfaces, for more extended periods of time, and for a larger array of sediment transport laws. The manner in which rill-like patterns evolve overtime, and whether the same rilling patterns first emerge on lower segments of a surface, also requires further investigation, as does the issue of how the nature of the initial surface determines the generation of channelized patterns. The preceding analysis, for example, focussed on relatively small but finite perturbations. It would be of value to investigate whether a large range of perturbations leads to similar patterns. It is also important to attempt an integration of numerical and analytical results concerning these equations.

As noted above, the family of models that we have introduced appear to be suitable for answering a number of issues in geomorphology. In particular, it is suggested that suitable generalizations of these models may be employed as an "experimental" facility for examining a broad range of geomorphic issues. It appears relatively straightforward, for example, to use numerical solutions to such equations in investigating characterizations of channelized flows in terms of variational principles. From such numerical solutions to (10) - (11), one may easily compute integrals of various functions of slope and velocity over drainage surfaces. By examining the behavior over time of integrals of a variety of candidate functions that may be minimized/maximized during the evolution of channelized flow systems, it may prove possible to identify specific variational principles that govern drainage basin evolution.

6. SUMMARY AND CONCLUSIONS

In summary, numerical solutions of equations of the form

$$-\alpha = \nabla \cdot \frac{\nabla(z + h)}{|\nabla(z + h)|} nh^{5/3}|\nabla(z + h)|^{1/2} \qquad (10)$$

$$\frac{\partial z}{\partial t} = \nabla \cdot \frac{\nabla(z + h)}{|\nabla(z + h)|} kq_w^\gamma |\nabla(z + h)|^\delta \qquad (11)$$

indicate that such equations provide a simple model of the emergence of channelized drainage patterns. In particular

1. the introduction of a free water surface, simple Manning-type flow resistance equations, and the assumption that water flows down the water surface gradient into models of the form (1) - (2) appears sufficient to model the emergence of stable patterns of channelized flow. In particular, it does not appear necessary in modeling such patterns to introduce other stabilizing mechanisms, such as gravity-induced sediment transport down the gradient of the land surface.
2. The cases of equations (10) - (11) that were examined indicate that well-structured, rill-like flow patterns emerge on an initially randomized planar and erodible surface after a relatively small amount of erosion has occurred. The patterns appear to emerge first in the

lower parts of the surface where the flow of water is deeper and more erosive.

3. The patterns that emerge arise as a result of both deposition and erosion of the surface materials.
4. The dominant initial patterns of rill-spacing on the surfaces are well-characterized by a relationship of the form

$$M_w = 2.19 \left[\frac{\gamma^{0.63} \delta^{0.06} \theta^{0.46} D^{0.16}}{I^{0.07}} \right] \quad (21)$$

5. The results have clear geomorphic significance for explaining both channelized drainage patterns and for variations in such patterns.

It is suggested that extensive investigations of generalizations of these equations and, in particular, those having the form (6) - (7) will be of great value in explaining a variety of geomorphic phenomena.

Acknowledgements. We wish to thank David Siegel and Leal Mertes for many helpful discussions, and David Furbish for a thoughtful review and helpful comments.

REFERENCES

Dingman, S. L., *Fluvial Hydrology,* W. H. Freeman and Company, New York, 1984.

Fennema, R. J. and M. H. Chaudhry, Explicit methods for 2-D transient free-surface flows, J. *of Hydraulic Eng., 116,* 1013-1034, 1990.

Julien, P. Y. and D. B. Simons, Sediment Transport Capacity of Overland Flow, *Am. Soc. for Agricultural Eng., 28,* 755-762, 1985.

Kirkby, M. J., A 2-Dimensional Simulation Model for Slope and Stream Evolution, in *Hillslope Processes,* edited by A. D. Abrahams, 203-222, Allen and Unwin, Winchester, MA, 1986.

Kramer, S. and M. Marder, Evolution of River Networks, *Physical Rev. Letters, 68,* 205-208, 1992.

Loewenherz, D. S., Stability and the Initiation of Channelized Surface Drainage: a Reassessment of the Short Wavelength Limit, *J. of Geophys. Res., 96,* 8453-8464, 1991.

Loewenherz-Lawrence, D. S., Hydrodynamic Description for Advective Sediment Transport Processes and Rill Initiation, *Water Resour. Res., 30,* 3203-3212, 1994.

Smith, T. R., A Derivation of the Hydraulic Geometry of Steady-State Channels from Conservation Principles and Sediment Transport Laws, *J. of Geol., 82,* 98-104, 1974.

Smith, T. R., Set-Determined Process and the Growth of Spatial Structure, *Geog. Analysis 8,* 354-375, 1976.

Smith, T. R. and F. P. Bretherton, Stability and the Conservation of Mass in Drainage-Basin Evolution, *Water Resour. Res., 8,* 1506-1529, 1972.

Todini, E. and Venutelli, M., Overland Flow: a Two-dimensional Modeling Approach, in *Recent Advances in the Modeling of Hydrologic Systems,* edited by Bowles, D. S. and P. E. O'Connell, Kluwer Academic Publishers, Netherlands, 153-166, 1991.

Willgoose, G., Bras, R. L., and I. Rodriguez-Iturbe, Results from a New Model of River Basin Evolution, *Earth Surf. Proc. and Landforms, 16,* 237-254, 1991.

Zhang, W. and T. W. Cundy, Modeling of Two-Dimensional Overland Flow, *Water Resour. Res., 25,* 2019-2035, 1989.

Terence R. Smith and George E. Merchant, Department of Computer Science and Department of Geography, University of California, Santa Barbara, CA 93106.

Microtopography of Hillslopes and Initiation of Channels by Horton Overland Flow

Thomas Dunne, Kelin X Whipple[1] and Brian F. Aubry[2]

Department of Geological Sciences, University of Washington, Seattle, Washington

On long hillsides one can examine systematic downslope changes in the ground surface as Horton overland flow gathers into depressions and eventually incises the surface to form channels. Microtopography plays an important role in this process. We have sampled the microtopography of two long hillslopes in a savanna region of southern Kenya, and defined the spectral characteristics of its roughness at various distances from the drainage divide. The microtopography is fractal with a dimension that decreases systematically downslope, and the overall roughness varies between and along hillslopes in response to: (i) the weathering characteristics of the underlying bedrock; (ii) the type and density of patchy vegetation, and (iii) the tendency for wash to incise the surface with increasing intensity as runoff discharge increases downslope. The downslope decrease in fractal dimension reflects the progressive development of low-frequency roughness, here referred to as "swaley" microtopography, which the wash develops even far upslope of the channel head. The statistical analysis of surface roughness motivates a discussion of the role of microtopography in the interaction between wash and diffusive sediment transport processes that ultimately determines the critical distance from the divide at which channels begin.

1. BACKGROUND

The formation of channels by runoff is a crucial process in fluvial geomorphology [*Dietrich and Dunne*, 1993], and yet it is little understood despite being a focus of concern in theoretical geomorphology for 50 years. The first formal theory of the mechanics of channel incision resulting from hillslope runoff, proposed by *Horton* [1945], revolutionized geomorphological analysis. It led to analyses of: climatically induced influences of erosion mechanics on topography [e.g. *Melton*, 1958]; lithological influences on topography [*Schumm*, 1956a]; and changes in the density of channel incision as a result of changing environmental conditions [e.g. *Strahler*, 1956].

[1]Now at St. Anthony Falls Hydraulic Laboratory, University of Minnesota, Minneapolis, MN 55414
[2]Now at Dames and Moore, Inc., 221 Main St., Suite 600, San Francisco CA 94105-1917.

Natural and Anthropogenic Influences in Fluvial Geomorphology
Geophysical Monograph 89
Copyright 1995 by the American Geophysical Union

In Horton's theory of channel incision, an irregular, diffuse sheet of overland flow incises the hillslope surface and separates into distinct rivulets wherever it exerts a shear stress greater than the shear resistance of the soil. In a rainstorm of a fixed intensity the critical distance from the drainage divide beyond which such incision occurs will vary directly with soil resistance and inversely with gradient, hydraulic roughness, and infiltration capacity. Slightly deeper parts of the sheet flow will exert a greater shear stress than the surrounding flow and therefore will become even deeper, capturing still more flow through cross-grading of the hillslope surface and micropiracy of flow from smaller channels, until most of the water drains into a few, relatively deep channels. Thus, wrote *Horton* [1945, p. 332], "Sheet erosion implies the formation of either a rilled or gullied surface", and he equated the critical distance at which sheetwash becomes erosive with the distance downslope at which channel incision begins.

This threshold view of channel incision proved attractive in soil conservation, where the goal became manipulation of the length, gradient, infiltration capacity, and shear resistance of fields to keep land surfaces within the range of conditions that discouraged rapid sheetwash erosion and channel formation. For example, *Schaeffer et al.* [1979] used a Hortonian balance of force and resistance to calculate the area of a "zero-order basin" required to scour a first-order channel head during a rainstorm with a 2-year recurrence interval in landscapes weakened by mining. Such a force balance is a useful approximation where the rate of sediment transport is negligible below some value of shear stress and increases strongly above it. The approximation is a pragmatic solution in the short-term view of design calculations if an "effective" soil resistance and a "dominant" or design storm are included, and for the convenience of stabilizing landscape models in which the emphasis is on other aspects of the interaction between hydrology and geomorphology [e.g. *Escobar and Rodriguez-Iturbe,* 1982; *Cordova et al.,* 1983].

However, field observations during rainstorms indicate that Horton's fundamental assumption equating the onset of channelization with the threshold of erosion is widely violated. During plot experiments one can observe that, even in the absence of raindrops and on low gradients, the critical shear stress required for eroding soil is exerted by flow depths of a few millimeters on even low gradients, confirming, for example, a Shields calculation for cohesionless sediment. Such conditions develop within several meters of the drainage divide (upper end of the plot) during rainstorms of only a few centimeters per hour [*Dunne and Aubry,* 1986]. On the same hillslopes, the heads of small channels (in the sense of concentrations of flow and sediment transport between definable banks that can be identified between rainstorms) [*Dietrich and Dunne,* 1993, p. 178] occur approximately 450-900 m from the divide. During natural rainstorms on irregular surfaces, both raindrop impact [*Moss et al.,* 1979] and flow convergence around microtopography [*Zhang,* 1990; *Zhang and Dunne,* 1991] mobilize soil at average shear stresses even lower than would be predicted from a laboratory measurement of the critical shear stress required for erosion by a sheet flow of uniform depth. Channelization should start even closer to the divide than predicted by Horton according to his threshold paradigm.

Thus, a simple force balance between one-dimensional shear stress and soil resistance does not govern the onset of channelization. *Montgomery and Dietrich* [1994] suggested a variant on this balance wherein an erosion threshold based on a critical soil resistance causes channel erosion to extend upstream to a limiting combination of drainage area and local gradient after which diffusive sediment transport processes partially fill the resulting depression. However, the mechanics of the threshold phenomenon remain unexplored, and it is quite clear from field observations and measurements of erosion rates that on some long hillslopes distributed erosion, commonly referred to as "sheet erosion" or "sheetwash", is widespread hundreds of meters upslope of distinct channels.

On many hillslopes the onset of channelization appears to be controlled by a competition between diffusive processes (rainsplash, soil creep, bioturbation) which act to fill depressions and advective processes (wash, channel flow) which tend to incise the surface [*Smith and Bretherton,* 1972; *Dunne,* 1980; *Dunne and Aubry,* 1986; *Loewenherz,* 1991]. *Smith and Bretherton* [1972] formulated the condition required for channelization of a steady-state (constant form) hillslope surface in terms of the continuity equation for sediment transport. Their generalized sediment transport formula allowed for both diffusive and advective processes, and therefore encapsulated the basis for evaluating the competition between these processes, as clearly explained by *Loewenherz* [1991]. They performed a stability analysis of the steady-state surface with respect to an infinitesimal microtopographic perturbation, and found that such perturbations (interpreted as channels) should grow on the concave parts of the profile where transport is dominated by the effect of wash [*Kirkby,* 1971], the intensity of which depends on local gradient and local water discharge raised to a power greater than 1.0.

Dunne [1980] elaborated the field interactions implied by the Smith and Bretherton mathematical analysis, emphasizing that under conditions of Horton overland flow the onset of channelization is driven downslope by diffusive rainsplash. This paper was interpreted inaccurately by *Loewenherz* [1991; *Loewenherz-Lawrence,* 1994] as an analysis of the wavelength selection, or stable channel spacing, a problem that *Smith and Bretherton* [1972] had raised without solution, and to which Loewenherz proposed an important partial solution. The interactions between rainsplash and wash were further elaborated by *Dietrich and Dunne* [1993, p. 195-198], and the result of altering the relative efficacy of wash and splash on the position and density of channels was illustrated in laboratory experiments by *Moss et al.* [1979], field experiments by *Dunne and Aubry* [1986], and land-surface changes resulting from the eruption of Mt. St Helens, Washington in 1980 [*Collins and Dunne,* 1986, p. 899-903].

It is an implication of these papers that the distinction made by *Dietrich and Dunne* [1993, p. 179] and by *Montgomery and Dietrich* [1994, p. 222] between rills and channels becomes irrelevant after sufficient time. These authors defined rills as ephemeral straight troughs, incised into a smooth, planar hillslope, which are effaced before they have evacuated enough sediment to increase the planform concavity of a hillslope and form a new valley.

They may, for example, be obliterated by frost action, as documented by *Schumm* [1956b]. However, if such a rill survives indefinitely, competing successfully for space and water against neighboring rills, it will become a valley-forming channel with distinct banks and properties scaled by its drainage area. We include these survivor rills in our definition of channels and explore the competition between diffusive and advective processes, their mutual interaction with the evolving microtopography, and the controls on the onset of channelization.

2. PURPOSE

This paper examines the microtopography of soil-mantled hillslope surfaces upslope of channel heads on two long hillsides eroded by Horton overland flow. The lengths of the hillslopes (700 m and 1450 m) and their unchannelled portions (approximately 450 m and 900 m) allow us to examine gradual changes that are not usually apparent around the heads of channels on short, intricately dissected hillslopes. Changes in microtopography along the hillslope profile are interpreted in light of the ideas, outlined above, about the way in which flow gathers to form channel heads and advective wash and diffusive rainsplash interact to set the critical distance at which this concentration occurs. We focus on the role of finite-scale effects which ultimately control the inception of channels and the long-term evolution of hillslope profiles. Such finite-scale effects are beyond the scope of perturbation analyses and may explain the discrepancy between theoretical predictions and observed lengths of unchannelized hillslopes.

A secondary purpose of the paper is to emphasize that microtopography exerts a crucial influence on the hydrology and hydraulics of Horton overland flow and on the resulting sediment transport. Microtopography has received little study by hydrologists and geomorphologists. For example, *Dunne et al.* [1991] demonstrated that the presence of microtopography with greater permeability on root-impregnated mounds than in depressions with less-dense root masses should force a downslope increase in spatially averaged infiltration capacity, even when all other influences such as soil texture, vegetation density, and antecedent wetness are constant along the hillslope profile. A consequence is that steady-state runoff increases with downslope distance at a rate that is less than linear. This result influences profile shape when it is inserted into the current hillslope evolution paradigm [*Hirano*, 1966; *Kirkby*, 1971; *Smith and Bretherton*, 1972], which assumes a linear downslope increase in specific water flux.

Emmett [1970], *Dunne and Dietrich* [1980], *Abrahams et al.* [1989] have also emphasized that microtopography causes cross-slope variations in flow depth and velocity. *Zhang and Dunne* [1991] used a numerical model of two-dimensional flow around microtopography to show that analysis of these cross-slope depth variations is required in order to parameterize local shear stresses in sediment transport equations. *Abrahams et al.* [1991] incorporated such effects into a soil erosion model of their plot measurements. For these reasons there is value in characterizing microtopography and its variation over a range of downslope distance, gradient, lithology, vegetation density and other controlling variables [*Mitchell and Jones*, 1978; *Preston*, 1966; *Stone and Dugundji*, 1965]. In addition, we emphasize that analysis of microtopography, such as our exploration of its spectral properties on scales of 0.2 - 30 m, can yield important insights into the physical processes responsible for the degradation of the surface and the simultaneous generation of the microtopography.

3. DESCRIPTION OF FIELD SITES

The field sites lie on the south side of a ridge of Precambrian Basement rocks, draining to the dry bed of Pleistocene Lake Amboseli on the northern boundary of Amboseli National Park, southern Kenya [*Western and Dunne*, 1979, Fig. 1]. One hillslope (Jonah's Basement) is underlain by schist and the other (Amphibolite Hill) by amphibolite gneiss. The rocks have weathered to poorly aggregated, kaolinitic, sandy clay loam, that form ferric luvisols ranging in depth from 0-1.0 m depending on local gradient variations between 0.01 and 0.22 [*Dunne and Aubry*, 1986, p. 38]. Where the soils are thin, gravel and cobbles from quartz veins or particularly coarse-grained parts of the amphibolite, emerge at the surface and spread downslope as thin aprons [*Western and Dunne*, 1979, Fig. 3]. More generally, the soil surface is covered by a discontinuous, poorly sorted lag of sand with a median grain size in the range 0.35-0.75 mm on Jonah's Basement and 0.4-1.8 mm on Amphibolite Hill [Figure 1]. Gravel occurs on the surface over the upper several hundred meters of both profiles, but is much more common on Amphibolite Hill. Detailed examination of the hillslope surfaces, especially during artificial rainstorms on experimental plots [*Dunne and Dietrich*, 1980; *Dunne and Aubry*, 1986; *Dunne at al.*, 1991] or immediately after rainstorms (Figure 2) reveals microtopography with an amplitude of up to 0.25 m. The downslope sequence of photographs in Figure 2 qualitatively illustrates progressive changes in the microtopography and the degree of flow concentration with distance from the divide.

Like many soil-mantled hillslopes in semi-arid, cratonal

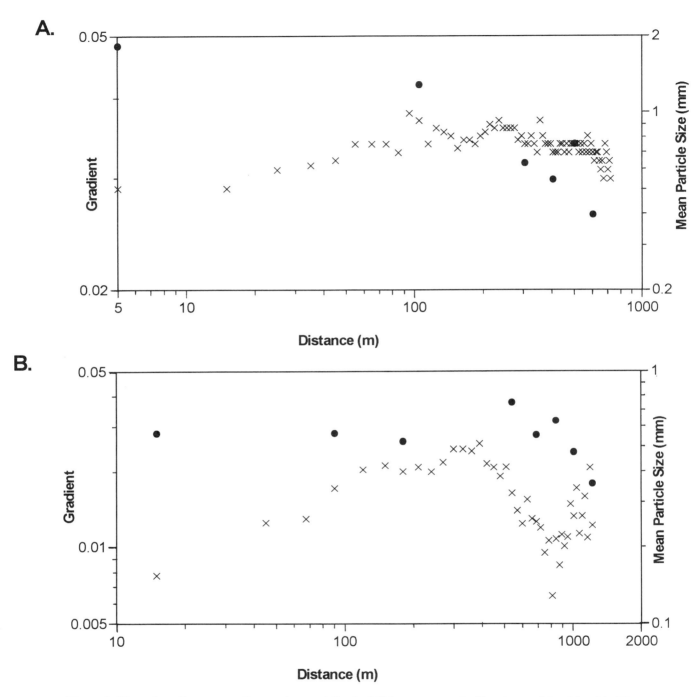

Figure 1. Plots of gradient versus distance (crosses) for the hillslopes surveyed with a tape and hand level in 10 m increments (Amphibolite Hill, A) and 30 m increments (Jonah's Basement, B). Circles represent the median grain size of sediment scooped from the soil surface.

environments, the hillslope profiles are long and overall gradients are low with only slight convexity and concavity. Figure 1 shows that Jonah's Basement has an upper convexity that begins with a gradient of 0.008 over the upper 30 m of the profile and extends to 400 m where the gradient is 0.026. Thereafter, the gradient decreases to 0.006 at 800 m downslope of which another convexity develops. The total relief is 20 m. It will be shown later

Figure 2. Photographs of microtopography highlighted by overland flow at different distances from the divide on Jonah's Basement. The photographs were taken within minutes of the end of a rainstorm that lasted approximately 20 minutes, and their locations were not recorded. However, they are shown in order of increasing distance from the divide. Photograph A was located approximately 200-300 m from the divide, and D was taken approximately 1100-1200 m from the divide. The linear feature crossing C is a path.

that channel heads occur between 780 m and 945 m on this profile. On Amphibolite Hill, the convexity begins with a gradient of 0.029 over the upper 10 m of the profile and increases to 0.038 at 100 m before decreasing to 0.030 at 725 m for a total relief of 25 m. On this profile the channel heads occur between 400 m and 500 m.

The 300 mm average annual rainfall of the area is erratic but usually distributed in two seasons (April - June and

November - December) in the form of short, intense rainstorms. The vegetation cover of grass and shrubs varies in density from 10 to 80 percent depending on the seasonal variation of rainfall and intense grazing pressure from wild herbivores and the domestic stock of nomads.

4. FIELD MEASUREMENTS

Microtopography was surveyed along 100 m-long transects normal to the visually estimated fall line at downslope intervals of 100 m (Amphibolite Hill) and

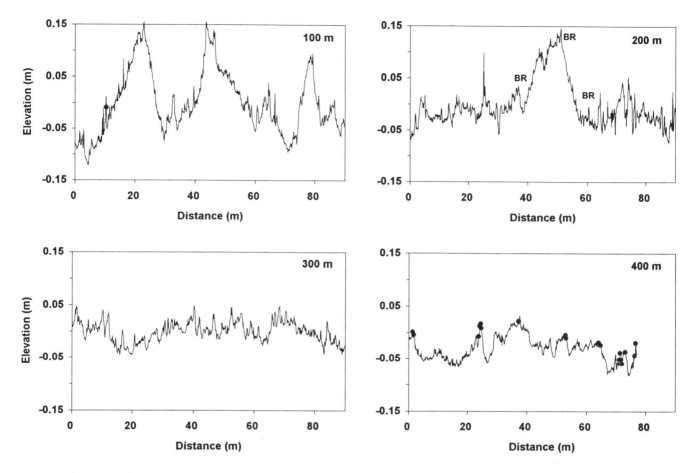

Figure 3. Microtopographic transects from Amphibolite Hill. Dots indicate the locations of bushes. BR denotes outcropping bedrock. Channels (C) begin between 400 and 500 m from the divide. The profile at 400 m was shortened because a termite mound was excised from it. The profile at 600 m was curtailed by a surveying error.

approximately 150 m (Jonah's Basement). A tape and self-leveling engineering level were used, and because of the hard soil surface, the short distance of sighting, and the constant instrument position elevations could be reproduced to within 1 mm by different operators interpolating the rod readings. The sampling interval was 0.1 m on Amphibolite Hill and 0.3 m on Jonah's Basement. The mean elevation for each profile was subtracted from each measured elevation and the results are presented in Figures 3 and 4.

During the surveys, it was easy to recognize a channel as a zone of concentrated sediment transport between definite banks [*Dietrich and Dunne*, 1993], quite distinct from the evidence of some concentrated wash marks in other depressions upslope of and between the channels. When the microtopographic profiles were plotted, these incisions coincided in almost every case with depressions that were distinctly lower than surrounding depressions. Channel depths averaged 0.06-0.07 m on Amphibolite Hill and 0.05 m on Jonah's Basement. The occurrence of bushes and the extent of bedrock outcrops were also recorded..

5. DATA ANALYSIS

The upper two transects of microtopography on Amphibolite Hill (Figure 3) include one or more protuberances with amplitudes of approximately 0.2 m standing distinctly above the rest of the profile. A bedrock outcrop was identified at one of them; the others are in a zone of thin rubbly colluvium upslope of the outcrop, and we interpret them as results of bedrock heterogeneities influencing regolith production and texture. Away from the bedrock outcrop and from a few termite mounds, and over most of the profiles at 200, 300, and 400 m, amplitudes are lower. Bushes occur on some mounds. There is a general downslope increase in the size of the depressions,

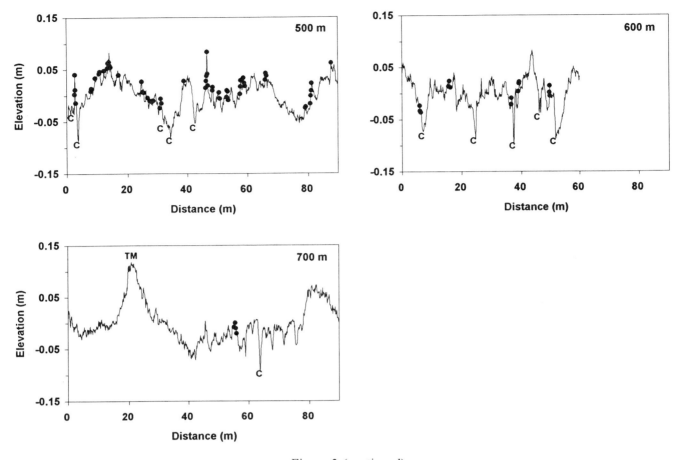

Figure 3 (continued)

and between 400m and 500 m channels begin within these depressions. The same trend occurs on Jonah's Basement (Figure 4) without the complication of the large-scale bedrock disruptions, there is a progressive downslope development of broad swales up to tens of meters wide in which channels eventually occur.

The root-mean-square (rms) deviations of these profiles yields a rough measure of their roughness. Each sample of elevations was approximately normally distributed, and on Amphibolite Hill the rms deviation decreased downslope from 0.061 m at the divide to 0.019 m at 300 m over the bedrock-affected portion of the hillslope before increasing monotonically downslope to 0.038 m. On Jonah's Basement, there was a more general increase from 0.015 m at the divide to 0.041 m at the foot of the slope, but again with a minimum of 0.012 m at 680 m. In both cases there is a roughening in the upper several hundred meters that reflects the influence of gravely colluvium and of bedrock at or near the surface. Downslope of this influence, the finer-grained soil surface attains its smoothest condition 100 m or more upslope of the onset of channel formation. Further downslope, the surface becomes progressively rougher again.

Since the rms elevation provides only an index of overall roughness without regard to scale, we attempted to quantify microtopographic roughness at various scales using spectral analysis methods described by *Press et al.* [1986]. For each profile of digital relief data we computed an estimate of the power spectral density function (SDF) using *Welch's* [1967] technique for ensemble averaging modified periodograms of overlapping sections of each profile via the fast Fourier transform (FFT). First, a crude estimate of the power spectral density function is obtained via direct application of the FFT to the entire data set:

$$x'_t = x_t * W_t \qquad (1)$$

$$FT\{x'_t\} = X'(\omega) \qquad (2)$$

$$\Phi(\omega) = (1/2\pi UN) * |X'(\omega)|^2 \qquad (3)$$

where

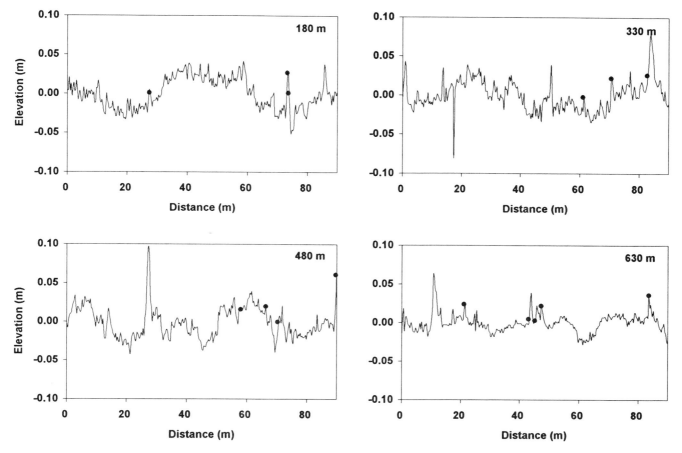

Figure 4. Microtopographic profiles from Jonah's Basement. The profile at 45 m was omitted because of space limitations. Dots indicate bushes. Channels (C) begin between 780 and 945 m from the divide.

x_t is the detrended data set
W_t is the window sequence (a standard Hanning cosine window was used)
x'_t is the tapered data set
FT{ } indicates the Fourier transform (via FFT)
$X'(\omega)$ is the transform of x'_t
$\Phi(\omega)$ is the raw periodogram
U is the mean power of the window sequence
N is the number of measurements in the data series

The use of a tapered cosine window function is necessary to reduce side lobe leakage problems inherent in spectral estimation of any finite data series; it has the benefit of minimizing loss of spatial information as well as minimizing degradation of the spectral signal. The mean power of the window sequence was used in (3) to compensate for the power loss associated with the windowing procedure.

The raw periodogram can be improved without loss of information through use of the ensemble averaging technique of *Welch* [1967]. If $\{X_k(t)\}$ is the set of K realizations of the random process under study (e.g. elevation), it is useful to sample the process repeatedly as if there were multiple survey lines at each station. The ensemble of the K samples of profiles with length N will return an improved estimate of the power spectrum:

$$\Phi(\omega) = \Sigma^K_{k=1} (1/KNU\, 2\pi) * |X'(\omega)|^2 \qquad (4)$$

An alternative approach, when multiple realizations are not available, is to subdivide a given series into K overlapping blocks of length L, and compute the ensemble average of the individual power spectra [*Welch*, 1967]. The method has the advantage of reducing the variance in the power estimates by a factor proportional to 1/K. The disadvantage is in the diminished number of points available for each computation, and in the loss of resolution at the low frequency end of the spectrum. However, most of the variance in the SDF estimate is at the high-frequency end

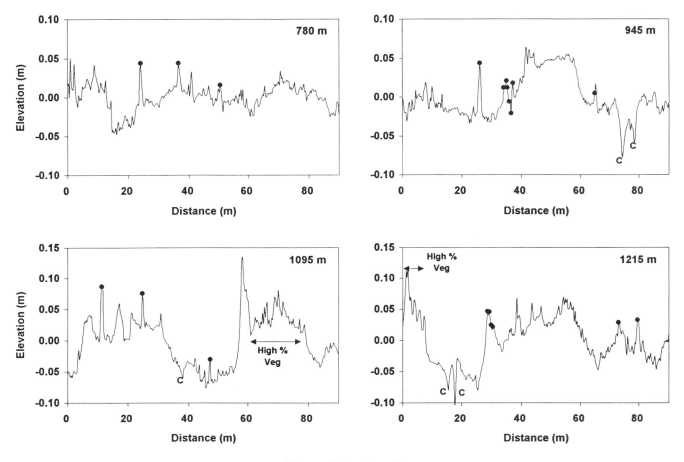

Figure 4 (continued)

of the spectrum, and in our application the ensemble averaging did not affect the long-wavelength information, but rather had the desired effect of dramatically reducing the variance at the high-frequency end of the spectral estimate. Still, with the wisdom of hindsight, it is apparent that much longer transects might have revealed other interesting features not discussed here, and we recommend to others the pursuit of this topic.

The results of this calculation are shown in Figures 5 and 6 as logarithmic plots of Φ against ω, the spectral density function (SDF), the form and slope of which provides a measure of the relative degree to which harmonics of various frequency are represented in the Fourier transform of the original data series. The data indicate a power-function relationship, at least over the frequency range 2-5 m^{-1} to 0.02-0.03 m^{-1}, or 0.2-0.5 m to 30-50 m. in terms of wavelength. Linear regression was used to estimate the slope of the SDF in log-log space.

Perhaps the best test of a spectral estimate lies in comparison with results obtained using several different computational schemes, which avoid some of the difficulties inherent in the Fourier approach to spectral analysis. We have computed spectra using: (i) the standard periodogram method; (ii) the ensemble averaging technique; and (iii) Burg's maximum entropy algorithm [*Press et al.,* 1986]. Although the various techniques yielded slightly different values of the SDF slope, each method yielded the same pattern of downslope changes in microtopographic characteristics as those evident in Figures 5 and 6, which is important to our later discussion. This consistency implies that our computations have not been seriously affected by difficulties such as aliasing and computational artifacts associated with the Fourier transform of spiky data sets.

An implication of the power-function form of the SDF is that the microtopography is fractal, or self-similar, at least over the wavelength range 0.2-0.5 m to 30-50 m, and that the scaling of the surface roughness is described by a single constant: the fractal dimension, a quality that is lacking in other roughness parameters such as rms height.

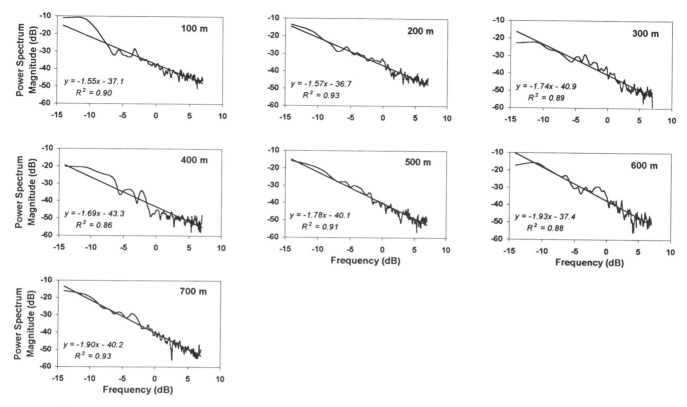

Figure 5. Periodograms for each transect on Amphibolite Hill computed using Welch's (1967) ensemble averaging technique. The power density or magnitude (m^3) and the frequency (m^{-1}) are both expressed in decibels (dB, 10 log$_{10}$ x). Thus the data are plotted in log-log space. The straight lines are best-fit regressions.

The fractal characteristic implies that there is no preferred wavelength in the microtopography. All wavelengths are represented but for each successive increment of wavelength the amplitude changes in a consistent, logarithmic manner. The fractal dimension (D) can be calculated from the slope (S) of the SDF in log-log space [*Mandelbrot*, 1983; *Brown and Scholz*, 1985]:

$$D = (5+S)/2 \qquad (5)$$

Figure 7 shows that although there is a rough parallelism between the power spectral density functions of the various profiles there are subtle differences between profiles in the relative representation of power at various frequencies. First, the SDFs for Amphibolite Hill plot higher than those for Jonah's Basement, reflecting the greater total power, and therefore roughness, of the former set. In fact, when all the data are combined there is a rough positive correlation between the rms elevation values of the profiles and the integrals under the regression lines of the periodograms, and with the vertical intercepts of the regression lines in Figures 5 and 6. The intercepts are generally greater for Amphibolite Hill than for Jonah's Basement, especially those on the left-hand side of the periodograms, whereas the intercepts on the high-frequency ends of the graphs are closer together, although still slightly lower for Jonah's Basement. The implication of this separation is that the microtopography of Amphibolite Hill exhibits more power at all wavelengths measured, especially the longer ones. The most probable explanations for this difference is that on Amphibolite Hill the underlying gneiss weathers to a shallower, gravely colluvium with a rougher surface than is the case on Jonah's Basement, and that Amphibolite Hill has a denser cover of bushes that stabilize larger mounds. The mounds grow for a longer time under bushes, which average about 10-12 years in age on this slope, and the bush and its grassy undercover may provide greater protection against erosion than does grass alone. In general, there also exists the possibility for enhanced bioturbation to raise the surface underneath bushes, although this process does not appear to have a large effect on these particular hillslopes [*Dunne*

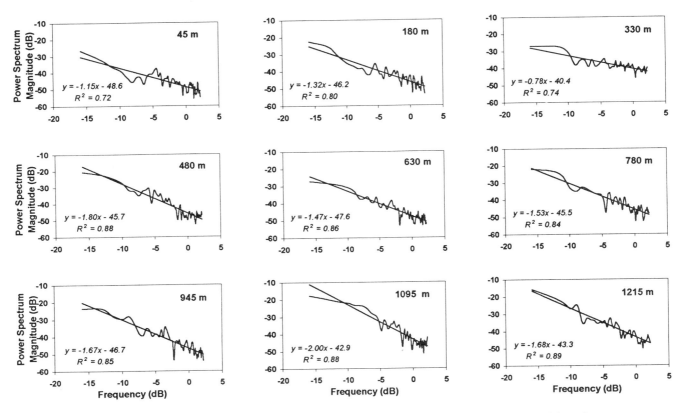

Figure 6. Periodograms for each transect on Jonah's Basement. See caption of Fig. 5 for elaboration.

et al., 1978, p.134].

The second interesting feature of the periodograms is the slope of each spectral density function, which reflects the distribution of power among the various frequencies and therefore provides a scaling measure for the roughness. On both hillslopes there is a general downslope increase in the (negative) slope of the SDF (Figure 8), with the trend being smoother on Amphibolite Hill. For example, the upper right panel in Figure 7 shows a progressive increase between 400 m and 600 m in the power at low frequencies relative to that at high frequencies, confirming the informal observations made about Figures 3 and 4. In other words, Figure 8 implies that the fractal dimension of the microtopographic profiles generally decreases downslope from 1.73 to 1.55 (Amphibolite Hill) and from 1.93 to 1.66 (Jonah's Basement). On Jonah's Basement, it is not clear whether two trends were measured: one on the upper 630 m; and another downslope of that point, but the general trend reflects the progressive increase in the representation of longer-wavelength roughness relative to the short-wavelength roughness. However, it is likely that the break in the trend at 330 m is due to a limitation in the procedure we have used for estimating the SDF slope, or is an artifact of the short record with a resolution of only 0.3 m. The result of our regression for the profile at 330 m (SDF slope >1; $D>2$) would clearly be in error if it were extrapolated without bound, since fractal dimensions in excess of two are physically impossible for a transect. An implication of this result and of the generally high fractal dimensions that we have calculated is that at higher frequencies than those resolved in our surveys the spectral power must decay more rapidly with spatial frequency than would be indicated by linear extrapolation of the regression lines in Figures 5 and 6. Otherwise, the topographic profile would approach a "space-filling" condition [*Brown and Scholtz*, 1985], which it does not (Figure 2). Our analysis is limited, therefore, to microtopographic features with wavelengths greater than 0.2-0.6 m.

6. INTERPRETATION

Several processes create microtopography on these hillslopes. The universal presence of high-frequency roughness elements over each hillslope suggests that they are continually being produced since simple, scale-dependent diffusion [*Andrews and Hanks*, 1985] would

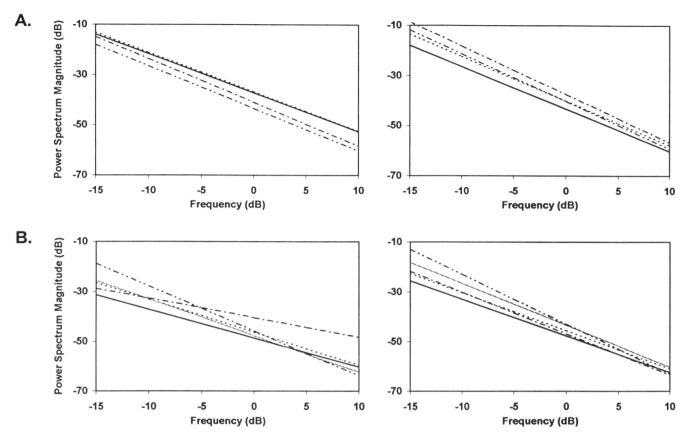

Figure 7. Superposition of linear regression fits to individual power spectra. Amphibolite Hill left-hand panel: solid line refers to the transect at 100 m, dotted line 200 m, dot-dash line at 300 m, and dash-double dotted line at 400 m. Right-hand panel: solid line at 400 m, dotted line at 500 m, dot-dash line at 600 m, and dash-double dotted line at 700 m. Jonah's Basement left-hand panel: solid line refers to the transect at 45 m, dotted line at 180 m, dot-dashed line at 330 m, dash-double dotted line at 480 m, and gray line at 630 m. Right-hand panel: solid line refers to transect at 630 m, dotted line at 780 m, dot-dash line at 945 m, dash-double dotted line at 1095 m, and gray line at 1215 m.

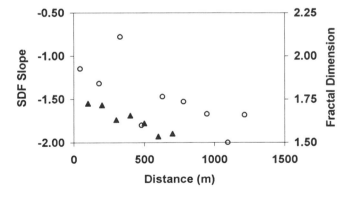

Figure 8. Downslope variation in the slope of the power spectral density function and the associated fractal dimension for Amphibolite Hill (triangles) and Jonah's Basement (circles). The position of one point at an apparent fractal dimension greater than 2 and the generally high computed fractal dimensions are discussed in the text.

quickly remove such small features, causing the SDF to decline sharply at high frequencies. The small-scale features (and smaller, grassed mounds not resolved by our surveys) survive because the development of microtopography is a continuous process of growth and degeneration. Even in the absence of vegetation, incision by wash tends to create microtopographic roughness, at least until the incision is balanced by rainsplash [*Dunne*, 1980; *Dunne and Aubry*, 1986; *Loewenherz-Lawrence*, 1994]. Near the upper ends of hillslope profiles, and on the convexities of some other parts of the profiles, outcrops of bedrock or of gravely colluvium roughen the surface [see *Western and Dunne*, 1979, Fig. 3 for a surveyed example], and the roughness scale must be generated by the weathering characteristics of the bedrock and the size of weathered debris. These fragments may be transported infrequently by flow concentrations, or receive only minor disturbance through bioturbation as they weather *in situ*.

On parts of the hillslope with deeper soil, the type, growth and longevity of plants influence the growth of mounds. Ground cover drastically lowers rainsplash fluxes [*Dunne*, unpublished measurements], slowing the rate of erosion of inter-rill areas, and even allowing them to grow if splashed sediment accumulates under the vegetation cover or if bioturbation beneath the vegetation reduces the bulk density, although these accumulation processes do not appear to be significant in the rapidly eroding environment of these study sites [*Dunne et al.*, 1978]. The mound itself can then divert sheetflow into neighboring microtopographic depressions, protecting the mound and deepening the trough. The vegetation also reinforces the mound against animal trampling. Small termitaria stabilize the surface locally. The pedestals produced by these biotic processes are up to 0.2-0.4 m wide and 0.05-0.10 m high, and are readily observed in photographs (Figure 2) and profiles (Figures 3 and 4) where the locations of bushes are indicated by dots. Other mounds are grassed and increase in amplitude for the same reasons, and still others are bare and susceptible to erosion by animal trampling and rainsplash erosion. Grassed patches with widths of 0.1-0.2 m and heights of 0.01-0.02 m are widespread, and although they influence sheetwash behavior [*Zhang and Dunne*, 1991], they are effectively lost in the noise at the high-frequency end of the roughness spectrum. After the death of the plant, rainsplash, bioturbation by soil fauna, and trampling by grazing animals reduce the amplitude of the protuberance.

These processes and their interaction may cause the ensemble of roughness elements to be stationary or to vary through time. For example, a long drought accompanied by intense cattle trampling probably smoothes the surface at some scales, while wet periods probably extend the influence of sheetwash, deepening the depressions between protected mounds, and even extending channel formation upslope. In this paper, we have analyzed the microtopography of two hillslopes at a time that followed neither an intense drought nor a particularly wet season.

We interpret the general downslope increase in the power at longer wavelengths (Figure 7) as the result of the progressive tendency for incision by wash with consequent flow concentration and cross-grading, and development of "swaley" microtopography (concave in cross-section, but unchannelized) upslope of the channel heads. At the upper end of Amphibolite Hill (Figure 3), the outcropping of bedrock and rubbly regolith associated with shallow depths to bedrock produces surface material that is immobile until weathered or is moved by animals and concentrated wash, rather than by rainsplash. On finer-grained regolith between or downslope of bedrock outcrops and rubble lags,

rainsplash smoothes the surface within the constraints imposed by the patchy vegetation. This diffusive process lowers the surface simultaneously with sediment removal by wash, which is preferentially concentrated in the lower parts of the microtopography and acts to incise the surface. On the upper parts of the hillslope, diffusive rainsplash stabilizes the surface against incision by wash because: (a) wash discharge and gradients are both low; (b) rainsplash occurs even during the early parts of a rainstorm before ponding of runoff; and (c) rainsplash occurs even during most rainstorms in which ponding does not occur. The right-hand sides of each panel in Figures 5 and 6, and of Figure 7 indicate that the high-frequency (>1 m^{-1}) part of the roughness spectrum persists more or less unchanged (within the limits of our ability to discriminate) along each hillslopes profile. It dominates the entire profile near the top of the hill, and further downslope it occupies interfluves between swales and even the interiors of some swales (Figures 3 and 4).

As the average discharge and duration of wash increase with increasing distance from the divide, the capacity of this process for incision [*Smith and Bretherton*, 1972; *Dunne and Aubry*, 1986; *Kirkby*, 1986] progressively overcomes the diffusive influence of rainsplash, even far upslope of the distance at which recognizable channel heads develop. Thus, even on the unchanneled parts of the hillslope the sequence of profiles exhibits increasing power at frequencies of less than 1 m^{-1} (i.e. wavelengths greater than 1m). As it progressively intensifies its influence on the net sediment flux, the wash alters the nature of the surface in a way that concentrates the dispersed flow and increases its specific sediment transport capacity while simultaneously decreasing its lateral extent. *Abrahams et al.* [1989] provide an erosion model that incorporates such effects, and summarize similar work. For example, in a one-dimensional sheetflow with a sediment transport capacity (q_s, expressed as volume transported per unit width per unit time) represented by an equation of the Du Boys type:

$$q_s = A(\tau - \tau_c)^n \quad (6)$$

$$\tau = f(\rho g h s) \quad (7)$$

A sheetflow of average depth 0.001m may be too shallow to transport material unless the flow is concentrated by microtopography [*Zhang and Dunne*, 1991]. In (6) and (7), is the effective bed shear stress, τ_c is the critical shear stress for the bed material, ρ is the density of water, h is the local flow depth within the microtopographic trough, and s is the gradient. The coefficient, A, and n (an exponent greater than 1.0) probably vary between bedload and suspended load transport. The increase in the maximum depth associated with topographic steering may be sufficient to render the flow competent for transport, and the cross-slope average sediment transport capacity will be the integration of q_s across the wetted perimeter. The critical shear stress threshold and the nonlinearity of (6) both increase the sensitivity of the sediment flux to small changes in flow-depth distribution. Thus, the sediment transport equation to be inserted in theoretical analyses of hillslope profile evolution [e.g. *Kirkby*, 1971] depends intimately on the nature of the microtopography. The microtopography of the surface outside of the inundated swales should also control the efficacy of rainsplash in supplying sediment to the wash and therefore should affect its capacity for incision. Our illustration of the downslope change in the microtopography suggests that it should therefore affect the evolution of the longitudinal profile of the entire hillslope, and of course, in turn be affected by this profile because the competition between diffusive rainsplash and incision by wash depends in part on hillslope gradient.

7. EVOLUTION OF MICROTOPOGRAPHY AND THE INITIATION OF CHANNELS

The tendency for the time-averaged efficacy of surface wash to increase downslope causes it to evacuate sediment from larger depressions, generating more roughness at all frequencies less than about 1 m^{-1} in the cases reported here. At least on Jonah's Basement, this occurs at all distances downslope of the divide, including the convex portions of the hillslope, contrary to the analyses of infinitesimal perturbations of sediment transport fields on a constant-form hillslope [*Smith and Bretherton*, 1972; *Loewenherz*, 1991; *Loewenherz-Lawrence*, 1994], which predict that convex slopes should not be perturbed away from the base state. After an infinitesimal perturbation of the sediment transport field in sheetwash, the surface erosion is dominated by finite-scale effects such as: (i) the lengths of splash trajectories; (ii) flow depth which influences the splash-wash interaction through detachment and transport; (iii) the influence of patchy vegetation, bioturbation, and trampling on mound development over years to decades; (iv) downslope variations of regolith texture; (v) downslope variations of vegetation type and density associated with spatial gradients of soil moisture caused by scale effects on the frequency, duration and intensity of infiltration and runoff; and (vi) convective accelerations in flow around microtopography.

Thus, microtopography that is influenced by these finite-scale processes as well as the simple additive interaction of rainsplash and wash forces convergence and divergence on the wash, altering its long-term (i.e. integrated over many rainstorms) sediment transport capacity at a given distance downslope (for a fixed gradient and rainfall climate). The inter-swale areas simultaneously evolve a spectrum of shapes (which may be steady or unsteady, as described above) mainly as a result of the interaction between the erosion processes and vegetation. This spectrum governs the sediment supply from the inter-swale to the neighboring depressions where the wash transports sediment. This high-frequency end of the microtopographic spectrum does not change much along the hillslope because it is governed mainly by the interaction of rainsplash and vegetation. In fact, the larger components of it are not even inundated by the increasing discharge of runoff downslope because the growing swales accommodate most of the increased flow.

Near the top of a hillslope, wash, though forced to converge by high-frequency microtopography that modulates its sediment transport capacity, also receives a sediment supply from that microtopographic spectrum, and this supply approximately balances the resulting transport capacity. Therefore incision of the small depressions relative to the microtopographic protuberances is non-existent or slow. As water discharge increases downslope from the divide (here we ignore temporarily the spatial effects on infiltration and runoff generation modeled by *Dunne et al.* [1991]), the transport capacity of wash increases faster than linearly with distance because of the form of (6), but the rate at which sediment is supplied by splash from the inter-swales remains constant because the high-frequency portion of the microtopography remains constant (Figure 7). The developing imbalance between sediment mobilization from the high-frequency microtopography and the transport capacity of wash in the depressions causes long-term, slow, diffuse evacuation of the lower-frequency depressions, further concentration of flow in the resulting swales, and a greater degree of exposure of the longer wavelength protuberances (> 1-3 m in the cases documented in Figures 3, 4, 5, 6, and 7). However, as these protuberances are progressively exposed through time they supply more sediment that eventually slows the incision of the swales to approximately the same rate as the lowering of the surrounding microtopography, and keeps it widespread and diffuse.

Eventually, however, the gradual enlargement of the swales and their gathering of runoff increases the area of exposed inter-swale, and therefore decreases the sediment supply by scale-dependent diffusion [*Andrews and Hanks*, 1985] from the longer inter-swale areas of high-frequency microtopography. A downslope increase in vegetation density caused by some of the factors referred to above may further decrease the lateral influx of sediment to the wash on some hillslopes, but this is not a general process. Thus, the lateral sediment supply decreases downslope while the transport capacity of wash, on a constant gradient, continues to increase faster than linearly. At some distance that still cannot be quantified, this imbalance occurs with sufficient intensity and frequency that the surface is lowered in the form of a trough distinctly below the floor of the surrounding swale and this trough is occupied by a zone of concentrated sediment transport, often with a strip of sandy bed material and small bedforms. The edges of the trough become steep (up to at least 20° in the present case) and distinct because the colluvium of the hillslopes studied is slightly cohesive. The cohesion is likely to be more effective in maintaining steep banks wherever the incision is faster because weathering has then had less time to reduce the cohesion and degrade the bank. In the absence of regolith cohesion, the survival of distinct banks presumably requires some other form of reinforcement such as plant roots. The inter-swale zones may eventually attain a form on which the diffusive processes can lower the microtopography at the same rate as the channel is lowered, but that equilibrium form will be associated with a distinct, degrading channel.

The growth of long-wavelength microtopographic roughness hundreds of meters upslope of channels heads results from both: (i) the interaction between diffusive and incising sediment transport processes during single runoff events, and (ii) their interaction integrated over time periods long enough to allow the removal of sufficient mass to create the swaley topography that we have measured. This latter integration should include the probability distribution of both types of transport and their interaction, including the relatively frequent rainstorms that generate rainsplash without runoff. Although there would be value in initially considering the simple condition of a bare slope such as a groomed field or construction site, in most natural environments the sheltering effects of vegetation and the diffusive effects of bioturbation would also need to be considered. Long-term monitoring of the spectral characteristics of microtopography and repeated mapping of channel heads on a hillslope-scale plot where rainfall intensity and runoff are simultaneously recorded might eventually yield a clearer view of the interactions which lead to channel formation.

8. SUMMARY

Field surveys and spectral analysis lead to the conclusion that microtopography on two long semiarid hillslopes subject to Horton overland flow is fractal with a dimension that generally decreases downslope. The overall roughness depends on: (i) weathering characteristics of bedrock and colluvium; (ii) the patchiness of vegetation; and (iii) the tendency for wash to incise the surface. The fractal dimension indicates the relative degree of roughness at different scales, and in particular that low-frequency microtopography becomes more strongly represented with increasing distance downslope. The fact that the microtopography is fractal over the observed range of scales indicates that the surface is degraded by a combination of diffusive and incising processes at all measured scales and on all parts of the hillslope. This observation is particularly relevant to the convex part of the hillslope where diffusive processes have long been thought to dominate the sediment transport and profile evolution. They do not dominate at the local scale of microtopographic mounds. Instead, the integrated effect of the interaction between diffusive and wash processes acts to prevent channelization and the evolution of a concave hillslope profile. Wash becomes competent to transport sediment within a few meters of the drainage divide, but immediately interacts with finite-scale microtopography imposed by biotic processes (in most cases). This microtopography forces the wash to develop a depth distribution that controls its transport capacity, and diffusive processes to release sediment from mounds into the wash. The net result of this interaction is that the sediment supply is high enough to prevent wash from incising channels on the upper portion of these long hillslopes.

Although there is no preferred wavelength on a fractal surface, for purposes of qualitative discussion one can arbitrarily distinguish in the field: flow concentrations (with wavelengths less than 1 m), microtopographic "swales" (with wavelengths greater than 1 m), and distinct channels, the spacing of which increases rapidly downslope from 50 m on Jonah's Basement and 10 m on the steeper Amphibolite Hill to more than 100 m at the base of each hillslope.

We have interpreted the downslope changes in these surface roughness components in the light of current theory and recent field observations of processes to propose a qualitative description of the interactions between wash and diffusive sediment transport that gradually alter the microtopography upslope of channel heads and then cause channel incision at some critical distance which arises from the interaction. In particular, we emphasize the finite-scale interactions that are driven by microtopography and the factors that control its evolution. The discussion does not yet produce a quantitative theory but indicates some of the non-local influences called upon by *Loewenherz-Lawrence* [1994] to stabilize channels and to locate the channel heads at some characteristic distance downslope. Our argument is distinct from that of Loewenherz-Lawrence in that we emphasize the importance of the finite-scale effects that are beyond the scope of her treatment.

Acknowledgments. The work was supported by grants from the National Science Foundation (EAR 80/8286 and 83/3172) and by a J. S. Guggenheim Foundation Fellowship. The paper was improved by suggestions from John Costa, Milan Pavich, Kenneth Potter, and Jonathan Stock.

REFERENCES

Abrahams, A .D., A. J. Parsons, and S-H. Luk, Distribution of depth of overland flow on desert hillslopes and its implications for modeling soil erosion, *J. Hydrol., 106,* 177-184, 1989.

Abrahams, A .D., A. J. Parsons, and S. H. Luk, The effect of spatial variability in overland flow on the downslope pattern of soil loss on a semiarid hillslope, Southern Arizona, *Catena, 18,* 255-270, 1991.

Andrews, D. J., and T. C. Hanks, Scarp degraded by linear diffusion: inverse solution for age, *J. Geophys. Res., 90,* 10,193-10,208, 1985.

Brown, S. R. and C. H. Scholtz, Broad bandwidth study of the topography of natural rock surfaces, *J. Geophys. Res., 90,* 12,575-12,582, 1985.

Collins, B. D., and T. Dunne, Erosion of tephra from the 1980 eruption of Mount St. Helens, *Geol. Soc. Amer. Bull., 97,* 896-905, 1986.

Cordova, J. R., I. Rodriguez-Iturbe, and P. Vaca, On the development of drainage networks, *Internat. Assoc. Hydrol. Sci. Pub. 137,* 239-249, 1983.

Dietrich, W. E. and T. Dunne, The channel head, in *Channel Network Hydrology,* edited by K. Beven and M. J. Kirkby, pp. 175-220, John Wiley and Sons Ltd., Chichester, U.K., 1993.

Dunne, T., Formation and controls of channel networks, *Prog. Phys. Geogr., 4 (2),* 211-219, 1980.

Dunne T. and B. F. Aubry, Evaluation of Horton's theory of sheetwash and rill erosion on the basis of field experiments, in: *Hillslope Processes,* edited by A. D. Abrahams, pp. 31-53, Allen and Unwin, London, 1986.

Dunne, T., and W. E Dietrich, Experimental study of Horton overland flow on tropical hillslopes, 2, Hydraulic characteristics and hillslope hydrographs, *Z. Geomorphol. Suppl. Bd. 35,* 60-80, 1980.

Dunne, T., W. E. Dietrich and, M. J. Brunengo, Recent and past rates of erosion in semi-arid Kenya, *Z. Geomorphol. Suppl. Bd. 29,* 215-230, 1978.

Dunne, T., W. Zhang, and B. F. Aubry, Effects of rainfall, vegetation, and microtopography on infiltration and runoff, *Water Resour. Res., 27,* 2271-2286, 1991.

Emmett, W. W., The hydraulics of overland flow on hillslopes, *U.S. Geol. Surv. Prof. Pap.,* 662-A, 68 pp., 1970.

Escobar, L. A.., and I. Rodriguez-Iturbe, A modeling scheme for the study of drainage density, *Water Resour. Res., 18* (4), 1029-1035, 1982.

Hirano, M., A study of a mathematical model of slope development, *Geogr. Rev. Japan, 39,* 324-336, 1966.

Horton, R. E., Erosional development of drainage basins: hydrophysical approach to quantitative morphology, *Geol. Soc. Amer. Bull., 56,* 275-370, 1945.

Kirkby, M. J., Hillslope process-response models based on the continuity equation, *Institute of British Geographers Spec. Pub. 3,* pp. 15-30, 1971

Loewenherz, D. S., Stability and the initiation of channelized surface drainage: A reassessment of the short wavelength limit, *J. Geophys. Res., 96,* 8453-8464, 1991.

Loewenherz-Lawrence, D. S., Hydrodynamic description for advective sediment transport processes and rill initiation, *J. Geophys. Res., 30,* 3203-3212, 1994.

Mandelbrot, B., *The Fractal Geometry of Nature,* 468 pp., W.H. Freeman, San Francisco, CA, 1983.

Melton, M. A., Correlation structure of morphometric properties of drainage basins and their controlling agents, *J. Geol., 66,* 442-460, 1958.

Montgomery, D. R. and W. E. Dietrich, Landscape dissection and drainage area-slope thresholds, in: *Process Models and Theoretical Geomorphology,* edited by M. .J. Kirkby, pp. 221-246, John Wiley and Sons, Chichester, U.K., 1994.

Mitchell, J. K., and B.A. Jones, Jr., Micro-relief and surface depression storage: changes during rainfall events and their application to rainfall-runoff models, *Water Resour. Bull, 14,* (4), 777-802, 1978.

Moss, A. J. P., P. H. Walker, and J. Hutka, Raindrop-stimulated transportation in shallow water flows: an experimental study, *Sedimentary Geol., 22,* 165-184, 1979.

Press, W. H., B. P. Flannery, S. A. Teukolsky, and W. T. Vetterling, *Numerical Recipes: The Art of Scientific Computing,* Cambridge University Press, Cambridge, 818 pp, 1986.

Preston, F. W., Two-dimensional power spectra for classification of landforms, in *Computer application in the earth sciences: colloquium on classification procedures,* edited by D. F. Merriam, University of Kansas, State Geological Survey, Computer Contribution No. 7, pp. 64-69, 1966.

Schaeffer, M., D. Elfrits, and D. J. Barr, Sculpturing reclaimed land to decrease erosion, *Proc. Symp. on Surface Mining, Hydrology, Sedimentology, and Reclamation,* Lexington, KY, pp. 99-108, 1979.

Schumm, S.A., The role of creep and rainwash on the retreat of badland slopes, *Amer. J. Sci., 254,* 693-706, 1956a.

Schumm, S.A., Evolution of drainage systems and slopes in badlands at Perth Amboy, New Jersey, *Geol. Soc. Amer. Bull., 67,* 597-646, 1956b.

Smith, T. R. and F. P. Bretherton, Stability and the conservation of mass in drainage basin evolution, *Water Resour. Res., 8,* 1506-1529, 1972.

Stone, R. O., and J. Dugundji, A study of microrelief -- its mapping, classification, and quantification by means of Fourier analysis, *Engineering Geology, 1,* 89-187, 1965.

Strahler, A. N., The nature of induced erosion and aggradation, in *Man's Role in Changing the Face of the Earth,* edited by W. L. Thomas, Univ. of Chicago Press, Chicago, pp. 621-638, 1956.

Welch, P. D., The use of fast Fourier transform for the estimation of power spectra: a method based on time averaging over short, modified periodograms, IEEE Trans. Audio Elec., AU-15, 2, 70-74, 1967.

Western, D. and T. Dunne, Environmental aspects of settlement site decisions among pastoral Maasai, *Human Ecology, 7(1),* 75-98, 1979.

Zhang, W., Numerical simulation of the hydrodynamics of overland flow including spatial variation in its characteristics, Ph. D. thesis, 231 pp., Univ. of Washington, Seattle, Aug., 1990.

Zhang, W., and T. Dunne, Numerical simulation of the effects of microtopography on hillslope soil erosion (abstract), *EOS Trans. AGU, 72,* 137, 1991.

Thomas Dunne, Kelin X Whipple, and Brian F. Aubry, Department of Geological Sciences, University of Washington, Seattle, WA 98195

Geomorphically Effective Floods

John E. Costa and Jim E. O'Connor

U.S. Geological Survey, Vancouver, Washington

Investigations of the hydrology and geomorphology of recent floods from the rapid failure of two small upland dams document the unusually large peak boundary shear stress and peak stream power per unit area for each flood. Downstream consequences to alluvial channels and floodplains, however, were minimal. Lack of geomorphic change is attributed to the short duration of the floods, which lasted about six and sixteen minutes each. Distribution of stream power over hydrographs of eight exceptional floods is constructed from channel geometry, discharge rating curves, and flood hydrographs; the resulting curve is defined as a stream-power graph. A stream-power graph gives a better portrayal of the potential for a flood to be geomorphically effective than simple statements of flow magnitude. From stream-power graphs, total energy expended over a flood hydrograph can be computed. Total flood energy may not be a sensitive measure of geomorphic effectiveness without consideration of channel and floodplain resistance. A conceptual model combining flow duration, peak stream power per unit area, flood energy, and alluvial and bedrock thresholds may represent the effectiveness of floods and can distinguish among such cases as (a) floods of long duration, moderate to large energy expenditure, but low peak stream power per unit area. These floods are ineffective in causing significant landform changes in alluvial or bedrock channels; (b) floods of medium to long duration, with medium to large total energy expenditure, and large peak stream power per unit area. These are believed to be the most effective geomorphic floods in any kind of channel because of the optimal combination of peak flood power, duration, and total energy expenditure; and (c) floods of very short duration, low total energy expenditure, but large peak stream power. These floods are also ineffective agents of geomorphic change in spite of record values of peak stream power per unit area because of their short duration, and resulting low energy expenditures.

1. INTRODUCTION

One fundamental underpinning of the science of geomorphology is that the form of the earth's surface is the consequence of past and present geophysical forces acting on the earth's landforms. In fluvial geomorphology, this notion led to the classic question of whether valleys and channels were primarily shaped by frequently-occurring moderate flows, and resulting small forces, or by rare and cataclysmic flows with corresponding large forces [*Wolman and Miller*, 1960].

The maximum discharge of a flood is commonly used as a measure of the potential of a flow to be an effective geomorphic agent, primarily because maximum discharge is routinely measured or computed and published for large floods. In general, the larger the discharge, sometimes indexed by drainage area or recurrence interval, the more change that is anticipated in the channel and valley. A dilemma for geomorphologists is the observation that floods of similar magnitude and frequency sometimes produce surprisingly dissimilar geomorphic results. Unfortunately, few quantitative hydraulic data on large floods have been presented to assess disparities in landform response. Recently, channel boundary shear stress and stream power per unit boundary area have been shown to be more useful concepts than discharge alone in assessing the potential of flood flows to affect landscapes [*Baker and Costa*, 1987].

The concept of "geomorphic work" is difficult to define precisely, partly because the issue is clouded by semantics.

Natural and Anthropogenic Influences in Fluvial Geomorphology
Geophysical Monograph 89
This paper is not subject to U. S. copyright. Published in 1995 by the American Geophysical Union

Geomorphic work in fluvial systems has been variously defined to be represented by the rate of sediment movement [*Wolman and Miller*, 1960], or as the mass of sediment transported through a vertical distance in unit time [*Caine*, 1976]. Depending on the interpretation of what constitutes geomorphic "work", and the relative magnitude of forces acting on the landform during the time of interest, slow and persistent processes that generate low forces may appear to predominate [*Wolman and Miller*, 1960; *Andrews*, 1994], or large and rare floods that generate large forces may be considered to be most significant [*Baker*, 1977].

The recognition that some really large, rare floods may not have long-lasting effects, or cause long-term changes in channel and valley morphology [*Costa*, 1974; *Moss and Kochel*, 1978; *Huckleberry*, 1994] led to the realization that the absolute magnitude or force of a geomorphic process is not the sole factor responsible for the resulting landforms, nor their perseverance. Other controlling factors include landsurface resistance [*Bull*, 1979; *Graf*, 1979; *Brunsden*, 1993], the frequency and ordering of effective processes [*Beven*, 1981], and the rate of recuperative processes following formative events [*Costa*, 1974]. This holistic view of what constitutes an effective event in geomorphology is well captured in the benchmark paper by *Wolman and Gerson* [1978]. There are, at present, no simple measures of flow and effect that have been used consistently in describing the interaction of floods and the landscape.

1.1 The significance of flood-flow duration

Our investigations of the recent floods from two small dam failures in Washington and Oregon inspired us to further consider the importance of flood-flow duration with respect to the geomorphic effectiveness of floods. Although these floods had extremely high instantaneous values of shear stress and stream power, they produced few or no geomorphic changes in downstream valleys or channels. The purpose of this paper is to document the role of flood duration as an obvious, but often ignored, critical factor. Flood duration can affect geomorphic response to large flows in several ways. Long flow duration may be necessary to saturate channel banks before they will fail, or aid in the wetting and subsequent expansion of floodplain soils, with concomitant reduction in shear strength. Some finite amount of time may be necessary to break down floodplain vegetation or erode through the cohesive, root-strengthened top strata, after which erosion of less-cohesive substrata can proceed more rapidly. Also, sediment entrained from hillslopes or channels requires time to be transported onto floodplain surfaces, especially if it travels as bedforms.

Flow duration, in addition to flow magnitude and frequency, stream power, resistance of the land surface, and the restorative and recuperative processes between effective events, determines whether a large discharge event is geomorphically effective. Flow duration can be a key to understanding how floods with lower values of peak discharge, shear stress, or stream power, can have greater geomorphic impact in some alluvial channels than floods with larger instantaneous values.

2. INSIGHT FROM FLOODS FROM THE FAILURE OF SMALL DAMS

Floods resulting from the failure of small dams in upland areas can offer a unique perspective into the influence of a high-magnitude event on steep channels and floodplains in small basins. Dam failures in upland areas involve a precisely known volume of water being introduced to a channel at a point location. Dam failures also offer a mechanism for the creation of floods far larger than possible from snowmelt or rainfall-runoff, and that may be unprecedented in the recent or past geological history of the basin [*Jarrett and Costa*, 1986]. Such small-dam failures occur frequently, and many go unreported in the literature. Recent documented examples include rainfall-induced failure of seven earthfill gravity dams in 1977 near Johnstown, Pennsylvania [*Hoxit and others*, 1982], and three earthfill gravity dams in 1989 at Fayetteville, North Carolina [*Mason and Caldwell*, 1992]. The recent failures of two small upland dams in Washington and Oregon present the opportunity to evaluate the role of stream power and flood duration on geomorphic effectiveness in downstream channels and floodplains. Both dams failed rapidly and nearly instantaneously released their stored water down small, steep, upland channels and floodplains.

2.1 The failure of Reservoir No. 3, Centralia, Washington

Reservoir No. 3 is a small concrete-lined water-supply reservoir for the city of Centralia, Washington (Figure 1). On Oct. 5, 1991, the bedrock hillslope under the southwest side of the reservoir suddenly failed, and instantaneously released 13,250 m^3 of water down a small steep valley that led to the eastern edge of the city of Centralia. Two houses were destroyed, four city blocks were flooded, and 400 people were evacuated (Figure 2) [*Costa*, 1994].

The reason for the failure is believed to have been a landslide in the silty sandstone bedrock beneath the reservoir, caused by some combination of (a) seepage from

COSTA AND O'CONNOR 47

Fig. 1. Location map of Centralia, Washington, and Porter Hill, Oregon, dam failures.

cracked and deteriorated concrete panel seams into the fractured bedrock foundation; (b) stress patterns caused by the quarterly draining and refilling of the reservoir; or (c) a recent increase of 0.6 m in the water level in the reservoir. Sedimentological characteristics of deposits, high-water mark distribution, transport of unbroken beer bottles, and landforms preserved on the valley floor indicated the dam-failure flood consisted initially of a debris flow that deposited coarse gravel and boulders along the channel and floodplain. The debris flow had an estimated volume of 1,800 m^3, and was immediately followed by a water flood that achieved a stage about 0.3-0.5 m higher than the debris flow.

A four-section slope-area indirect discharge estimate was made on Oct. 10, 1991, five days after the dam failure, at a site 275 m below the emptied reservoir (Figure 3). Scour and deposition, a steep channel slope of 0.09, and uncertain roughness coefficients all contribute to some uncertainty in the final peak-discharge estimate of 71 m^3/s. An official for the city of Centralia, responsible for the operation of the reservoir, reported that the reservoir drained in three to five minutes. At a constant discharge rate of 71 m^3/s, it would take 3.1 minutes to drain the reservoir.

Several pieces of data about the dam-failure and resulting flood, such as reservoir volume, reports of drainage time, peak discharge calculations, and average velocity of the flood, allow construction of a flood hydrograph. Using the average peak-flow velocity of 4.2 m/s calculated for the slope area reach, it would take 1.1 minutes for the flood to travel 275 meters from the reservoir to the measurement site. If a triangular-shaped hydrograph is assumed, considering the 13,250 m^3 reservoir volume and the 71 m^3/s peak discharge, the duration of the flood past the slope-area site would be about 6.2 minutes. Consequently, after about 7.3 minutes from the time of the reservoir failure, the flood had passed the indirect-discharge measurement site, and moved into the city (Figure 4).

2.2 *Failure of Porter Hill dam near Roseburg, Oregon*

A private landowner constructed several small earthen dams to collect spring discharge on the flanks of Porter Hill in southwestern Oregon. The dams blocked an unnamed tributary into Olalla Creek, which flows into Lookingglass Creek and eventually into the South Umpqua River. The Porter Hill dam is the largest of these dams, and is located in the NW1/4, SE1/4, sec. 32, T28S, R7W (Tenmile Quadrangle, Oregon). Porter Hill is underlain by rhythmically bedded sandstone and siltstone that has been folded, faulted, and weathered [*Baldwin*, 1974] (Figure 1).

The Porter Hill dam was 5.8 m high, about 20 m wide, and stored an estimated 15,000 m^3 of water at the time of failure. The earthen dam was constructed of local clayey residuum. The exact date of the dam failure is unknown, but it is believed to have failed on or about February 27, 1993 (John Falk, Oregon State Dam Safety Coordinator, personal communication, April 15, 1993). The dam apparently failed during a rainstorm when a large slump on the downstream face of the dam opened a breach with a top width of about 20 m (Figure 5). The large slump led to a near instantaneous failure of the dam, and the release of about 15,000 m^3 of water down a steep upland valley.

Peak discharge from the dam failure was estimated to have been about 30 m^3/s at a location about 150 m downstream from the dam, using the slope-conveyance method (Figure 6). If a triangular-shaped hydrograph is assumed, considering the 15,000 m^3 reservoir volume and the 30 m^3/s peak discharge, the duration of the flood past the slope-conveyance site would be about 16.6 minutes. A reconstructed hydrograph for the flood is shown in Figure 4. Data for the two dams are summarized in Table 1.

Fig. 2. Airphoto of the failure of Reservoir No. 3, Centralia, Wash. (from Costa, 1994).

3. STREAM POWER, GEOMORPHIC WORK, AND CHANNEL CHANGES

Stream power per unit boundary area (ω) expressed in watts per square meter (W/m^2), is computed from

$$\omega = \gamma QS/w$$

where γ is specific weight of the fluid (9800 N/m^3 for clear water), Q is discharge, S is energy slope, and w is water-surface width. Peak stream power per unit boundary area at the sites of indirect-discharge estimates for the Centralia flood, 3,300 W/m^2, and for the Porter Hill flood, 2,900 W/m^2, are among the largest values ever documented for historic flows [*Baker and Costa*, 1987]. The historic floods in other basins that generated similar or smaller values of peak stream power per unit boundary area all were considered geomorphically effective according to our inspections, and the authors' reports.

In contrast, in spite of the magnitudes of the peak stream power of the two dam-failure floods in Oregon and Washington, the erosional effects on downstream alluvial channels were unimpressive. Characteristics of the original channel at Centralia are not known. The floodplain is about 20 m wide and bounded by a bedrock ridge on one side, and roadfill on the other. The slope is about 0.09, and the surface is grass-covered and regular, with three or four widely-spaced large trees. Floodplain sediment consists of gravel and cobbles in a silt and clay-rich matrix. During the flood, the original channel in the small valley was enlarged, and a 1.5-m headcut formed. The 20-m-wide floodplain was entirely inundated by about one meter of water flowing at about 4.2 m/s, but neither the floodplain nor floodplain vegetation were destroyed or greatly modified (Figure 3). Most of the visible change in the floodplain is attributable to deposition of coarse sediment from the preceding debris flow, not the water flood.

At Porter Hill, the floodplain slopes at about 0.10 and consists of open forest and moss-covered stumps. The surface is covered with leaf litter, ferns, and a few fallen trees. Floodplain sediment consists of poorly-drained and unstratified gravel, silt, and clay. Following the flood at Porter Hill, moss was still intact on the upstream side of trees below high-water marks, and ferns and leaf litter were virtually undisturbed. It was nearly impossible to tell that a large flood, 10 m wide, and 1 m deep, had recently

Fig. 3. Photograph of the floodplain below Reservoir No. 3, Centralia, Wash., where there was only minimal damage from the flood.

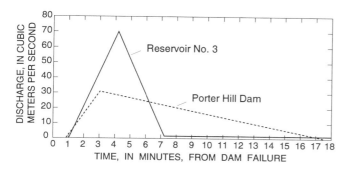

Fig. 4. Reconstructed triangular hydrographs for the Centralia and Porter Hill floods.

passed at about 3 m/s (Figure 6).

During large floods along high-energy fluvial systems, floodplains can become unraveled, severely eroded, and inundated by coarse gravel and debris [*Nanson*, 1986; *Jarrett and Costa*, 1986]. A floodplain that is not ravaged by an extreme flood is the exception, and requires explanation. Some studies that documented the lack of erosion and modifications to channels and floodplains accompanying large floods attributed the lack of land surface disruption to insufficient stream power [*Nanson and Hean*, 1985] or extraordinary stabilization of surfaces by vegetation [*Zimmerman and others*, 1967]. These explanations refer to thresholds of landscape resistance that must be overcome by the flow for it to be effective. The Centralia and Porter Hill floods generated peak stream power values that were likely capable of surpassing resistance thresholds offered by the alluvial valleys and floodplains.

We infer that the lack of disruption of the valley floor despite extraordinarily large peak stream power acting against the floodplain was chiefly a consequence of the very short duration of high stream power during the flood. In both dam-failure floods, high flows were not sustained, and the entire hydrographs passed the study areas within about 16 minutes (Figure 4). The maximum flood power, while large enough to greatly surpass landscape resistance thresholds, lasted for only a small fraction of that time and was not effective in breaking down floodplain vegetation and eroding channels and floodplains. Thus high-energy floods, of very short duration, may cause little geomorphic

Fig. 5. Photograph of breach and slump failure of the Porter Hill dam, near Roseburg, Oregon.

change. We hypothesize that other floods in the same basins with smaller peak stream power values, but longer duration, could precipitate significant and perhaps permanent changes in channels and floodplains.

3.1 Changing stream power over a flood hydrograph

If flow duration is believed to be an important factor in the ability of floods to alter landforms, then it is important to know the temporal and spatial distribution of stream power throughout a flood. For flooded locations not at gaging stations, indirect discharge methods are commonly used to calculate peak discharge associated with high-water marks. With data derived from these investigations, only instantaneous peak stream power for the flood can be ascertained and reported [*Costa*, 1987]. Peak stream power is useful in evaluating flood competence [*Costa*, 1983; *Williams*, 1983], but is not the sole factor in evaluating whether a flood may be geomorphically effective. Time-integrated flood power, computed over a hydrograph and combined with some quantitative measures of landscape resistance, such as shear strength of river channel banks and floodplains, may be more useful to evaluate potential for geomorphic effectiveness.

There are no widely applicable procedures to quantify landscape modification accomplished by a given flood. Valuable qualitative descriptions have been used in some studies [*Kochel*, 1988; *Miller*, 1990; *Miller and Parkinson*, 1993], but data requirements for a more rigorous quantitative analysis of landform modification preclude our use of anything but a simple two-class scheme at this time. The amount of geomorphic alteration is assigned a qualitative value of small or extreme. *Small* disruption represents sites where floodplains are inundated, but with little or no erosion of the floodplain surface. Channel scour and erosion are local, drainage patterns remain similar (e.g. a meandering stream still meanders), deposition and sedimentation are restricted to small, local areas, and the valley floor and channel have minimal changes. *Extreme* disruptions occur in areas where the entire floodplain and channel are substantially affected by erosion or deposition. New channels may be formed or the floodplain may be entirely eroded. Extensive areas of deposition may occur on uneroded floodplain or in newly eroded areas during the flood recession. Bedrock, if present, may have been eroded, and the stream channel and stream pattern may be completely realigned. We recognize that deposition of sediment from large floods can significantly alter alluvial

Fig. 6. Photograph of the channel and floodplain about 100 m downstream from the Porter Hill dam. Note the complete lack of any identifying evidence that a large flood had ever passed this location. The tape marks the maximum flood stage.

TABLE 1. Dam and flood characteristics of Reservoir No. 3, and Porter Hill Dam.

Feature	Reservoir No. 3 Centralia, Wash.	Porter Hill near Roseburg, Oreg.
Type of dam	Concrete-lined	Earthen
Date built	1914	early 1990s
Date failed	1991	1993
Height (m)	5.2	5.8
Volume (m^3)	13,250	15,000
Flood depth (m)	1.0	1.0
Slope	0.09	0.10
Discharge (m^3/s)	71	30
Peak ω (w/m^2)	3,300	2,900

forms, but our analysis focuses on the erosional thresholds of extreme floods.

For comparison to the Centralia and Porter Hill dam-failure floods, we analyzed floods from three thoroughly studied historic large-dam failures, one cloudburst-rainfall flood along a sand-bed stream in Colorado, and the flood of record along the Mississippi River (Table 2). We also include two well-known paleofloods, or series of paleofloods, the Missoula and Bonneville floods, which resulted from large natural-dam failures about 10,000 - 15,000 years ago along the Columbia and Snake Rivers.

These floods were selected because (a) we have done field work at all sites except the Mississippi River, and are familiar with the flood effects; (b) the floods were unequivocally large and capable of generating substantial hydrodynamic forces; and (c) we had the appropriate data for our analysis for each flood. All available hydraulic data, except that from the Mississippi River, are from indirect-discharge measurements. Gaging stations, where present, were destroyed during the floods. Indirect-discharge data increase the likely error, but our attempt is to present a concept and approach that can be further verified with better data.

Field investigations and descriptive reports indicate which floods were effective geomorphic floods, and which were not. Little or no change occurred to channels or floodplains associated with the floods from the two small upland dams at Centralia and Porter Hill discussed

TABLE 2. Hydraulic, energy, and geomorphic data for ten well-documented floods that demonstrate different kinds of stream-power graphs

Flood	Peak stream power (W/m^2)	Mean stream power (W/m^2)	Duration (s x 10^3)	Energy expended per unit area (joules x 10^3)	Geomorphic impact	Kind of power-graph	Reference
Centralia, Wash.	3300	1650	0.38	620	Small	C	Costa, 1994
Porter Hill, Oreg.	2900	1450	1.0	1500	Small	C	This report
Plum Creek, Colo.	630	110	68	3900	Extreme	B	Osterkamp and Costa, 1987
Roaring River, Colo.	4300	1200	7.2	8500	Extreme	B	Jarrett and Costa, 1986
Rubicon River, Calif.	6100	3600	22	29,000	Extreme	B	Scott and Gravlee, 1968
Teton Dam, Id.	17,200	3400	29	109,000	Extreme	B	Ray and Kjelstrom, 1978
Mississippi River, Ark.	12	6	1200	21,600	Small	A	Baker and Costa, 1987
Bonneville Flood, Burley Basin, Id.	300	150	11,200	1,700,000	Small	A	O'Connor, 1993
Bonneville Flood, Rock Creek, Id.	90,000	20,000	11,200	220,000,000	Extreme	B	O'Connor, 1993
Missoula Flood, Columbia River Gorge, Oreg. and Wash.	60,000	8100	430	3,500,000	Extreme	B	Benito and O'Connor, 1991

previously, nor to the Mississippi River floodplain or the Snake River alluvial floodplain at the Burley Basin in Idaho. The Mississippi River and Bonneville paleoflood in the Burley Basin were similar in that wide alluvial floodplains and flat channel gradients prevented peak or average stream power per unit area from exceeding erosion thresholds. The other floods all caused severe and widespread channel and floodplain erosion, channel modifications, and erosion of bedrock, where present. These floods exceeded alluvial or bedrock erosion thresholds, and were clearly effective geomorphic agents.

3.2 Calculations of total energy expenditure using time-integrated stream power per unit area

The average energy per unit area (Ω) that is expended over the duration of a flood can be represented by:

$$\Omega = \int \gamma QS/w \, dt$$

where γ is specific weight of the fluid (9800 N/m^3 for clear water), Q is discharge in m^3/s, S is energy slope, w is water-surface width, and t is time in seconds. We have numerically calculated Ω for seven large, well-documented historical floods, and two paleofloods (Table 2), by evaluating reported measurements of valley cross-sections, the flood hydrograph, and a stage-discharge curve. Limitation of the data sources are discussed below.

Following floods, hydrographs are constructed in a variety of ways. The ideal situation is to have a stream gage properly operating throughout the flow. In other situations hydrographs can be constructed from peak-discharge measurements, observations of duration, and assumptions about hydrograph shape [e.g. *Costa*, 1994] (Figure 7). For dam-failure floods, downstream hydrographs can be constructed from reservoir draw-down rates, or dam-break models [e.g. *Jarrett and Costa*, 1986]. Cross-sections of channels and floodplains are nearly always made during surveys following floods [*Williams and Costa*, 1988] (Figure 8). They are required for determining the hydraulic variables necessary to calculate discharge. The primary problem with cross-section accuracy results from possible scour or deposition during the flood, and the

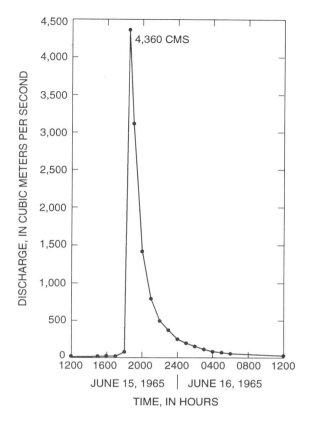

Fig. 7. Hydrograph for the Plum Creek, Colo. flood of June, 1965 (from Osterkamp and Costa, 1987).

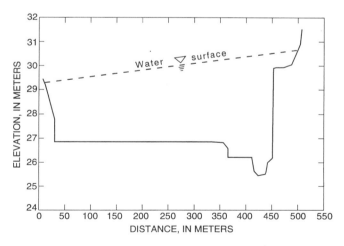

Fig. 8. Cross-section used to construct a flood rating curve for Plum Creek, Colo. flood of June, 1965.

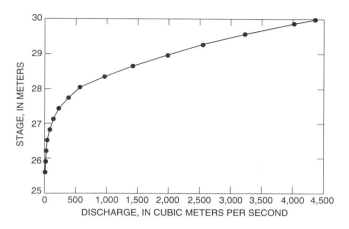

Fig. 9. Stage-discharge curve for the Plum Creek flood of June, 1965.

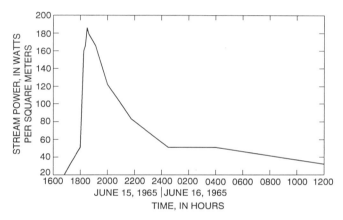

Fig. 10. Stream-power graph showing the distribution of stream power over time for the Plum Creek flood of June, 1965.

consequent uncertainty in the exact location of alluvial boundaries during the flood peak. Judicious selection of cross-section locations can minimize possible errors. Stage-discharge curves (Figure 9) can be constructed for individual cross-sections by computing hydraulic conveyance for different elevations in the cross-section. The primary source of error in this method as we have applied it is the assumption that the flow resistance is constant at all stages. This is clearly untrue, but roughness values generally change by less than a factor of two with increasing stage once flows are overbank [*Hicks and Mason*, 1991]. Using a hydrograph, cross-sections, and a stage-discharge relation, it is possible to construct a curve of the distribution of stream power per unit area throughout the flood. We refer to these plots as "stream-power graphs" (Figure 10). From these graphs, it is possible to integrate the area under the curve to derive the total amount of energy expended by a flood per unit area, as well as an average value of stream power for the flood (Table 2). These data can then be compared to observations and measurements of the magnitude of channel and floodplain disruptions generated by a flood.

Data in Table 2 are not exhaustive, but the information demonstrates the kinds of data required to compute the energy expended per unit area by a flood. The absolute value of flood energy expended per unit area, or the average flood power, may provide no clear differentiation of effective and ineffective floods. Floods with a relatively low average stream power and expended energy can produce catastrophic impacts on alluvial channels and floodplains, such as during the Plum Creek, Colorado flood in 1965 [*Osterkamp and Costa*, 1987]. Other floods like the Centralia, Washington, and Porter Hill, Oregon floods, with five times the peak stream power, and over ten times the average stream power of the Plum Creek flood, can cause only minimal changes. Likewise, long-duration floods on the Mississippi River are capable of generating large values of total energy, but minimal geomorphic changes, because the peak stream power per unit area is too low to exceed resistance thresholds of its channels and floodplain. Apparently, effective floods require some optimal combination of stream power, duration, and energy expenditure. This optimal combination depends on the floodplain and channel resistance thresholds, and the hydrologic characteristics of a particular fluvial system. More data like those in Table 2 will help clarify this important problem. In the next section we propose a model to guide these investigations.

4. EFFECTIVE FLUVIAL EVENTS: A MODEL TO INCLUDE FLOW DURATION

The ability to compute the distribution of stream power per unit area of a flood throughout the hydrograph, combined with consideration of potential landsurface resistance thresholds, allows us to construct a conceptual model of geomorphically effective floods (Figure 11). Three hypothetical stream-power graphs are plotted in Figure 11. Curve A represents a flood of long duration but very low peak stream power. Total energy generated by the flood at a particular site, represented by the area under the stream-power graph, may be large. But in spite of a large total energy expenditure, and long flow duration, peak stream power never rises above the threshold required to significantly disrupt alluvial channels and floodplains. There has been some effort to identify minimum thresholds of critical stream power and boundary shear stress for alluvial systems, but far more work in a variety of environments is required [*Magilligan*, 1992; *Prosser and Slade*, 1994]. Great floods along large, low-gradient rivers such as the Mississippi River flood of 1927, which

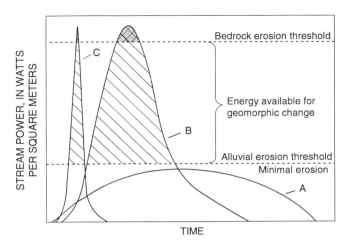

Fig. 11. Conceptual stream-power graphs used to document geomorphic effectiveness of different kinds of floods.

generated peak stream power per unit area of about 12 W/m^2, would be representative of curve A.

Curve B represents large floods that generate high values of peak stream power per unit area, and have moderate to long duration. Average flood stream power per unit area is high, and total energy expended by the flood is large. Peak stream power per unit area can be great enough to generate processes capable of eroding some bedrock boundaries, such as cavitation or macroturbulence [*Baker and Costa*, 1987; *O'Connor*, 1993]. Tremendous changes in alluvial channels are possible, even total unraveling of floodplains, because of the large energy expenditure represented by the area under the stream-power graph above the alluvial threshold. Area above the bedrock threshold represents the amount of energy available to erode and effectively modify bedrock flood-channel boundaries. Floods represented by curve B are likely to be the most geomorphically effective fluvial events in any landscape, and would include exceptional floods like the Rubicon River and Teton River dam-failure floods, and colossal paleofloods like the Missoula and Bonneville floods [*O'Connor and Baker*, 1992; *O'Connor*, 1993].

Stream-power graph C represents floods that generate high values of instantaneous peak stream power per unit area, but are short-lived. The energy represented by the area under the stream-power graph above the alluvial threshold is small, and these floods are impotent to accomplish any significant amount of geomorphic change, even though instantaneous peak stream power per unit area may be among the highest values documented and well above landscape resistance thresholds. Total energy represented by the area under the curve above the bedrock threshold is also small, and the flood engenders little or no perceptible change in alluvial or bedrock channels, even though instantaneous peak stream power per unit area may be comparable to other floods that eroded and shaped bedrock channels. These kinds of stream-power graphs are representative of flash floods in small basins that rise quickly and are gone in a matter of minutes. Such floods are generally caused by cloudburst rainstorms or the rapid failure of natural or constructed dams such as Reservoir No. 3 in Centralia, Washington, or the Porter Hill dam in Oregon.

5. CONCLUSIONS

Floods are a fascinating phenomenon that may or may not be effective agent in shaping the channels and floodplains through which they flow. We have attempted to demonstrate that it is possible to quantify approximately the amount of energy in a flood available for geomorphic work. Construction of stream-power graphs from channel geometry and flood hydrographs shows how stream power varies at a cross-section throughout a hydrograph, and allows computation of the total geomorphic energy expended by a flood. A conceptual model combining flood duration, stream power per unit area, and thresholds for alluvial and bedrock channel erosion can predict geomorphic effectiveness and distinguishes between cases where (a) some floods with long duration and large total energy expenditure along alluvial channels may not be effective channel or floodplain-disrupting events; (b) some floods with very large peak instantaneous stream power per unit area, but low total energy expenditure, may also not be effective channel or floodplain-disrupting events; and (c) floods with a combination of high peak instantaneous stream power, sufficient flood-flow duration, and large total energy expenditure are able to alter significantly the land surface, and become geomorphically effective floods.

While we believe the preceding analysis is a valuable way to conceptualize and perhaps predict channel and floodplain changes, it only addresses half of the force versus resistance equation. We suspect that quantifying landscape resistance and erosion thresholds will prove to be much more difficult than quantifying the hydraulic forces. One conclusion is sure: floods will continue to provide a bounty of questions and opportunities for generations of present and future students.

Acknowledgments. The inspiration and example of Reds Wolman helped focus our thinking about fluvial processes, especially floods. Questions he formulated nearly four decades ago continue to challenge us. We also thank Andy Miller for careful and thoughtful comments that helped reshape an early draft, and Robb Jacobson, Andy Miller, Michael Church, and Peter Wilcock for their insightful reviews. Jasper Hardison assisted in the calculation of hydraulic parameters.

REFERENCES

Andrews, E. D., Marginal bed load transport in a gravel bed stream, Sagehen Creek, California: *Water Resour. Res.*, 30, 2241-2250, 1994.

Baker, V. R., Stream channel response to floods with examples from central Texas: *Geol. Soc. Am. Bull.*, 88, 1057-1071, 1977.

Baker, V.R. and J. E. Costa, Flood power: in Mayer, L. and Nash, D., (Eds), *Catastrophic Flooding*, Allen & Unwin, Boston, pp. 1-21, 1987.

Baldwin, E. M., Eocene stratigraphy of southwestern Oregon: *Oregon Dept. of Geol. and Mineral Ind. Bull.* 83, 40 pp., 1974.

Benito, G. and J. E. O'Connor, Hydraulics and geomorphic features of Late Pleistocene Missoula flooding in the Columbia River Gorge (abstract): *Geol. Soc. Am. Abst. with Prog.*, 23, 5, A207, 1991.

Beven, K. J., The effect of ordering on the geomorphological effectiveness of hydrological events: *Inter. Assoc. Sci. Hydr. Pub.* 132, 510-526, 1981.

Brunsden, Denys, Barriers to geomorphological change: in Thomas, D.S.G. and Allison, R.J., (Eds), *Landscape Sensitivity*, John Wiley and Sons, Inc., Chichester, pp. 7-12., 1993.

Bull, W. B., Threshold of critical power in streams: *Geol. Soc. Am. Bull.*, 90, 453-464, 1979.

Caine, Nel, A uniform measure of subaerial erosion: *Geol. Soc. Am. Bull.* 87, 137-140, 1976.

Costa, J. E., Response and recovery of a Piedmont watershed from Tropical Storm Agnes, June, 1972: *Water Resour. Res.* 10, 106-112, 1974.

Costa, J. E., Paleohydraulic reconstruction of flash-flood peaks from boulder deposits in the Colorado Front Range: *Geol. Soc. Am. Bull.* 94, 986-1004, 1983.

Costa, J. E., A comparison of the largest rainfall-runoff floods in the United States, with those of the Peoples Republic of China: *J. Hyd.*, 96, 101-115, 1987.

Costa, J. E., Multiple flow processes accompanying a dam-break flood in a small upland watershed, Centralia, Washington: U.S. Geol. Surv. *Water-Res. Invest. Rep.* 94-4026, 20 pp., 1994.

Graf, W. L., The development of montane arroyos and gullies: *Earth Surf. Proc.*, 4, 1-14, 1979.

Hicks, D. M., and P. D. Mason, *Roughness characteristics of New Zealand rivers*: Water Res. Surv., Wellington, New Zealand, 329 pp., 1991.

Hoxit, L. R., R. A. Maddox, C. F. Chappell, and S. A. Brua: Johnstown-western Pennsylvania storm and floods of July 19-20, 1977: U.S. Geol. Surv. *Prof. Pap.* 1211, 68 pp.,

1982.

Huckleberry, G., Contrasting channel response to floods on the middle Gila River, Arizona: *Geol.*, 22, 1083-1086, 1994.

Jarrett, R. D. and J. E. Costa, Hydrology, geomorphology, and dam-break modeling of the July 15, 1982 Lawn Lake and Cascade Lake Dam failures, Larimer County, Colorado: U.S. Geol. Surv. *Prof. Pap.* 1369, 78 pp., 1986.

Kochel, R. C., Geomorphic impact of large floods: review and new perspectives on magnitude and frequency: in Baker, V.R., Kochel, R.C., and Patton, P.C., (Eds), *Flood Geomorphology*, John Wiley and Sons, Inc., N.Y., pp. 169-187, 1988.

Magilligan, F. J., Thresholds and their spatial variability of flood power during extreme floods: *Geomorph.* 5, 373-390, 1992.

Mason, R. R., and W. S. Caldwell, The storm and flood of September 15, 1989, in Fayetteville, North Carolina: U.S. Geol. Surv.*Water-Res. Invest. Rep.* 92-4097, 26 pp., 1992.

Miller, A.J., Flood hydrology and geomorphic effectiveness in the central Appalachians: *Earth Surf. Proc. and Landforms*, 15, 119-134, 1990.

Miller, A.J. and D. J. Parkinson, Flood hydrology and geomorphic effects on river channels and flood plains: the flood of November 4-5, 1985, in the South Branch Potomac River Basin of West Virginia: U.S. Geol. Surv. *Bull.* 1981, pp. E1-E96., 1993.

Moss, J. H. and R. C. Kochel, Unexpected geomorphologic effects of the Hurricane Agnes storm and flood, Conestoga drainage basin, southeastern Pennsylvania: *J. Geol.*, 86, 1-11, 1978.

Nanson, G.C., Episodes of vertical accretion and catastrophic stripping: a model of disequilibrium flood plain development: *Geol. Soc. Am. Bull.* 97, 1467-1475, 1986.

Nanson, G.C. and D. Hean, The West Dapto flood of February 1984: rainfall characteristics and channel changes: *Australian Geog.*, 16, 4, 249-258, 1985.

O'Connor, J.E., Hydrology, hydraulics, and geomorphology of the Bonneville Flood: Geol. Soc. Am. *Spec. Pap.* 274, 82 pp., 1993.

O'Connor, J. E., and V. R. Baker, Peak discharges from Glacial Lake Missoula: *Geol. Soc. Am. Bull.*, 104, 267-279, 1992.

Osterkamp, W.R. and J. E. Costa, Changes accompanying an extraordinary flood on a sand-bed stream: in Mayer, L. and Nash, D. (Eds), *Catastrophic Flooding*, Allen & Unwin, Boston, pp. 201-224, 1987.

Prosser, I.P. and C.J. Slade, Gully formation and the role of valley-floor vegetation, southeastern Australia: *Geol.*, 22, 1127-1130, 1994.

Ray, H.A. and L. C. Kjelstrom, The flood in southeastern Idaho from the Teton Dam failure of June 5, 1976: U.S. Geol. Surv. *Open-File Rep.* 77-765, 48 pp., 1978.

Scott, K. M., and G. C. Gravlee, Flood surge on the Rubicon River, California-hydrology, hydraulics, and boulder transport: U.S. Geol. Surv. *Prof. Pap.* 422-M, 40 pp., 1968.

Williams, G. P., Paleohydrological methods and some examples from Swedish fluvial environments, I - Cobble and boulder deposits: *Geograf. Annaler*, 65A, 227-243, 1983.

Williams, G. P., and J. E. Costa, Geomorphic measurements after a flood: in Baker, V.R., Kochel, R.C., and Patton, P.C., (Eds), *Flood Geomorphology*, John Wiley and Sons, Inc., N.Y., pp. 65-77, 1988.

Wolman, M.G. and R. Gerson, Relative scales of time and effectiveness of climate in watershed geomorphology: *Earth Surf. Proc.*, 3, 189-208, 1978.

Wolman, M. G., and J. P. Miller, Magnitude and frequency of forces in geomorphic processes: *J. Geol.* 68, 54-74, 1960.

Zimmerman, R.C., J. C. Goodlett, and G. H. Comer, The influence of vegetation on channel form in small streams: *Inter. Assoc. Hydrol. Sci. Pub.* 75, 225-275, 1967.

John E. Costa and Jim E. O'Connor, U.S. Geological Survey, 5400 MacArthur Blvd., Vancouver, WA 98661

Valley Morphology and Boundary Conditions Influencing Spatial Patterns of Flood Flow

Andrew J. Miller

Department of Geography, University of Maryland Baltimore County, Baltimore, Maryland

Large floods in narrow, steep mountain valleys are more likely to leave a lasting imprint on the landscape than floods of comparable magnitude in broad, low-gradient valleys. Longitudinal variations in valley width and channel orientation are potentially more important than average width in determining location and severity of flood impacts. In order to explore the hydraulic implications of locally-varying boundary conditions, a set of eight finite-element meshes representing alternative versions of valley topography was designed for use with numerical flow models. A two-dimensional depth-averaged flow model (HIVEL2D), based on the conservation form of the shallow-water equations, was used to simulate flood flows routed through each design reach. Results, presented in the form of velocity fields, water-surface elevations, and calculated Froude number and shear stress distributions, indicate that valley expansions, channel bends along straight valleys, and flow obstructions may have a profound impact on the pattern of flow and geomorphic consequences. Maximum calculated shear stress on the floodplain along the outside bank at a channel bend is comparable to maximum channel shear stress along a straight, narrow canyon reach. Maximum shear stress on the floodplain where a valley expansion coincides with a channel bend is as much as three times greater than the maximum channel shear stress along the canyon reach and 5-7 times greater than the maximum floodplain shear stress along a constant-width valley with a straight channel. Comparison of the predicted spatial distributions of hydraulic parameters with photointerpretation of flood impacts suggests that hydraulic modeling may be useful for predicting spatial patterns of erosion and deposition resulting from a catastrophic flood.

1. INTRODUCTION

M.G. Wolman first questioned the relative importance of floods as geomorphic agents after observing the modest impacts of an outstanding flood on the Connecticut River in 1955 [*Wolman and Eiler,* 1958]. In one of the most frequently cited papers in all of geomorphic literature, *Wolman and Miller* [1960] argued that frequently recurring flow events of moderate size accomplish more cumulative geomorphic work than rare, large-magnitude floods. Another landmark paper, published in the same year by Hack and Goodlett, cited evidence from the central Appalachians in suggesting that rare, large-magnitude floods might in fact have a dominant impact on some landscapes. *Wolman and Gerson* [1978] later focused attention on the broader concept of geomorphic effectiveness, the capacity of a geomorphic event to create a new landform or a change in an existing landform that persists over time scales of decades to centuries.

Large unit discharges are necessary but not sufficient conditions for geomorphic effectiveness. Geomorphic constraints imposed by geologic setting have a critical influence on flood hydraulics, and effective floods are therefore more likely to occur in steep, narrow valleys than in broad, low-gradient valleys [*Baker,* 1977; *Nanson,* 1986; *Baker and Costa,* 1987; *Kochel,* 1988; *Jacobson, et al.,* 1989; *Miller,* 1990; *Magilligan,* 1992; *Costa and O'Connor,* this volume]. Most research on this question involves calculation of hydraulic parameters, including boundary shear stress and stream power per unit boundary area. Several of the authors listed above have sought to define threshold values of these parameters to distinguish effective floods from ineffective floods.

For most geomorphically effective floods, hydraulic parameters must be reconstructed from indirect evidence [*Williams and Costa,* 1988]. Even where such a flood affects a gaged stream, the gage in most cases is inundated by rising flood waters or provides an inadequate record because peak flood stage exceeds the maximum for the calibrated stage-discharge relationship. Reconstruction of peak discharge for assessment of hydraulic parameters is accomplished most frequently using either the slope-area method [*Benson and Dalrymple,* 1967] or the standard step method [*Chow,* 1959; Hydrologic Engineering Center, 1982]. Both methods use surveyed high-water marks and cross-sections in calculating solutions for the one-dimensional energy equation. Cross-sectional average values of boundary shear stress and unit stream power typically are calculated as

$$\tau = \rho g R s$$

and

$$\omega = \rho g Q s w^{-1}$$

where ρ is the density of water (1000 kg m^{-3}), g is the acceleration due to gravity (9.808 ms^{-2}), R is hydraulic radius, s is energy gradient, Q is discharge in m^{3}s^{-1}, and w is cross-section width in m.

These parameters are helpful in discriminating effective floods from ineffective floods, but the explanatory power of most hydraulic reconstructions is limited by a relative lack of temporal and spatial resolution. For example, reconstruction of discharge and unit stream power at peak stage does not tell us how flow conditions varied before and after peak stage, nor does it tell us how long peak-flow conditions persisted. *Costa and O'Connor* [this volume] ask whether flood duration is a significant determinant of geomorphic effectiveness and suggest that integrating stream power over the duration of the flood hydrograph may provide useful insights.

In this paper I consider the role of spatially-varying boundary conditions and flow patterns as determinants of geomorphic effectiveness. Slope-area measurements are typically made along straight reaches of uniform or nearly uniform width, and therefore cannot account for significant longitudinal variations in boundary conditions that may have affected flood hydraulics elsewhere along the valley. Hydraulic reconstructions based on the standard step method can account for longitudinal variability, but cannot predict lateral variations within a cross-section or the spatial pattern of flow vectors. In order to explore these

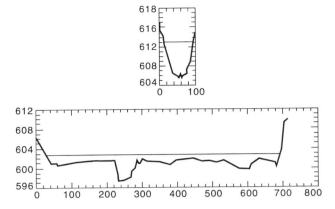

Fig. 1. Two cross-sections of the valley of the South Fork South Branch Potomac River at Fame, upstream of Moorefield, West Virginia. The upper cross-section is located approximately 2 km upstream of the lower cross-section. Water levels indicated are based on surveyed high-water marks from the November 1985 flood. Elevation datum is arbitrary.

features of the flow field it is necessary to use two-dimensional flow models [*Miller,* 1994]. Following initial presentation of observations motivating this approach, I use a two-dimensional flow model to explore a series of alternative spatial patterns of flood hydraulics associated with varying boundary conditions.

1.1 Spatially-varying Boundary Conditions and Geomorphic Effectiveness

Evidence that spatially-varying boundary conditions may have a critical influence on the location and severity of flood impacts is provided by observations from two historic floods: the November 1985 flood along the South Branch Potomac River and its tributaries [*Miller and Parkinson,* 1993], and the Hurricane Camille flood that struck the area around Nelson County, Virginia in August 1969 [*Williams and Guy,* 1973; *Johnson,* 1983].

The two cross-sections illustrated in Fig. 1 were surveyed at sites along the South Fork South Branch Potomac River in West Virginia that are only about 2 km apart. Drainage area at this location is between 300 and 400 km^2, and peak discharge during the record flood of November 1985 probably was on the order of 1500 to 2000 m^{3}s^{-1} (based on interpolation between indirect discharge calculations from U.S. Geological Survey gage sites upstream and downstream). In the aftermath of the flood the most dramatic examples of channel and floodplain erosion in this valley and adjacent valleys were observed at sites where flow emerged from a constriction [*Miller and Parkinson,* 1993].

The longitudinal trend in valley width along an extended

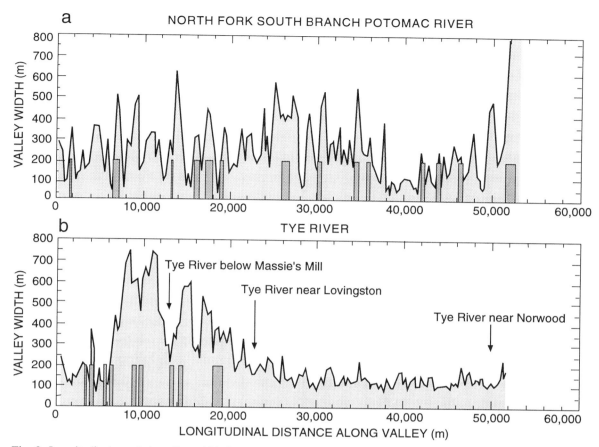

Fig. 2. Longitudinal trends in valley width for two rivers; light shaded bars superimposed on the line graphs indicate locations affected by extensive floodplain scour and channel widening. Downstream is to the right. (a) North Fork South Branch Potomac River upstream of the South Branch Potomac River confluence; flood damage caused by the November 1985 flood. (b) Tye River upstream of the James River confluence; flood damage caused by the Hurricane Camille flood of August 1969.

reach of the North Fork South Branch Potomac River is illustrated in Fig. 2a. An indirect discharge estimate of 2550 m^{3}s^{-1}, based on U.S. Geological Survey field surveys following the flood, was made near the downstream end of this reach at a location with a drainage area of 813 km^2 [*Miller and Parkinson,* 1993]. Variations in valley width along the reach appear to have influenced the spatial distribution of flood damage: shaded bars superimposed on the line graph indicate locations along the valley where a significant portion of the floodplain was destroyed in the November 1985 flood by stripping away of alluvial deposits or dissection by anastomosing erosion channels. Among the 14 sites identified, 11 are located at or just below valley expansions.

A similar analysis was performed for the Tye River, which was one of the two rivers most heavily impacted by the Hurricane Camille flood. As Fig. 2b shows, the upstream reaches of the Tye exhibit valley-width variations similar to those observed along the North Fork South Branch Potomac River; whereas the narrower downstream reaches show modest width variations. Peak discharge estimates based on indirect slope-area measurements by U.S. Geological Survey personnel at three locations along the Tye River following Hurricane Camille (Table 1) were the highest discharges ever recorded for comparable drainage areas in this part of the U.S. [*Miller,* 1990]. Shear stress and unit stream power values were calculated using results from these slope-area measurements (Table 1). The approximate locations of these measurements are indicated in Fig. 2b.

The values of stream power calculated for the two indirect-discharge measurements along the upper Tye suggest considerable local variability in flood hydraulics, possibly as a result of local variations in energy gradient

Table 1. Hydraulic Parameters Based on Slope-area Measurements for Three Sites along the Tye River
Hurricane Camille Flood, August 19-20, 1969

Drainage area (km^2)	Discharge (m^{3}s^{-1})	Average top width (m)	Energy gradient	Channel gradient	Shear stress (Nm^{-2})	Unit stream power (Wm^{-2})
Tye River below Massie's Mill, Virginia						
170.9	1910	285	0.0056	0.0040	120	344
Tye River near Lovingston, Virginia						
238.3	2180	238	0.0006	0.0029	27	45
Tye River near Norwood, Virginia						
932.4	5490	151	0.0022	0.0025	184	692

under the controlling influence of valley constrictions and expansions. The indirect discharge measurement farther downstream near the mouth of the Tye River yields a value of unit stream power twice as high as the larger of the two upstream values. The downstream increase in peak discharge, together with the decline in average valley width, makes it likely that peak cross-sectional average unit stream power at most locations along the lower Tye would probably have been higher than either of the values calculated for the upper Tye.

The longitudinal distribution of the filled bars on Fig. 2b indicates virtually no significant floodplain erosion along the lower Tye, despite the extremely high discharge and high unit stream power estimated for this reach. Significant erosion was observed only along parts of the upper Tye, where valley width was much more variable. If we assume that reach-average or cross-section-average values of unit stream power can be used to define hydraulic thresholds for geomorphic effectiveness [e.g. *Miller*, 1990; *Magilligan* 1992], the contrast between the upper and lower Tye is puzzling and does not appear to be explicable by assuming a difference in floodplain erosion resistance. This was one of the observations motivating the two-dimensional modeling approach described below.

1.2 Topographic irregularities and channel orientation

The association between valley expansions and floodplain erosion along the upper Tye River in the Camille flood (Fig. 2b) is less obvious than along the North Fork South Branch Potomac River in the 1985 flood (Fig. 2a). The width variations documented in Fig. 2b are based on photointerpretation of the width of the inundated area, which in virtually all cases is synonymous with the distance between one valley wall and the other. It may be, however, that complex topography of the valley floor creates effective constrictions at locations that are not identified as such in the longitudinal plots shown here.

If the valley cross-section is occupied by multiple surfaces of differing elevation, the nature of the flow cross-section may change with rising stage, and flow over different surfaces may follow different trajectories. A case in point is the Tye River near Roseland (Fig. 3), which is located a short distance upstream of the Lovingston slope-area measurement site. At this location there are at least three different topographic levels on the valley floor. On the west side of the valley, the floodplain and two terraces are separated by several scarps oriented almost perpendicular to the channel. Peak discharge at this site during Hurricane Camille was approximately 2300 m^{3}s^{-1} from a drainage area of 230 km^2; flood waters covered the floodplain and both terrace levels. The flow cross section as measured by the total width of valley inundated shows no constriction. However, at the middle of the reach indicated in Fig. 3, most of the flow evidently was confined between the first terrace scarp and the east wall of the valley. This portion of the cross section is only 85 m wide, forming a partial or effective constriction. The erosion pattern observed immediately downstream appears similar to erosion patterns observed in other settings where valley constrictions are more clearly defined. The two-dimensional distribution of flow vectors associated with this geomorphic impact would not be predicted by a one-dimensional analysis of trends in valley width.

Channel plan form may inhibit or enhance flood effectiveness. When a flood fills the entire valley to

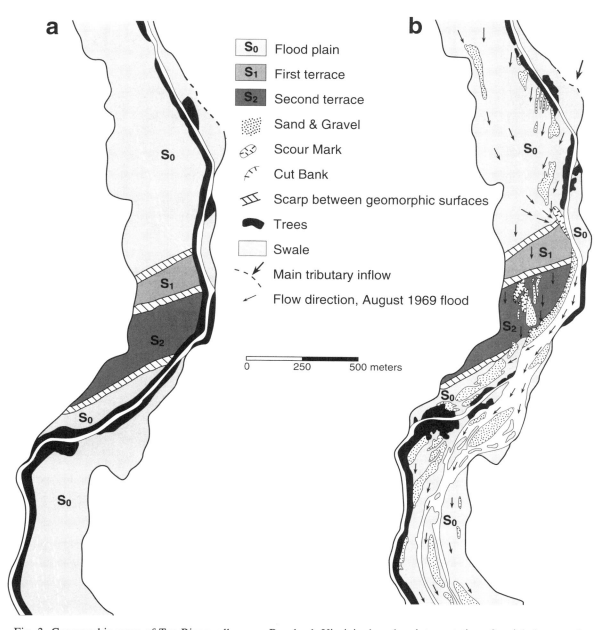

Fig. 3. Geomorphic maps of Tye River valley near Roseland, Virginia, based on interpretation of aerial photographs taken before and after the Hurricane Camille flood. S_0, S_1, and S_2 are geomorphic surfaces interpreted as the floodplain and the first and second terraces above the floodplain. (a) January 1967. (b) August 1969. Arrows represent flow direction interpreted from alignment of erosional and depositional features in post-flood aerial photographs.

sufficient depth, the valley in effect becomes the channel and the valley walls have a tendency to steer the flow, particularly along the narrower reaches. As long as the orientation of the channel is parallel to the valley walls and the width of the valley does not change very much, there is relatively little opportunity for development of the kind of hydraulic discontinuity that generates erosion features around flow obstructions or at topographic steps. The response of the lower Tye River to the Hurricane Camille floods illustrates this quite well.

On the other hand, at valley crossings where the steering influence of the valley walls is transverse to the orientation of the channel, a significant component of flood flow crosses onto the floodplain. Prominent erosion and deposition features have an opportunity to develop at such locations. Where a valley crossing is located at the mouth of a constriction there is an opportunity for catastrophic unraveling of the valley floor (Fig. 4).

2. SIMULATION MODELING OF FLOOD HYDRAULICS

Floods are highly turbulent, three-dimensional, time-varying phenomena. Although flood flow should obey the basic principles of hydrodynamics, any attempt at numerical modeling necessarily involves simplifying assumptions.

However, floods of the type described here occur so rarely that scientists seldom have the opportunity of observing them in progress. Further, it is not at all clear that collection of detailed information on the spatial pattern of water-surface profiles, velocities, sediment transport rates, roughness parameters, and time-varying boundary conditions would be feasible. Under these circumstances, the use of a numerical model may provide insights that might not otherwise be available. These may in turn be useful in designing experiments to test hypotheses.

In the remainder of this paper I report on some simulation experiments that were designed to explore how different spatial patterns of boundary conditions affect flood hydraulics. Following the same strategy as *Miller* [1994], I have chosen a set of idealized valleys that are variants on the same basic template. For each example I have constructed a finite-element mesh and simulated flow patterns for selected values of peak discharge using a two-dimensional flow model. The results are compared with observations of the geomorphic impacts of real floods.

2.1 Flow Model

Models of the type utilized in this study are based on a vertically-integrated form of the full three-dimensional Reynolds equations for turbulent flow, known as the shallow-water equations [*Samuels*, 1989]. The assumption is made that vertical motions are negligible by comparison with horizontal motions and gravitational acceleration. Numerical solutions to the flow equations yield water depth and x- and y-components of velocity or unit discharge for each node in the flow field.

Because preliminary results indicated that supercritical-subcritical flow transitions may occur in some regions of the simulated flow fields, the model HIVEL2D was used in this study. HIVEL2D was designed specifically for simulation of flow fields containing supercritical and

Fig. 4. Photographs showing the valley of the North Fork South Branch Potomac River at a site 3 km downstream of Seneca Rocks, West Virginia. (a) May 1980. (b) November 1985.

subcritical flow regimes and transitions between them, and has yielded satisfactory results in simulating physical modeling results from a variety of test cases [*Stockstill and Berger*, 1994]. The shallow-water equations are expressed in conservation form, e.g. derivatives in the continuity and momentum equations are taken with respect to x- and y-components of unit discharge rather than velocity [*Abbott*, 1979]. The conservation forms yield solutions that are more stable in the vicinity of discontinuities such as hydraulic jumps and vertical steps [*Whitlow and Knight*, 1992; *Samuels*, 1989].

The governing continuity and momentum equations [see *Stockstill and Berger*, 1994, for a complete listing] are solved using a Newton-Raphson iterative method. The variables are discretized using a Petrov-Galerkin finite element formulation, which is quite stable and allows accurate representation of unstable features such as shock waves. HIVEL2D is an unsteady flow model using a finite-difference scheme to solve for the temporal derivatives. Steady-flow problems, such as those presented here, still require iteration over enough time steps to reach a steady flow solution.

Inflow and outflow boundary conditions are required. Subcritical outflow boundaries require specification of a tailwater elevation and subcritical inflow boundaries require specification of x- and y-components of unit discharge or velocity; supercritical inflow boundaries additionally require depth to be specified. A partial slip condition is enforced along the sides, e.g. bed resistance coefficients are also applied along the sidewall for elements along the mesh boundary [*Stockstill and Berger*, 1994].

2.2 Finite-element Mesh Construction

HIVEL2D can simulate flow over complex topography, represented by a set of quadrilateral and triangular elements with x, y, and z coordinates provided for each node in the mesh. The model is compatible with meshes created by the FastTABS graphical pre- and post-processor [*Brigham Young University*, 1994], which was used in this study.

The basic template for all of the finite-element meshes in this study had the following characteristics:
- Trapezoidal channel, 2 m deep and 35 m wide
- Valley attaining maximum width of 135 m
- Constant downstream gradient of 0.005 applied to both channel and floodplain, with no transverse gradient (i.e. flat floodplain, no levees or overflow channels)
- Constant Manning's n of 0.035 applied to both channel and floodplain along broad reaches.

Channel and valley dimensions and gradient, though idealized, fall within the range observed along the valleys of the South Branch Potomac River [*Miller and Parkinson*, 1993]. Although it is possible to assign different values of the resistance coefficient to different areas of the mesh, my intention is to focus exclusively on the influence of morphology and, to the extent possible, to set aside the role of spatially varying roughness.

All meshes were designed with an entrance reach and an exit reach far enough away from flow regions of interest to avoid any direct influence by upstream or downstream boundary conditions. Tailwater elevations in all simulations were set at levels close enough to normal depth that the flow field converged to normal depth within a short distance upstream of the outflow boundary. Although all examples illustrated here cover longitudinal distances of 500 to 1000 m, the simulations actually covered longitudinal distances of 1925 to 2325 m.

The following variants on the basic template were designed for purposes of comparison:
- Straight channel without floodplain, constant width of 35 m (mesh 1)
- Straight channel running along valley centerline, constant valley width of 135 m (mesh 2)
- Straight channel running along valley centerline with valley expansion from 35 m to 135 m (mesh 3)
- Straight channel running along valley centerline, channel bends to right and continues along valley wall; constant valley width of 135 m (mesh 4)
- Valley expansion from 35 m to 135 m, channel remains straight until 400 m downstream of expansion and then bends to right (mesh 5)
- Same as mesh 5, but with several elements removed from mesh along river bank to simulate an obstruction or barrier to flow (mesh 5b)
- Valley expansion from 35 m to 135 m, channel bends to right at beginning of expansion and runs along valley wall (mesh 6)
- Same as mesh 6, but with several elements removed from mesh along river bank to simulate an obstruction or barrier to flow (mesh 6b)

The relevant regions of meshes 1 through 6 are illustrated in Fig. 5. Comparison of flow fields routed through these meshes allows us to isolate the effects of individual morphologic factors or combinations of factors on the spatial pattern of flow and to explain the associated potential for severe geomorphic impacts.

At present HIVEL2D does not have an algorithm capable of simulating wetting and drying with rising and falling water levels; therefore mesh design must ensure that all wet elements under initial conditions remain wet throughout the flow simulation. Steep walls normally present in a constricted mountain valley were therefore excluded from these simulations. The walls would normally be expected to reduce flow velocity along the margins. In order to eliminate numerical instabilities associated with unrealistically high velocities along the boundary at the exit from the constriction, a single row of elements along the boundary in the constricted reaches and immediately downstream of the constrictions was assigned a Manning coefficient of 0.13. In addition, the bed profile along the channel bend in meshes 4 through 6 was made to slope downward toward the outer or concave bank, in order to simulate effects normally induced by topographic steering of flow through a bend.

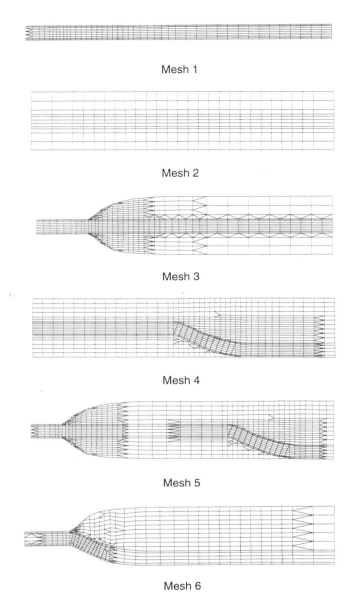

Fig. 5. Finite-element meshes compiled for simulation of flood flows, reflecting alternative valley and channel morphology. All examples shown are 700 m long; each mesh has extended entrance and exit reaches that are not illustrated.

2.3 Sidewall Roughness

As was indicated above, HIVEL2D assumes a partial slip condition. Thus elements along the flow boundary experience flow resistance both at the bed and along the outside wall, even though the wall is not defined as a topographic feature in the mesh. Other, related models based on the non-conservation form of the shallow-water equations, such as RMA2 [*Thomas and McAnally*, 1990], assume a slip condition along the boundary such that the resistance coefficients are applied only along the bed. This difference is of no practical importance in a broad flow field of modest depth; but it affects the results in a long, constricted reach with deeper flow, as is simulated in mesh 1. Application of a high value of Manning's n along a narrow row of elements on the boundary helps to compensate for the slip condition in RMA2 and gives somewhat more realistic results for mesh 1 than the same set of parameter choices in HIVEL2D. In the ensuing discussion I have chosen to use RMA2 results for mesh 1, and to compare these results with HIVEL2D simulations using an n value of 0.13 along the margins of the constricted reach in all meshes.

2.4 Calculation of Shear Stress

Bed shear stress is calculated in this class of flow models using a drag-force formulation, i.e. shear stress is expressed as the product of a drag coefficient and a squared velocity term. The drag coefficient can be defined as

$$C_D = \rho g n^2 h^{-1/3}$$

where ρg is the unit weight of water, n is Manning's n, and h is water depth.

Because this coefficient incorporates an n^2 term, it is clear that the specification of roughness coefficients has a powerful influence on the calculated distribution of shear stress. Because Manning's n is used as a parameter to account for many forms of flow resistance (including, for example, the retarding influence of standing vegetation), it is not clear how the calculated value of shear stress should be partitioned between bed shear stress and cumulative resistance to flow resulting from other sources such as bed forms, vegetation, and flow obstructions. This question is of key importance, as most sediment transport formulae are essentially power functions of bed shear stress [*Shen and Julien*, 1993], but sediment transport presumably is influenced only by stress that actually impinges on the bed. The decision to choose a fixed value of Manning's n for the entire flow field was made in order to analyze the pattern of flood hydraulics independent of this confounding factor. Note that although I have assumed an n value of 0.130 for the row of elements along the valley walls in the constriction, I have used an n value of 0.035 in calculating shear stress values for all points in the flow field. The elevated values of Manning's n along the margins were chosen as surrogates for steep, often forested valley walls that were not explicitly included in the mesh. The flow resistance that would actually be exerted along a considerable surface area of the valley walls is here allocated to a narrow strip only a couple of meters wide. A more detailed exploration of the shear stress distribution along the walls of a narrow canyon may be accessible through the use of physical modeling experiments.

3. RESULTS

Flow simulations were carried out for discharges of 300, 500, and 1000 m^3s^{-1} routed through each of the meshes described above. The 300 m^3s^{-1} simulations for meshes 4 and 5 developed a small area of instability with extremely high velocities at shallow depths, owing to the fact that HIVEL2D lacks an algorithm for wetting and drying of the surface with rising and falling stage. Although the model did in fact converge and the solutions appeared reasonable outside of this small area, the results for these two simulations were rejected and are not presented here.

I have used model output values of u- and v- velocity components and water depth to calculate values of bed shear stress (using equation (3) above) and Froude number, calculated as

$$F = (u^2+v^2)^{0.5}(gh)^{-0.5}$$

for each node. Froude number is significant in that transitions between supercritical and subcritical flow are generally accompanied by highly turbulent conditions and very high rates of energy expenditure.

3.1 Comparison of Mesh 1 and Mesh 2

The flows routed through both mesh 1 and mesh 2 are essentially one-dimensional simulations: they have straight valley and channel patterns, uniform longitudinal gradients with the water surface parallel to the bed, and no significant longitudinal variation in velocity, shear stress, or Froude number. Calculation of cross-sectional average values of unit stream power by applying equation (2) to the two meshes (Table 2) shows that for any discharge value the results vary only as a result of the difference in valley width; thus mesh 1 consistently has a unit stream power that is 3.9 times the value calculated for mesh 2, suggesting a much greater potential for accomplishing geomorphic work. However the geomorphic work actually accomplished in a flood is presumably more closely associated with local variations in the spatial distributions of hydraulic parameters than with cross-sectional averages. If we calculate instead the value of unit stream power for the

Table 2. Unit Stream Power Comparisons, Mesh 1 vs. Mesh 2

		Discharge in m^3s^{-1}		
		300	500	1000
		Wm^{-2}	Wm^{-2}	Wm^{-2}
Mesh 1	Cross-section average	420	700	1400
Mesh 2	Cross-section average	110	180	360
Mesh 1	Channel center	520	840	1650
Mesh 2	Channel center	410	540	790

center of the channel by substituting the point value of unit discharge (product of depth and velocity) for the average value of unit discharge (Qw^{-1}) in equation (2), the discrepancy between the two meshes is smaller but increases with discharge, from a ratio of 1.3 at 300 m^3s^{-1} to a ratio of 2.1 at 1000 m^3s^{-1}.

As the remainder of this paper is concerned with detailed two-dimensional patterns of flood hydraulics rather than cross-section averages, subsequent analyses focus on velocity, Froude number, and shear stress; point values of unit stream power differ from point values of shear stress but follow the same general spatial distributions and therefore add little useful information to our discussion.

A more detailed comparison of the modeling results for these two cases, illustrated in cross-section (Fig.6), yields the following observations: Water depth increases more steeply with discharge for mesh 1, as mesh 2 has a floodplain that can accommodate flow spilling out of the channel banks and mesh 1 does not. Water depth over the floodplain in mesh 2 is 0.4 m in the 300 m^3s^{-1} simulation and increases to about 1.6 m in the 1000 m^3s^{-1} simulation; therefore relative submergence of the floodplain compared with the channel increases from 0.17 to 0.44, and the disparity in velocity between the floodplain and the channel is less pronounced at higher discharge. At 300 m^3s^{-1}, most of the discharge in mesh 2 is confined to the channel, and peak velocity and shear stress attained in the confined valley of mesh 1 are only 10% and 15% higher than in mesh 2. As floodplain submergence increases at higher discharges, the difference between the confined channel of mesh 1 and the broader valley of mesh 2 becomes more apparent: peak velocity at 1000 m^3s^{-1} is 36% greater in mesh 1 than in mesh 2, and peak shear stress is 56% greater.

For both of these examples, Froude number along the channel is relatively insensitive to increases in discharge.

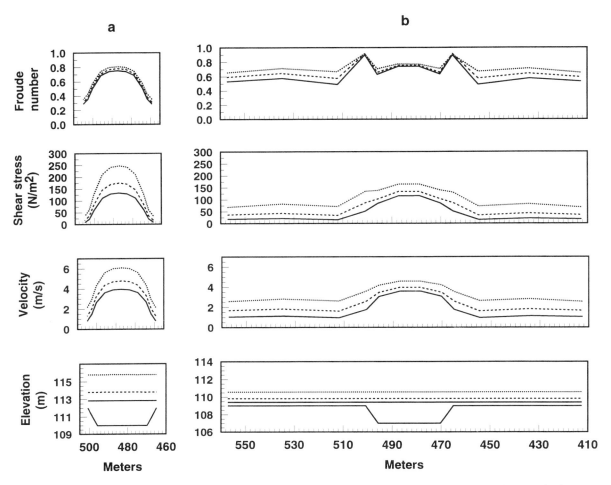

Fig. 6. Comparison of simulation results from typical cross-sections of mesh 1 and mesh 2. In both cases the bottom diagram shows the channel bed and the water surface for discharges of 300 m^{3}s^{-1}, 500 m^{3}s^{-1}, and 1000 m^{3}s^{-1}. In all diagrams the 300 m^{3}s^{-1} result is plotted as a solid line, the 500 m^{3}s^{-1} result is plotted as a dashed line, and the 1000 m^{3}s^{-1} result is plotted as a dotted line. (a) Mesh 1; results based on simulation using RMA2 with Manning's n = 0.130 along the edges of the flow field. (b) Mesh 2; results are based on simulation using HIVEL2D.

Froude number does increase somewhat on the floodplain, but flow remains subcritical throughout the mesh. Flow approaches a critical condition only along the channel banks in mesh 2, attaining a maximum value of 0.91 where the sudden lateral transition in flow depth is steeper than the lateral transition in velocity. Contour maps of the velocity, shear stress, and Froude-number distributions for mesh 2 (Fig.7) indicate that the most intense hydraulic conditions are confined to the channel or the channel banks. From the information available it cannot be determined whether the predicted shear stress is sufficient for disruption of the floodplain, but it appears that the most likely geomorphic effect even at extremely high discharge is moderate widening of the channel.

3.2 *Comparison of Mesh 1 and Mesh 3*

Although hydraulic conditions in the narrow valley illustrated by mesh 1 are more intense than in the broader valley of mesh 2, the transition that occurs at a valley expansion is accompanied by hydraulic conditions more extreme than those observed in either of these cases. Perspective views plotted for three different discharges (Fig. 8) indicate that the water surface in the mesh 1 simulations retains a constant gradient parallel to the bed, even as depth and velocity increase with discharge (Fig. 6a). The valley expansion in mesh 3 allows flood flows to diverge outward from the channel (Fig. 9), and the water surface drops steeply in the throat of the expansion just

Shear stress (N/m2)

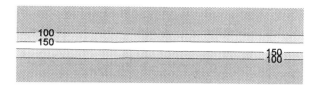

Froude number

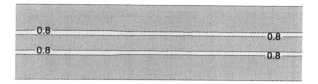

Velocity (m/s)

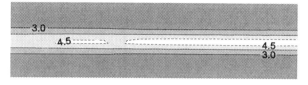

Elevation (m)

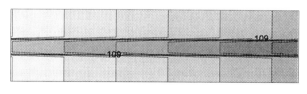

Fig. 7. Contour maps showing spatial distribution of hydraulic parameters in comparison to topography for a portion of mesh 2. Discharge is 1000 m^{3}s^{-1} and flow is from left to right. Contour intervals are 0.5 m for topography, 1.0 ms^{-1} for velocity, 0.2 for Froude number, and 50 Nm^{-2} for shear stress. Dashed lines indicate contour levels marking half contour intervals. Scale is in meters.

upstream of where flood flow spreads out over the valley floor (Fig. 8). The gradient at this location also becomes much steeper as discharge increases.

The lack of recirculating eddies flanking the channel along the expanding reach (Fig. 9) is attributable to the fact that channel and valley slope are constant and the tailwater elevation was set to produce normal flow. Other simulation experiments show eddy development when a jet enters stagnant water or when ponding induced by a downstream control causes an adverse water-surface gradient along the floodplain margins.

The steep gradient in the throat of the expansion leads to an increase in velocity, Froude number, and shear stress (Fig. 10). At a discharge of 300 m^{3}s^{-1}, flow remains mostly subcritical, with Froude numbers greater than 1 appearing only in narrow parallel strips along the channel banks where shallow flood flow spills out and over the floodplain. Peak velocity occurs in the center of the channel at the beginning of the expansion and decays rapidly downstream; elevated velocity and shear stress are mostly confined to the channel, with contour lines oriented parallel to the channel banks. Peak shear stress is nearly 50% higher than the value calculated for the same discharge along mesh 1.

At higher discharge, the steeper water-surface gradient and steeper increase in velocity lead to a transition from subcritical to supercritical flow that extends from the beginning of the expansion for a considerable distance downstream and covers the entire width of the channel as well as a portion of the floodplain. Peak shear stress is almost twice the value calculated for the same discharge along mesh 1. The location of maximum velocity shifts slightly downstream, and the shape of the mapped velocity, Froude-number, and shear-stress contours all indicate a bulb-shaped area characterized by extremely turbulent flow (Fig. 10). Examination of this pattern suggests the likelihood of dramatic channel widening with a bulb-shaped form beginning at the exit from the constriction and extending a short distance downstream; the steep drop in shear stress on the periphery of this zone should lead to deposition of coarse splay deposits rimming the eroded area. This is precisely the type of feature illustrated in Fig. 11.

3.3 Comparison of Meshes 2, 4, and 5

A bend in the channel along a straight valley also creates perturbations in the water surface and in the spatial pattern of hydraulic parameters associated with flood conditions. Although the results may be less dramatic than those discussed in the previous example, significant geomorphic impacts may still occur at high flow. Examination of vector plots (not reproduced here) for flow simulations of this situation indicates that much of the flow in the channel will continue to follow the channel outline, but that some component of the flow in the channel will follow a trajectory parallel to the valley walls, crossing onto the floodplain as the channel bends away from its upstream orientation.

Two meshes were designed to simulate flows associated with this configuration. Mesh 4 represents a straight valley

68 VALLEY MORPHOLOGY, BOUNDARY CONDITIONS, AND FLOOD-FLOW PATTERNS

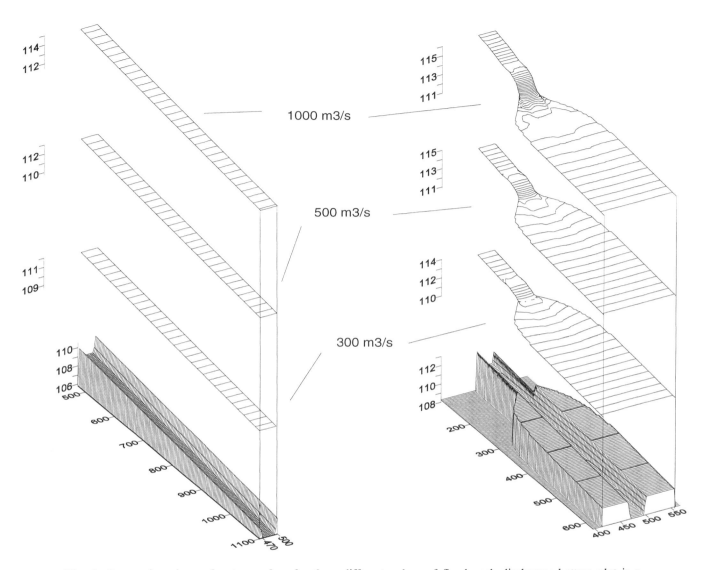

Fig. 8. Perspective views of water surface for three different values of flood peak discharge; bottom plot is a perspective projection of mesh topography. All plots have the same vertical scale. All scale units are in meters. (a) [left,] Mesh 1. (b) [right] Mesh 3.

of constant width, so that the only factor contributing to perturbations in the flow field is the bend in the channel itself. Mesh 5 has a bend several hundred meters downstream of a valley expansion; the intention of this experimental design was to ascertain whether an upstream expansion might "train" the flow in the channel and enhance the amplitude of the disturbance downstream in the vicinity of the bend.

Perspective views of the water surface for both meshes at a discharge of 1000 m^{3}s^{-1} (Fig. 12) indicate the development of superelevation in the bend, with a depression in the water surface along the inside of the bend and a ridge along the outside bank. The maximum difference in head across the channel in the mesh 4 simulation is 0.50 m at a discharge of 1000 m^{3}s^{-1} and 0.42 m at a discharge of 500 m^{3}s^{-1}. A slight trough in the water surface is observed on the floodplain just downstream of the ridge along the outside bank and a slight mound in the water surface over the channel persists for a short distance downstream. The same features are slightly more pronounced in the mesh 5 simulation than in the mesh 4 simulation, with a maximum head difference of 0.64 m at 1000 m^{3}s^{-1} and 0.45 m at 500 m^{3}s^{-1}.

Contour maps of the velocity, Froude number and shear

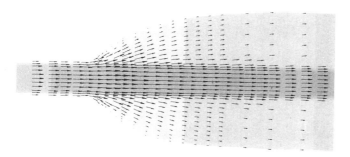

Fig. 9. Vector plot showing flow pattern from a portion of the $1000\ m^3s^{-1}$ simulation for mesh 3. Channel (darker shading) width is 35 m.

Fig. 11. Photograph showing catastrophic channel widening along the South Branch Potomac River at the exit from Petersburg Gap, downstream from Petersburg, West Virginia. Flow is from left to right.

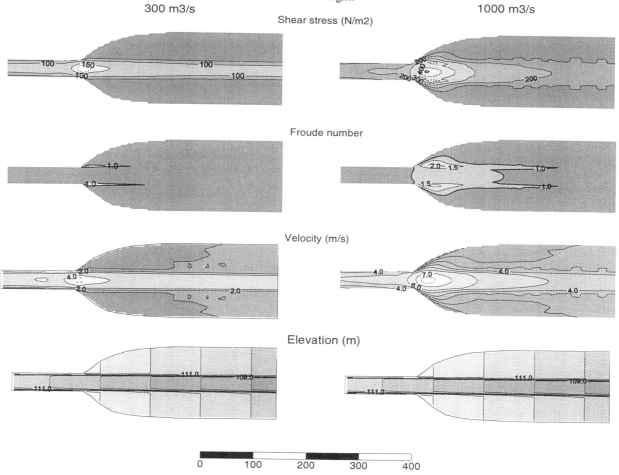

Fig. 10. Contour maps showing spatial distribution of hydraulic parameters in comparison to topography for a portion of mesh 3. Flow is from left to right. Contour intervals are 0.5 m for surface topography, 1 ms^{-1} for velocity, 0.5 for Froude number, and 100 Nm^{-2} for shear stress. Dashed lines indicate contours marking half contour intervals. Scale is in meters.

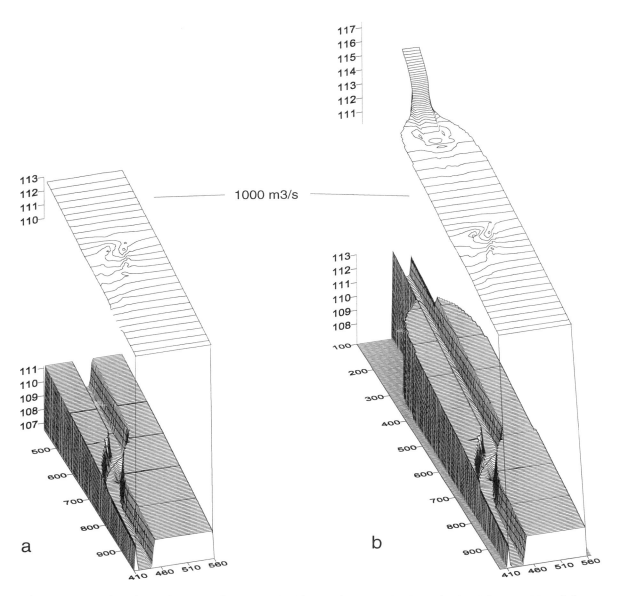

Fig. 12. Perspective views of water surface and underlying mesh topography for a simulated flow of 1000 m³s⁻¹. All plots have the same vertical scale. All scale units are in meters. (a) Mesh 4. (b) Mesh 5.

stress distributions for the same region of these two meshes at 1000 m³s⁻¹ (Fig. 13) indicate a local velocity maximum in the channel approaching the bend, a slight deceleration in the bend itself, and another local peak on the floodplain as flow leaving the channel accelerates over the step formed by the bank. A region of higher-velocity flow extends about 200 m downstream along the floodplain before the velocity field relaxes toward background levels. The region of accelerating velocity on the floodplain is also a local area of supercritical flow and experiences higher shear stresses than those calculated for the core flow in the channel immediately upstream. Recall that supercritical flow did not occur anywhere in the channel or floodplain for either the mesh 1 or mesh 2 simulations. Peak values listed in Table 3 indicate that floodplain shear stress maxima for mesh 4 and mesh 5 are 2.0 and 2.3 times as large as the maximum floodplain shear stress for the same discharge in the mesh 2 simulation, and are respectively about 90% and 100% of the maximum channel shear stress in the mesh 1 simulation.

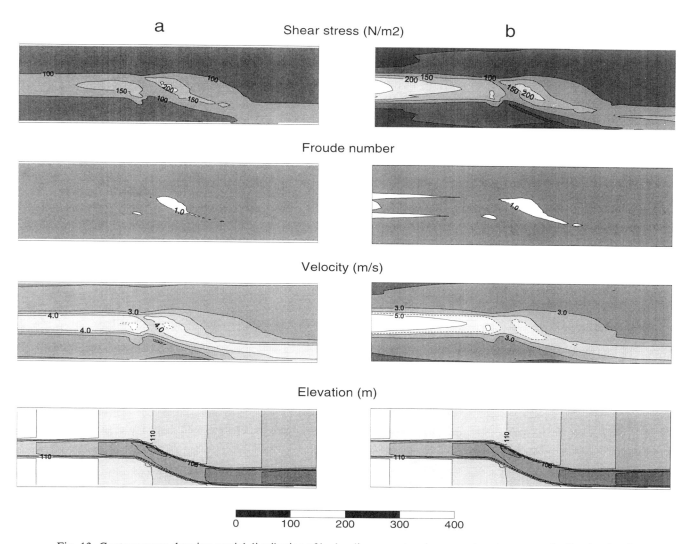

Fig. 13. Contour maps showing spatial distribution of hydraulic parameters in comparison to topography for simulated flow of 1000 m³s⁻¹. Flow is from left to right. Contour intervals are 0.5 m for surface topography, 1 ms⁻¹ for velocity, 0.5 for Froude number, and 50 Nm⁻² for shear stress. Dashed lines indicate contours marking half contour intervals. Scale is in meters. (a) Mesh 4. (b) Mesh 5.

Comparing the mesh 4 and mesh 5 results in Fig. 13, the velocity field in mesh 5 shows the effects of a steady deceleration of channel flow emerging from the expansion upstream (the expansion itself is not illustrated in the figure), with highest velocities and shear stress values at the upstream limit of the area shown. The area of high-velocity, supercritical flow is slightly larger and the maximum shear stress on the floodplain at the outside of the bend is slightly higher than in the mesh 4 simulation, indicating that the acceleration of flow in the expansion continues to exert a modest influence on hydraulic conditions at a distance of 3 to 4 valley widths or 10 to 12 channel widths downstream of the expansion. The extent of this influence undoubtedly would be affected by changing the gradient, the depth of the channel, or the width of the valley. The channel cross-section in Fig. 1b, for example, is about the same width but twice as deep as the channel simulated here and might therefore be expected to retain a significant component of high-velocity flow for a longer distance downstream of the constriction.

3.4 *Flow around Obstructions: Comparison of Meshes 5 and 5b*

Although a detailed exploration of the impact of spatially varying roughness on flood hydraulics is beyond the scope

Table 3. Maximum Shear Stress for Simulated Flood Flows

Mesh	Location	Discharge in m³s⁻¹		
		300	500	1000
		Nm⁻²	Nm⁻²	Nm⁻²
1	channel	135	175	260
2	channel	115	135	165
	floodplain	50	95	115
3	channel	200	285	505
	floodplain	75	165	465
4	floodplain	----	190	225
5	floodplain	----	195	265
5b	floodplain	----	195	330
6	floodplain	170	335	650
6b	floodplain	----	360	785

of this paper, the role of individual roughness elements or groups of roughness elements cannot be ignored entirely. Observations following the November 1985 flood showed that floodplain scour commonly occurred in association with flow around local obstructions or over topographic steps, even in areas where the floodplain surface was otherwise undisturbed [*Miller and Parkinson*, 1993]. Because of the enormous amount of floating debris that may be entrained in a large flood, groups of roughness elements such as clumps or rows of trees sometimes trap enough debris to form barriers of considerable height and breadth, and these may be considered effectively impermeable even if a small amount of flood flow leaks through or splashes over the top of the barrier. Of course such obstructions may form and break up during the course of a flood, leaving behind no evidence of their presence. Nevertheless a consideration of the spatial distribution of hydraulic parameters on the floodplain requires consideration of the potential impacts of such barriers on the flow field, even if they are transient and do not survive the flood.

In order to investigate this phenomenon I created mesh 5b by making a simple adjustment to mesh 5, deleting several elements on the floodplain along the outer bank at the channel bend. Flood flows routed through this mesh were diverted around the obstruction and it was possible to compare the resulting flow field with the unobstructed flow field without making additional assumptions about the value of Manning's n. The results are influenced by the shape and placement of the obstruction; this example serves merely to illustrate the type of effect that may occur. A more detailed explanation of flow phenomena in the immediate vicinity of the obstruction would require a 3-dimensional approach in order to simulate the vortices that roll up around the obstruction and associated vertical accelerations in the flow field.

The predicted effect of the obstruction on the shape of the water surface is quite impressive (Fig. 14). For a discharge of 1000 m³s⁻¹, maximum runup of the water surface on the upstream side of the obstruction was 0.9 m, with a corresponding drop of 1.0 m in the lee of the obstruction. The maximum difference between the front and back of the obstruction was 2.2 m.

The obstruction is oriented at an oblique angle to the flow approaching it from the channel. Flow diverted around it accelerates across the floodplain on the upstream side and along the edge of the channel on the downstream side. A low-velocity recirculating eddy forms in the lee, extending approximately two channel widths downstream (Fig. 15). Although the three-dimensional pattern of secondary currents in the vicinity of such an obstruction cannot be simulated by HIVEL2D, a computer animation of the vector field suggested that even in the 2-d simulation there would be some entrainment of fluid into the eddy on the upstream side and leakage of fluid out of the eddy near the downstream edge of the obstruction. If the barrier were to survive a flood, accumulation of sediment in its wake would be expected. An example of this kind of deposit in the wake of a tree flanked by a debris jam is illustrated in *Miller and Parkinson* [1993], Fig. 42.

A comparison of the spatial patterns of velocity, Froude number and shear stress for meshes 5 and 5b in the vicinity of the obstruction (Fig. 16) provides further evidence of the strong perturbations induced in the flow field that have implications for geomorphic impacts. Similar perturbations induced by bridge piers in stream channels are commonly associated with scour problems. The highest velocity predicted by the mesh 5b simulations is focused along a narrow pathway just to the left and downstream of the obstruction, and maximum shear stress occurs at the same location. There is a modest 27% predicted increase in maximum shear stress over the maximum for the unobstructed flow field (Table 3), but the focusing of the flow along a narrow pathway across the floodplain might well result in more extensive dissection of the floodplain than is indicated by this difference. It is also worth pointing out that in a real flood there could be multiple obstructions

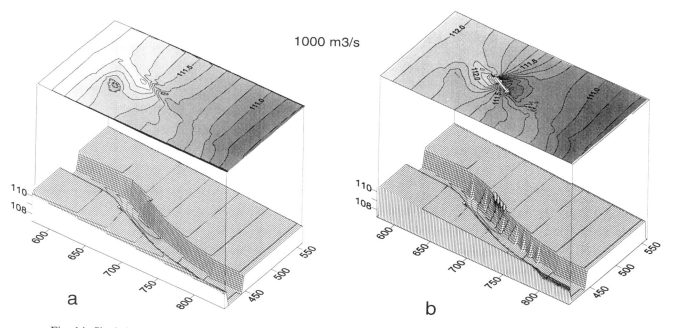

Fig. 14. Shaded contour maps of water surface for 1000 m³s⁻¹ simulation, superimposed on perspective view of mesh topography. All scale units are in meters. (a) Mesh 5. (b) Mesh 5b. White rectangle in center of plot is a flow obstruction created by removing several elements from the mesh.

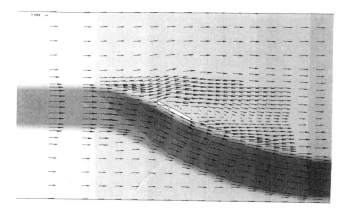

Fig. 15. Vector plot showing flow pattern from a portion of the 1000 m³s⁻¹ simulation for mesh 5b. Note low-velocity recirculating eddy in the lee of the obstruction. Channel (darker shading) width is 35 m.

constraining the flow to a greater extent than is shown in this example.

The spatial pattern of shear stress illustrated for mesh 5b shows a strong resemblance to the pattern of floodplain dissection along the South Fork South Branch Potomac River at the Adamson dairy farm upstream of Moorefield, West Virginia (Fig. 17). In fact mesh 5b is intended in part to represent a simplified form of the valley configuration at this site. The cross-sections shown in Figure 1 were surveyed upstream of the location shown in Fig. 17 and the channel remains almost perfectly straight for nearly 3 km from the upstream constriction until it bends to the right as shown. No attempt was made to simulate the precise configuration of features present at the site for this paper. The actual channel is more deeply entrenched than in the mesh and is bordered on both sides by lines of trees that may have helped to confine the floodwaters along a narrow alignment; this may help to explain the extent of the damage observed. Furthermore the large dairy barn shown in Fig. 17 is in almost the same position with respect to the channel as the obstruction in mesh 5b and probably played much the same role at the height of the flood.

3.5 Combined Effects of a Valley Expansion and a Bend in the Channel: Meshes 6 and 6b

Mesh 6 was designed to simulate the types of flow conditions occurring where a channel bend coincides with a valley expansion. A perspective view of the water surface for three different discharges (Fig. 18) reveals the same steep water surface gradient in the throat of the expansion as in mesh 3 (Fig. 8). This gradient is superimposed on a pattern of superelevation in the channel bend similar to those illustrated for meshes 4 and 5 (Fig. 12). With

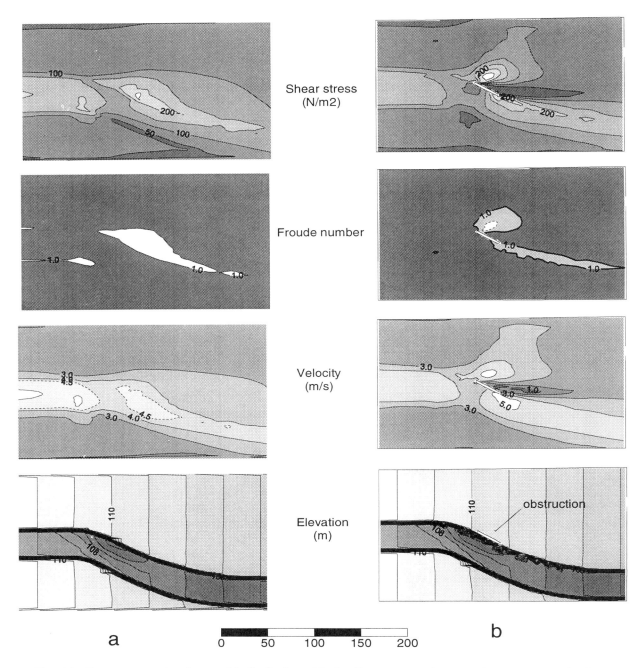

Fig. 16. Contour maps showing spatial distribution of hydraulic parameters in comparison to topography for a simulated flow of 1000 m³s⁻¹. Flow is from left to right. Contour intervals are 0.5 m for surface topography, 1 ms⁻¹ for velocity, 1.0 for Froude number, and 50 Nm⁻² for shear stress. Dashed lines indicate contours marking half contour intervals. Scale is in meters. (a) Mesh 5. (b) Mesh 5b.

increasing discharge the water-surface gradient in the throat of the expansion becomes steeper and the perturbations of the water surface associated with the channel bend become more pronounced. For a discharge of 1000 m³s⁻¹, the difference in water-surface elevation between the ridge on the outside of the bend and the trough on the inside of the bend is as much as 2.0 m. The water surface also drops by nearly the same amount from the crest of the ridge along

Fig. 17. Photographs showing the valley of the South Fork South Branch Potomac River at Fame, upstream of Moorefield, West Virginia. Cross-sections shown in Fig. 1 were surveyed a short distance upstream of this location. (a) May 1980. (b) November 1985.

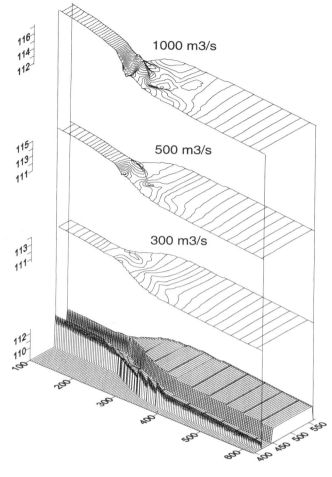

Fig. 18. Perspective views of simulated water surface over mesh 6 for three different values of flood peak discharge; bottom plot is a perspective projection of mesh topography. All plots have the same vertical scale. All scale units are in meters.

the outside of the bend to the bottom of the depression on the floodplain about 75 m downstream. The spatial pattern of flow vectors clearly indicates that a substantial component of flow crosses onto the floodplain as it emerges from the constriction and accelerates as it begins to flow across the floodplain (Fig. 19). The resulting distribution of hydraulic parameters is illustrated for two different discharges in Figure 20.

Many of the features illustrated resemble those previously described for the other mesh simulations. However, this set of simulations is distinguished by more intense hydraulic conditions and a more complex water surface. Maximum floodplain shear stress at 1000 m^3s^{-1} is 5.7 times

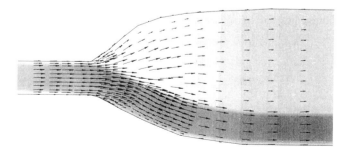

Fig. 19. Vector plot showing flow pattern from a portion of the 1000 m^3s^{-1} simulation for mesh 6. Channel (darker shading) width is 35 m.

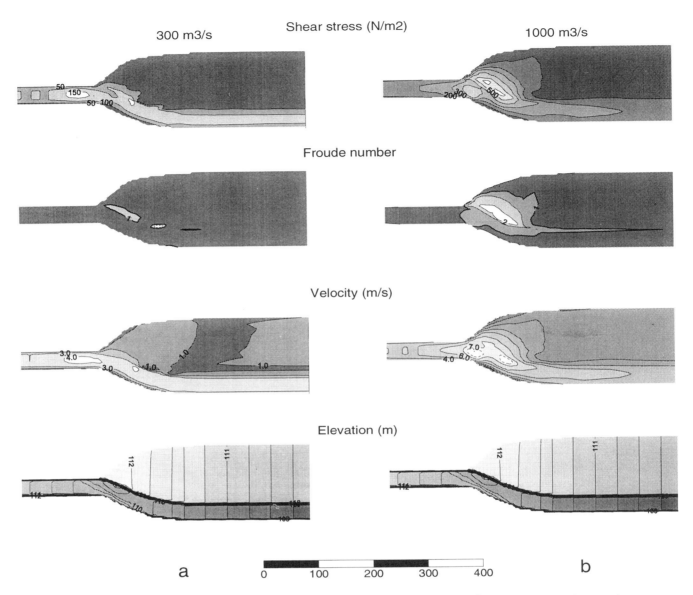

Fig. 20. Contour maps showing spatial distribution of hydraulic parameters in comparison to topography for a portion of mesh 6. Flow is from left to right. Contour intervals are 0.5 m for surface topography, 1 ms^{-1} for velocity, 0.5 for Froude number, and 100 Nm^{-2} for shear stress. Dashed lines indicate contours marking half contour intervals. Scale is in meters. (a) Simulation for 300 m^{3}s^{-1} discharge. (b) Simulation for 1000 m^{3}s^{-1} discharge.

as large as the value derived for mesh 2 and 35% larger than maximum floodplain shear stress just downstream of the valley expansion in mesh 3; and is also 2.5 times as large as the maximum channel shear stress derived for mesh 1 and 29% larger than the maximum channel shear stress derived for mesh 3. Predicted Froude numbers exceed 2.0 on the channel bank in the direct path of flow emerging from the constriction. The strong potential for catastrophic erosion is obvious.

Introduction of a barrier similar to the one in mesh 5b produces even more extreme hydraulic conditions (mesh 6b, Fig. 21). For a discharge of 1000 m^{3}s^{-1}, the maximum difference in head from one side of the obstruction to the other is 3.6 m and the maximum difference in head from one side of the stream to the other is 3.1 m. It is unlikely that any natural or man-made obstruction not made of solid rock or reinforced concrete would persist for long if exposed to a discharge of this magnitude at such a location.

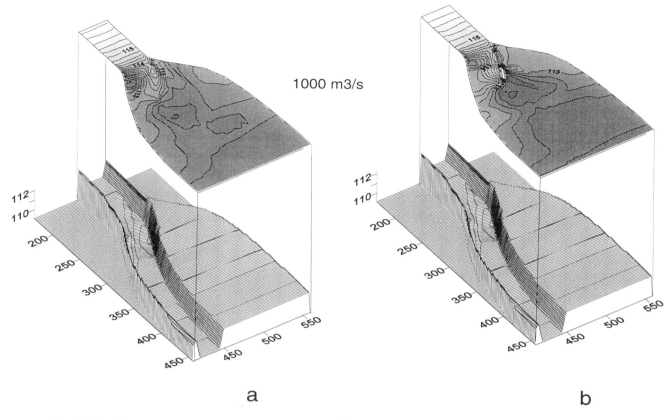

Fig. 21. Shaded contour maps of water surface for 1000 m^{3}s^{-1} simulation, superimposed on perspective view of mesh topography. All scale units are in meters. (a) Mesh 6. (b) Mesh 6b. White rectangle in center of plot is a flow obstruction created by removing several elements from the mesh.

Comparison of the velocity, Froude number, and shear stress distributions for meshes 6 and 6b reveals a trend similar to the trend observed in the comparison of meshes 5 and 5b, with a narrower thread of high-velocity flow bringing higher shear stress values in a more tightly focused area of the floodplain (Fig. 22). Maximum floodplain shear stress occurs on the channel bank just downstream of the obstruction, attaining a value of 780 Nm^{-2}, which is 21 % higher than the maximum in mesh 6, 55 % higher than the maximum channel shear stress in mesh 3, 3 times as large as the maximum channel shear stress in mesh 1, and nearly 7 times as large as the maximum floodplain shear stress in mesh 2. As was true in the case of mesh 5b, a different configuration of roughness elements might generate even more extreme conditions. The spatial patterns of flood hydraulics illustrated in Fig. 22 are consistent with the general pattern of erosion and deposition shown in Fig. 4, although the configurations shown in meshes 6 and 6b were not explicitly modeled on the site shown in the photograph.

3.6 Comparisons of Maximum Shear Stress

A summary of maximum shear stress as a function of discharge for each mesh is provided in Fig. 23. If we take the maximum channel shear stress from mesh 1 as a standard, it is evident that maximum channel shear stress along the straight uniform-width valley of mesh 2 is not very sensitive to changes in discharge by comparison. Maximum shear stress on the floodplain in mesh 2 is less than in the channel and is less than half of the channel maximum for mesh 1.

For meshes 4, 5, and 5b, however, maximum floodplain shear stress is comparable to the maximum channel shear stress in mesh 1, suggesting that a bend in the channel may induce perturbations in the flow field causing local shear stress peaks comparable to those that are found along a straight, narrow canyon. An obstruction in the path of the flow as it crosses the floodplain generates even higher local shear stresses and is associated with a steeper increase in maximum shear stress as a function of discharge.

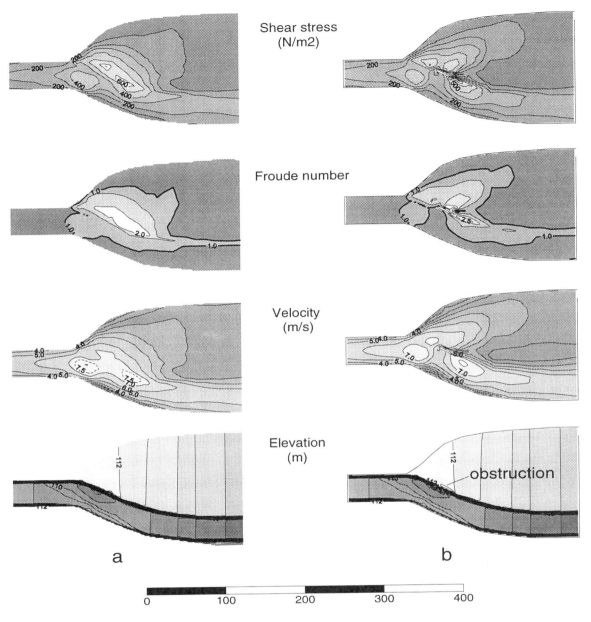

Fig. 22. Contour maps showing spatial distribution of hydraulic parameters in comparison to topography for a simulated flow of 1000 m^{3}s^{-1}. Flow is from left to right. Contour intervals are 0.5 m for surface topography, 1 ms^{-1} for velocity, 0.5 for Froude number, and 100 Nm^{-2} for shear stress. Dashed lines indicate contours marking half contour intervals. Scale is in meters. (a) Mesh 6. (b) Mesh 6b.

Shear stress increases much more steeply with discharge at a valley expansion than along either a straight canyon of constant width or along a broader valley (Fig. 23b); this is consistent with the increase in water-surface gradient at the expansion that is illustrated in Figs. 8 and 17. At the highest discharges the maximum shear stress for mesh 3 is nearly double the maximum for mesh 1. Along mesh 3 the highest shear stress values are found in the channel, but with increasing discharge the floodplain maximum approaches the channel maximum. The highest shear stress values overall, and the most dramatic increase in the maximum shear stress as a function of discharge, are observed on the floodplain in the mesh 6 and 6b simulations.

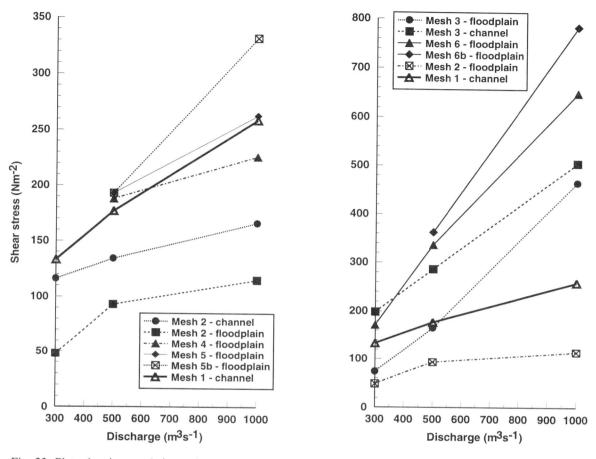

Fig. 23. Plots showing trends in maximum shear stress as a function of discharge for flow simulations in this study. Note difference in vertical scale between left and right plots.

4. DISCUSSION

The results presented above demonstrate the influence of longitudinal variations in channel and valley morphology on the spatial pattern of flood hydraulics and the resulting potential for catastrophic erosion. The concept of geomorphic effectiveness is generally discussed in conjunction with the concept of a resistance threshold, and several authors have proposed values of unit stream power or boundary shear stress that are associated with major channel widening and floodplain erosion or with mobilization of boulder deposits. *Magilligan* [1992], citing *Baker and Costa* [1987], *O'Connor et al.* [1986], and *Miller* [1990], selects unit power and shear stress values of 300 Wm^{-2} and 100 Nm^{-2} as reasonable minimum thresholds.

However, as these values are based on cross-sectional averages derived from solutions to the one-dimensional energy equation, they are not strictly comparable to the values compiled from simulation results in the present study and therefore cannot be used to map erosion and deposition zones.

4.1 *Relating Flow Patterns to Patterns of Erosion and Deposition*

For any site affected by the extreme velocities and shear stresses discussed here, the spatial distribution of erosion and deposition must be understood in the context of a local balance between sediment transport capacity and sediment load. If sediment transport is a power function of shear stress, a sharply increasing shear-stress distribution along the direction of flow should result in net erosion along the flow path as long as (a) shear stress exceeds the threshold for removing cohesive barriers such as turf or other forms of vegetation cover, and (b) there is a supply of erodible sediment available along the flow path.

If we assume this is the case, and that the flood has an opportunity to entrain material in amounts at or close to its sediment transport capacity, then a sharp decrease in shear stress along the same flow path should lead to deposition of a substantial amount of sediment, forming a rim of coarse splay deposits surrounding the eroded region. Such features were widely observed in the aftermath of the November 1985 flood and are illustrated in Figs. 4 and 11 and in many other photographs published in *Miller and Parkinson* [1993].

Zones of catastrophic erosion associated with valley expansions were generally limited to a short reach extending no more than 1-3 multiples of the downstream valley width from the expansion. It appears likely that a map of the eroded area and the deposition zone could be closely approximated by mapping shear-stress distributions based on hydraulic modeling of initial boundary conditions. This is not necessarily what would be expected, as the pattern of erosion might lead to a progressive change in the boundary conditions, with subsequent erosion and deposition responding to the altered conditions rather than the initial conditions. This observation indicates that either (a) the controls on flow patterns exerted by valley morphology at these sites are so strong that progressive growth of erosion features cannot occur beyond a limited distance; or (b) the flood peak was of such short duration that there was insufficient time for continued evolution and progressive growth of erosion and deposition features.

In effect, option (b) describes a scenario in which the flood rises to peak stage very rapidly, sculpts a new set of landforms within the reach most strongly influenced by initial boundary conditions, and then drops rapidly again, preserving these landforms. The question of how long flood waters must continue to flow above some threshold discharge and stage in order to accomplish this geomorphic impact is unresolved. The dambreak floods described by *Costa and O'Connor* [this volume], lasting only a few minutes, may be too short. A period of flood flow above the threshold level lasting more than a couple of hours probably is more than sufficient. Further investigation of this question and related questions about flood dynamics may be pursued with the help of flume experiments.

Acknowledgements. This paper is a long-overdue installment in a series based on investigations that began in the aftermath of the November 1985 flood. It is offered in tribute to M. Gordon Wolman, who has served as a source of inspiration to his students and colleagues, many of whom contributed to this volume. Work carried out with funding from NSF Grant SES-8722442 [Miller and Wolman, co-PIs] led directly to many of the ideas discussed in this paper. Discussions over the past several years with M. G. Wolman, Robb Jacobson, Douglas Parkinson, John Costa, Jim Smith, Diane Mas, Bob Jarrett, and others have also been helpful in formulating the ideas laid out here. Thoughtful reviews of the manuscript by John Costa and Hugh Mills served to improve the final product, and John Costa in his capacity as lead editor deserves special credit for shepherding the entire project through to completion. I thank Alan Howard and John Harper for the use of aerial photographs showing the Tye River valley following the Hurricane Camille floods. Rob Wallace of the Engineering Computer Graphics Laboratory at Brigham Young University provided much-needed assistance with the FastTABS program. Charlie Berger and Richard Stockstill of the Waterways Experiment Station were gracious in providing access to and assistance with the HIVEL2D model, and Barbara Donnell, also at WES, provided help with several versions of RMA2 used in this and other studies. Any errors that may have occurred in applying the models is purely my responsibility. Paul Danckaert of UMBC Academic Computing Services served as UNIX guru and provided critical assistance that made use of the models and analysis and presentation of the results possible. Initial work on mesh compilation was carried out with the help of James Burd and with financial assistance from a FY 1995 Designated Research Initiative Fund award from the University of Maryland Baltimore County.

REFERENCES

Abbott, M. B., Computational Hydraulics, *Elements of the Theory of Free Surface Flows*, 324 pp., Pitman, London, 1979.

Baker, V. R., Stream channel response to floods with examples from central Texas, *Geol. Soc. Am. Bull., 88,* 1057-1071, 1977.

Baker, V. R. and Costa, J. E., Flood power, in *Catastrophic flooding,* edited by L. Mayer and D. Nash, pp. 1-21, Allen & Unwin, Boston, Mass., 1987.

Benson, M. A. and Dalrymple, T., General field and office procedures for indirect discharge measurements, *Techniques of Water Resources Investigation,* U.S. Geol. Surv., Book 3, ch. A1, 1967.

Brigham Young University, *FastTABS Hydrodynamic Modeling Reference Manual*, 64 pp., Engineering Computer Graphics Laboratory, 1994.

Chow, V. T., *Open-channel hydraulics*, McGraw-Hill, NY, 680 pp., 1959.

Costa, J. E. and O'Connor, J. E., Geomorphically effective floods, this volume.

Hack, J. T., and Goodlett, J. C., Geomorphology and forest ecology of a mountain region in the Central Appalachians, *U.S. Geol. Surv. Prof. Paper, 347,* 66 pp., 1960.

Hydrologic Engineering Center, HEC-2 Water Surface Profiles: User's Manual, U.S. Army Corps of Engineers, Davis, CA, 1982.

Jacobson, R. B., Miller, A. J., and Smith, J. A., The role of catastrophic geomorphic events in central Appalachian landscape evolution, *Geomorph., 2,* 257-284, 1989.

Jarrett, R. D., Hydraulics of high-gradient streams, Journal of Hydraulic Engineering, *Am. Soc. of Civil Engineers, 110 (11),* 1519-1539, 1984.

Johnson, R. A., Stream channel response to extreme rainfall events: the Hurricane Camille storm in central Nelson County, Virginia, M.A. thesis, 109 pp., University of Virginia, Charlottesville, VA, 1983.

Kochel, R. C., Geomorphic impact of large floods: review and new perspectives on magnitude and frequency, in *Flood Geomorphology*, edited by V. R. Baker, R. C. Kochel, and P. C. Patton, pp. 169-187, John Wiley & Sons, NY, 1988.

Magilligan, F. J., Thresholds and the spatial variability of flood power during extreme floods, *Geomorph., 5*, 383-390, 1992.

Miller, A. J., Debris-fan constrictions and flood hydraulics in river canyons: some implications from two-dimensional flow modelling, *Earth Surf. Proc. and Landforms, 19*, 681-697, 1994.

Miller, A. J., Flood hydrology and geomorphic effectiveness in the central Appalachians, *Earth Surf. Proc. and Landforms, 15*, 119-134, 1990.

Miller, A. J. and Parkinson, D. J., Flood hydrology and geomorphic effects on river channels and flood plains: the flood of November 4-5, 1985, in the South Branch Potomac River Basin of West Virginia, *U.S. Geol. Surv. Bull., 1981-E*, 96 pp., 1993.

Nanson, G. C., Episodes of vertical accretion and catastrophic stripping: a model of disequilibrium flood plain development, *Geol. Soc. Am. Bull., 97*, 1467-1475, 1986.

O'Connor, J. E., Webb, R. H., and Baker, V. R., Paleohydrology of pool-and-riffle pattern development: Boulder Creek, Utah, *Geol. Soc. Am. Bull., 97,* 410-420, 1986.

Samuels, P. G., Some analytical aspects of depth averaged flow models, in Hydraulic and Environmental Modelling of Coastal, Estuarine and River Waters, Proceedings of the International Conference held at the University of Bradford, 19-21 September 1989, edited by R. A. Falconer, P. Goodwin, and R.G. S. Matthew, pp. 411-418, Gower Technical, Aldershot, England, 1989.

Shen, H. W. and Julien, P. Y., Erosion and sediment transport, in *Handbook of Hydrology*, edited by D. R. Maidment, pp. 12.1-12.61, McGraw-Hill, NY, 1993.

Stockstill, R. L. and Berger, R. C., HIVEL2D: a two-dimensional flow model for high-velocity channels, Technical Report REMR-HY-12, 53 pp., U.S. Army Corps of Engineers Waterways Experiment Station, Vicksburg, Mississippi, 1994.

Thomas, W. A. and McAnally, W. H., Jr., User's Manual for the Generalized Computer Program System: Open-Channel Flow and Sedimentation, TABS-2, U.S. Army Corps of Engineers Waterways Experiment Station, Vicksburg, Mississippi, 1990.

Whitlow, C. D. and Knight, D. W., An investigation of the effect of different discretizations in river models and a comparison of non-conservative and conservative formulation of the De St. Venant equations, in *Hydraulic and Environmental Modelling: Estuarine and River Waters*, Proceedings of the Second International Conference on Hydraulic and Environmental Modelling of Coastal, Estuarine and River Waters, v. 2, edited by R. A. Falconer, K. Shiono, and R. G. S. Matthew, pp. 115-126, Ashgate, Aldershot, England, 1992.

Williams, G. P. and Costa, J. E., Geomorphic measurements after a flood, in *Flood Geomorphology*, edited by V. R. Baker, R. C. Kochel, and P. C. Patton, pp. 65-77, John Wiley & Sons, NY, 1988.

Williams, G. P. and Guy, H. P., 1973, Erosional and depositional aspects of Hurricane Camille in Virginia, 1969, *U. S. Geol. Surv. Professional Paper, 804*, 80 pp.

Wolman, M. G. and Eiler, J. P., Reconnaissance study of erosion and deposition produced by the flood of August 1955 in Connecticut, *Am. Geophys. Union Trans., 39*, 1-14, 1958.

Wolman, M. G. and Gerson, R., Relative scales of time and effectiveness of climate in watershed geomorphology, *Earth Surf. Proc., 3,* 189-208, 1978.

Wolman, M. G. and Miller, J. P., Magnitude and frequency of forces in geomorphic processes, *J. Geol., 68,* 54-74, 1960.

Andrew J. Miller, Department of Geography, University of Maryland Baltimore County, Baltimore, MD 21218

Morphology and Processes of Valley Floors in Mountain Streams, Western Cascades, Oregon

G. E. Grant and F. J. Swanson

U.S. Department of Agriculture, Forest Service, Pacific Northwest Research Station, Corvallis, Oregon

Development of valley floor landforms and channel morphology in headwater streams is strongly influenced by processes external to the channel. The distribution of landforms on valley floors of two fifth-order mountain streams in western Oregon exhibit distinctive reach-to-reach variations, as defined by a valley floor width index: the ratio of approximate width of Holocene valley floor (surfaces less than or equal to 3 m in height) to active channel width. This variation corresponds to the distribution of bedrock outcrops and hillslope landforms, including large landslides and alluvial fans, which constrain the channel. Lowest valley floor width index values (less than or equal to 1.3) occur where the channel is incised in bedrock or where landslides or alluvial fans encroach upon the valley floor, commonly forcing the channel against bedrock in the opposite valley wall. Highest valley floor width index values (greater than or equal to 4.0) occur upstream of these constrained reaches where valley floors aggrade and little bedrock is exposed. Widths of valley floor landforms of different origins do not increase uniformly with increasing valley floor width; active channel widths remain relatively constant while reaches with greater valley floor width generally exhibit greater widths of floodplains created by fluvial processes and/or debris flows. Hillslope and tributary processes also influence the channel gradient and determine the size and density of boulders in the channel. Highest boulder densities and steepest channels occur where channels are bordered by alluvial fans, terraces containing debris-flow deposits, and active or inactive large landslides. Reaches differ in their susceptibilities to reworking by fluvial and non-fluvial processes. The distribution of landforms of fluvial origin varies in relation to reach-averaged shear stresses generated during large floods. The pattern of landforms of debris-flow origin is determined by the valley floor geometry, location of debris-flow producing tributaries, and the longitudinal sequence of reaches. Changes in valley floor morphology in mountain streams occur rapidly and episodically during infrequent, intense floods, in contrast to the more gradual, fluvial reworking of valley floor sediment in low-gradient alluvial streams.

1. INTRODUCTION

The morphology and dynamics of valley floor environments in mountain landscapes differ markedly from their lowland counterparts but have received comparatively little attention. In contrast to wide, low-gradient streams, where valley floor morphology reflects the predominance of fluvial processes, morphology of narrow, high-gradient streams is dominated by the interplay of fluvial processes with processes and landforms external to the channel. The latter include landslides, debris flows, alluvial fans, and bedrock outcrops. Interpreting the geomorphic, sedimentologic, and ecologic patterns of valley floor environments requires appreciation for how hillslope processes and landforms influence channel and valley floor morphology.

Despite the importance of hillslope processes as modifiers of valley floor morphology in mountain landscapes, there has been little attempt to examine this subject systematically, although aspects of this problem have been explored. Landslides, debris flows, and alluvial fans were recognized as important agents modifying valley bottoms by *Hack and Goodlett* [1960] in their landmark paper on

Natural and Anthropogenic Influences in Fluvial Geomorphology
Geophysical Monograph 89
This paper is not subject to U. S. copyright. Published in 1995 by the American Geophysical Union

geomorphology of Appalachian watersheds. More recent work has emphasized the role of bedrock controlling channel and valley floor morphology [*Baker*, 1984; *Lisle*, 1986; *O'Connor et al.*, 1986; *Baker and Pickup*, 1987; *Ashley et al.*, 1988; *Baker and Kochel*, 1988; *Kelsey*, 1988; *Jacobson et al.*, 1989; *Grant et al.*, 1990; *Miller*, 1990a] and debris fans as prominent landforms affecting channel planform and gradient [*Kieffer*, 1985; *Miller*, 1990b]. Many workers have documented the importance of mass movements leaving persistent impacts on mountain valleys during exceptional floods [e.g., *Tricart*, 1962; *Stewart and LaMarche*, 1967; *Scott and Gravlee*, 1968; *Nolan and Marron*, 1985; *Kochel*, 1988; *Jacobson et al.*, 1989; *Miller*, 1990a, b), and there has been some effort to describe the sedimentologic and rheologic characteristics of debris flow and debris flood deposits on valley floors [*Costa and Jarrett*, 1981; *Jarrett and Costa*, 1986; *Costa*, 1988; *Carling*, 1989].

The purpose of this investigation is to examine how variation in valley floor morphology in mountain landscapes corresponds to hillslope processes' and to contrast the dynamics of valley floors in mountain and lowland streams. Drawing from field studies of 5-km lengths of fifth-order valley floor in two streams typical of the western Cascade Range of Oregon, we examine the pattern and areal extent of valley floor landforms in relation to bedrock outcrops and hillslope processes. We hypothesize that hillslope processes control the distribution of valley floor landforms by determining the type, extent, and energetics of geomorphic processes operating on the valley floor. Our interpretations of geomorphic dynamics of montane valley floors are contrasted with the type and rate of change observed in lower-gradient, meandering river systems.

Hierarchical classifications of stream networks and watersheds are increasingly being utilized in riverine research and management [*Rosgen*, 1985, 1994; *Frissell et al.*, 1986; *Kishi et al.*, 1987; *Grant et al.*, 1990; *Montgomery and Buffington*, 1993] and have been employed to classify streams ecologically [*Kani*, 1944, 1981; *Mizuno and Kawanabe*, 1981; *Gregory et al.*, 1991; *Hawkins et al.*, 1993). By this view, drainage basins are composed of lengths of stream and associated valley floor environments at several spatial scales that are more or less homogenous with respect to controlling variables of geology, substrate, hydraulics, and hillslope influences (Figure 1). This approach is useful for expressing scales and sources of variation in watershed morphology that are not well captured in the classic continuum view of stream and valley morphology as controlled primarily by discharge [*Leopold and Maddock*, 1953; *Leopold et al.*, 1964; *Vannote et al.*, 1980; *Bhowmik*, 1984].

Within this hierarchical scheme, the channel unit scale (e.g., pools, riffles; see Figure 1) has been extensively studied for decades in both low- and high-gradient streams [e.g., *Richards*, 1976; *Keller and Melhorn*, 1978; *Grant et al.*, 1990; *Takahashi*, 1990]. The reach scale, the subject of this paper, has received much less attention, in part because interpreting reach morphology requires considering both fluvial and extra-fluvial influences. We develop this concept by examining how the morphology and distribution of valley floor landforms reflect hillslope processes. Understanding the distribution and dynamics of reaches is important for interpreting drainage basin structure and long-term behavior, predicting channel and valley floor conditions in unsampled watersheds, and designing effective stream and riparian management strategies.

We distinguish reaches as segments of valley floor separated by distinct breaks in valley floor width' and examine the correspondence between reaches and the type and degree of constraint on the width of the valley floor imposed by hillslope processes. Major agents of constraint in this study included bedrock outcrops, large landslides, and large alluvial fans. Bedrock outcrops are passive constraining agents. Active, slow-moving landslides and accreting alluvial fans, the latter growing by periodic accumulation of debris flow and bedload deposits delivered from tributary streams, may actively constrict valley floors and channels. In this study, reaches typically ranged from several hundred meters to several kilometers in length.

2. STUDY SITES

At the lower ends of the study sites, Lookout Creek and French Pete Creek drain 67 and 83 km^2, respectively, of moderate to steeply sloping land in the western Cascade Range of Oregon (Table 1, Figure 2). Elevations of the two basins range from 410 to more than 1600 m. Bedrock is a mixture of Tertiary volcaniclastic rocks and lava flows cut by scattered dikes [*Peck et al.*, 1964; *Swanson and James*, 1975a; *Priest et al.*, 1983; *Sherrod and Smith*, 1989]. Landforms have been sculpted by fluvial, glacial, and mass movement processes. The latter include shallow, rapid movements of soil on hillslopes (debris slides); rapid movements of alluvium, colluvium, and organic matter down stream channels (debris flows); and large, slow-moving landslides (earthflows) [*Swanson and James*, 1975a, b]. Streamflow regimes are characterized by high winter flows from November through April, with peak flows occurring primarily during rain-on-snow events [*Harr*, 1981]. The largest recorded peak flow for Lookout Creek was 185 m^3/s in December 1964, a storm that

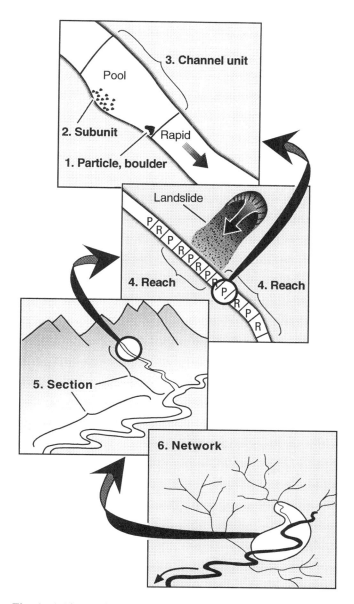

Fig. 1. A hierarchical model for organizing stream systems and associated landforms.

produced extensive debris slides and flows in tributaries. French Pete is ungaged but also experienced high flows and mass movements during this storm. This storm, which was regional in extent, left prominent landforms along the valley bottoms of both streams (described below) as well as many other watersheds throughout the Pacific Northwest [*Waananen et al.*, 1971]. A prolonged summer drought from July through September results in a very constant low flow discharge of approximately 0.015 m^3/s km^2 for both watersheds.

3. IDENTIFICATION OF VALLEY FLOOR LANDFORMS

The distribution of landforms in the valley floors of these two basins was examined by detailed mapping. The mapping was done with tape, compass, and clinometer using a baseline established along the channel and transects spaced at 100-m intervals extending from the channel to the base of the valley wall. We mapped geomorphic landforms as small as 10 m^2 and with as little as 0.5 m difference in elevation from adjacent landforms. Landforms were described by type (defined below) and height above low water level, which remained essentially constant throughout the month-long mapping period. Dense forest cover and the very coarse substrate of cobbles and boulders precluded stratigraphic analysis. Interpretation of age and origin of landforms was based on surficial morphology, age of vegetation, and other features such as orientation and degree of decay of woody debris accumulations, many of which dated from the 1964 flood.

The valley floors are composed of (1) the active channel, (2) secondary channels, (3) a progression of floodplains and terraces generally increasing in height away from the channel, and (4) alluvial fans (Figures 3, 4). These are described in the same order below.

The active channel is the area still inundated at summer low flow plus the adjacent unvegetated channel shelf [in the sense of *Hupp*, 1982; *Osterkamp and Hupp*, 1984] and gravel bars, plus secondary channels fed perennially from the main channel. These landforms are typically less than 0.5 m above low flow and are inundated by flood flows several times each year. Transport of coarse bedload and woody debris over these surfaces repeatedly prunes vegetation (mainly willow and red alder) and maintains a shrubby growth form on exposed bars.

Intermittent, ephemeral, and abandoned channels comprise a separate landform: the secondary channels. These channels typically are incised into terrace or floodplain landforms, and some have abundant woody debris and perennial vegetation within the channel. Some may originally have formed as primary channels but now carry discharge from a tributary or spring.

Floodplains are vegetated landforms less than or equal to 3 m above the low flow water surface and sloping in the down-valley direction parallel with the main channel. They include both coarse-textured fluvial deposits and boulder-berms resulting from debris-flow runout or fluvial deposition due to abrupt expansion of the valley floor [*Carling*, 1989]. In these two streams, the 3-m cutoff represents the approximate maximum height of Holocene-age deposits, based on field evidence and paleohydraulic

TABLE 1. Geomorphic Characteristics of the French Pete and Lookout Creek Study Sites and Watersheds

	French Pete	Lookout
Length of study site (m)	4950	5130
Drainage area (km^2)		
Upstream end	60.0	47.0
Downstream	84.0	60.0
Mean channel slope (m/m)	0.038	0.022
Mean active channel width (m)	18.1	21.5
Mean basin elevation (m)	1300	1200
Mean annual precipitation (mm)	2500	2500
Mean annual discharge (m^3/s)	3.5	3.6
Peak discharge (m^3/s)	95.*	190.**

*From regional flood frequency analysis; estimated recurrence interval = 100 years [*Harris et al.*, 1979].
**From gage record; estimated recurrence interval = 100 years [*Waananen et al.*, 1971].

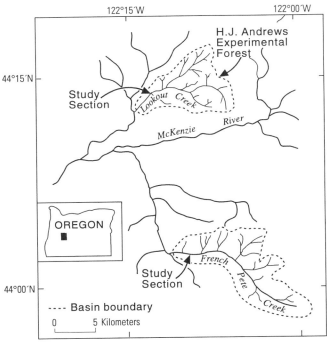

Fig. 2. Location of Lookout Creek and French Pete Creek study sections. Vertical lines on streams show upstream and downstream boundaries.

reconstruction [*Swanson and James*, 1975b; *Grant et al.*, 1990]. We defined floodplains using this criterion to reflect the extent of valley floor that potentially has been reworked during the Holocene and because of the difficulty in interpreting floodplains from surficial morphology, stratigraphy, or vegetation alone. Mountain stream floodplains differ from lowland floodplains in that the relationship in lowland floodplains among surface height, frequency of inundation, and age is obscured in mountain streams by debris-flow deposition and woody debris dams that create young surfaces substantially higher than older ones [*Wolman and Leopold*, 1957; *Hack and Goodlett*, 1960; *Costa and Jarrett*, 1981; *Kochel et al.*, 1987]. For example, some surfaces only 1 m above low flow bear 500-year-old forest, whereas debris flows in 1964 created fresh surfaces more than 2 m above the present low flow water surface. In analyzing floodplains, we distinguished between those that formed during or after the 1964 flood and those older than 1964, based on the age of vegetation.

Terraces are valley floor features greater than 3 m but less than 10 m above the low flow water surface. Terraces underlain entirely by alluvium, strath terraces cut in bedrock, and alluvial deposits overlapping bedrock are all found in the study area. High terrace remnants greater than 10 m above the low flow channel are also present but rare and are considered valley wall features (Table 2). Ages of terraces are poorly known, but all predate the 500-year-old forest growing on them. Most terraces have surficial deposits of Mazama ash (6600 yrs B.P.) and are therefore probably late Pleistocene in age [*Swanson and James*, 1975a; *Gottesfeld et al.*, 1981].

Alluvial fans, 140 to 66,000 m^2 in size, have developed at the junctions of tributary streams and the mainstem

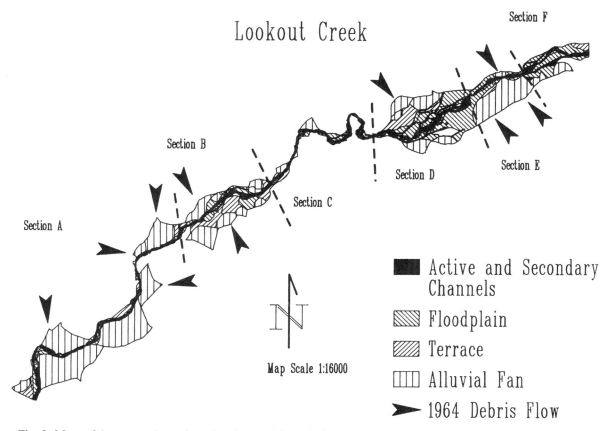

Fig. 3. Maps of the two study sections showing reach boundaries and locations of channels, floodplains, terraces, and alluvial fans as defined in text. Debris flows that occurred in the December 1964 storm also are shown. Bedrock outcrops occur where the stream flows against the valley wall. (a) Lookout Creek. (b) French Pete Creek.

where the valley floor is or was wide enough to accommodate fan deposits supplied by tributaries. Both active and inactive fans are present (Figure 4). Tributary streams on active fans generally are incised less than 4 m below the fan surface, and channels flowing over the fan may have the form of discontinuous gullies. Bedload and debris flows from tributary basins can be deposited on the fan front or margins, adding to fan volume. Inactive fans are those deeply incised by the tributary stream and cut at the toe by the mainstem, leaving fan remnants 5 m or more above the present channels. These fan fragments are inactive in the sense that they no longer accumulate material transported as bedload or debris flow from tributaries; so sediment and debris flows pass directly to the mainstem channel. Some fan fragments in the study areas have been inactive since at least the time of deposition of Mazama ash and possibly much longer [*Swanson and James,* 1975a].

4. SPATIAL PATTERN OF VALLEY FLOOR LANDFORMS

4.1 *Reach Delineation*

To analyze the effects of external controls on valley floor geometry, we first delineated reaches based on the width of the valley floor. Two questions motivated this analysis. First, how do different external controls influence the width of the valley floor, and second, how does the distribution of landforms of different types change in relation to those controls? The width of the valley floor was measured on the maps at 50-m intervals along and perpendicular to the stream. We expressed the width of the valley at each station using a *valley floor width index* (VFWI), defined as VFWI = $(W_{ac} + W_{fp})/W_{ac*}$, where W_{ac} and W_{fp} are the widths of the active channel and floodplain

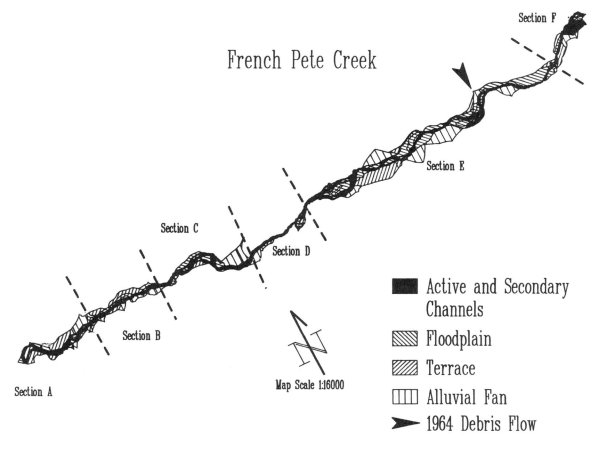

Figure 3 (continued)

landforms, respectively, and W_{ac*} is the average width of the active channel for all reaches at each site (Figure 4). The VFWI therefore provides a measure, in channel widths, of the variation between narrow and wide valley floors. Only the active channel and floodplain widths were considered in defining the VFWI in order to emphasize that portion of the valley floor that has been most active during Holocene time. The VFWI is potentially useful as a dimensionless index in comparing the geomorphic behavior of a wide range of reaches and streams. Its inverse, for example, has been used to discriminate between reaches experiencing scour or deposition during floods [*Wolman and Eiler,* 1958].

The valley floor width index was calculated as a 200-m running average for each site. We defined reaches as lengths of valley floor at least 300-m long bounded by changes of 25% or more in the smoothed valley floor width index (Figure 5a, b). On this basis, we delineated six reaches in French Pete and five in Lookout Creek. An additional reach (LE) was delineated in Lookout Creek where a large alluvial fan complex constricts the channel, even though the length criterion was not strictly met in this case (Figure 3a, b; Tables 2, 3). Reach identification is explained in the caption to Figure 6.

4.2 Controls on Valley Floor Width

Variation in reach widths results from terraces, alluvial fans, and valley wall landforms, including earthflows, bedrock outcrops, high terraces, and colluvium (Table 2; Figure 3a, b). Some reaches are dominated by a single type of constraint, i.e., valley walls (reach FC) or terraces (reach LD). Most reaches have at least two factors controlling valley floor width, such as valley walls and fans (reaches FD, FE, LE), terraces and fans (reaches FB, LA, LB), or terraces and valley walls (reaches FA, FF, LC, LF). Bedrock outcrops along the valley wall are important constraints in reaches FA, FD, FE, FF, LA, LC, and bedrock also underlies the strath terraces constraining the valley floor in reaches LA and LB. Earthflow complexes

TABLE 2. Reach Characteristics and Constraining Agents in French Pete and Lookout Creek Study Sites.

Reach*	Length (m)	Agents of Constraint**	Sinuosity (m/m)	Channel Gradient (%)	Valley Floor Width Index (m/m)	Channel Perimeter in Bedrock (%)	Average boulder density (No./100 m)	Fans (%)	Terraces (%)	Valley Wall (%)	Active-active (%)	Active-passive (%)	Passive-passive (%)
FRENCH PETE CREEK													
FA	900.0	Terraces, valley wall (bedrock)	1.2	4.2	2.2	5.9	22.4	19	60	21		19	81
FB	400.0	Fans, terraces	1.0	5.5	1.3	2.2	34.7	47	38	15		47	53
FC	600.0	Valley wall (high terrace)	0.9	3.9	2.9	6.0	13.7	18	14	68		18	82
FD	700.0	Valley wall (earthflow, bedrock, fans	1.2	3.3	1.3	9.5	18.4	27	10	63		90	10
FE	2000.0	Valley wall (high terrace), fans	1.1	3.5	4.0	10.1	13.0	25	21	53		25	75
FF	350.0	Terraces, valley wall (bedrock)	1.3	3.3	1.8	16.0	2.7	3	62	35		3	97
TOTAL/SITE AVERAGE	4950.0		1.1	3.8	2.8	8.8	16.6					32.5	67.5
LOOKOUT CREEK													
LA	1800.0	Fans, strath (bedrock) terraces	1.1	1.9	1.2	44.1	5.4	43	40	17		80	20
LB	690.0	Terraces, fans	1.0	2.1	2.8	13.1	6.9	29	51	20		67	
LC	1130.0	Valley wall (earthflow, bedrock)	1.2	2.8	1.3	23.6	12.2	11	31	58	33	75	25
LD	690.0	Terrace, valley wall (colluvium)	1.3	2.2	6.9	0.5	1.1	16	71	13		16	84
LE	375.0	Fans, valley wall (colluvium)	1.0	2.2	2.8	1.6	6.5	39	28	33	25	75	
LF	445.0	Terrace, valley wall (colluvium), fans	1.0	2.0	5.5	4.9	0.1	26	44	30		25	75
TOTAL/SITE AVERAGE	5130.0		1.1	2.2	2.6	23.3	6.2				6.5	65.7	31.4

*Reaches are listed in order from downstream to upstream.
**Terrace and valley wall constraining agents are further modified by the dominant underlying material or process.

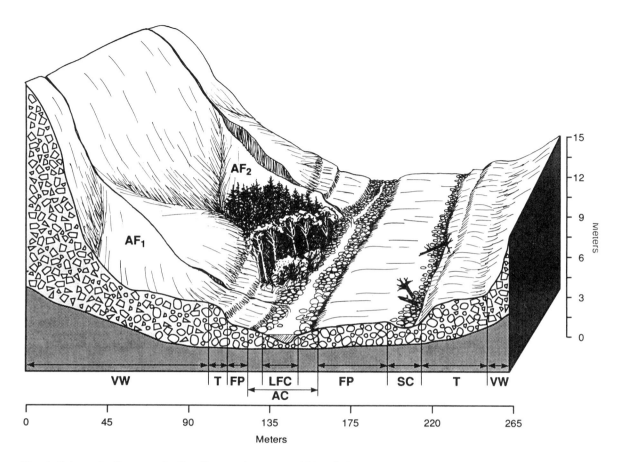

Fig. 4. Schematic diagram of valley floor surface types. Abbreviations used: LFC = low-flow channel; AC = active channel; FP = floodplain; SC = secondary channel; T = terrace; AF_1 = unincised, active alluvial fan; AF_2 = incised, inactive fan; VF = valley wall.

encroach on the valley floor in reaches FD and LC. Narrowest valley floors occur where fans (FB, LA) or earthflows (FD, LC) encroach upon the valley floor, forcing the channel against bedrock in the opposite valley wall. Widest valley floors (FE, LD) occur immediately upstream of the most actively constricting earthflow reaches. Little bedrock is exposed in the wide reaches; in general, there is a negative relation between valley floor width and percent of channel margin length comprised of exposed bedrock. This is presumably due to burial of bedrock in wide, alluviated reaches (Table 2).

In general, there are three classes of valley floor-valley margin interaction: active-active, where the valley floor is bounded by two active constraining processes (i.e., earthflows, active alluvial fans) on either side of the channel; active-passive, where an active constraining process pushes the stream against a passive constraint (i.e., terraces, bedrock, inactive fans, colluvium); and passive-passive, where no active constraint is occurring. The active-active type is quite rare and found in only two reaches (LB and LE) comprising 6% of channel length in Lookout Creek. Most of French Pete Creek falls in the passive-passive category (68%), while most of Lookout Creek is in the active-passive category (66%) (Table 2), implying that channel-adjacent hillslope processes may currently be more active in Lookout Creek than in French Pete Creek.

4.3 *Distribution of Landforms by Reach*

The proportions of valley floor occupied by different landforms varies by reach. We consider the 12 sampled reaches as a population along a continuum of valley floor width and examine how landforms of different types contribute to increasing valley floor width (Figure 6, Table 3).

Active and secondary channels. The width of the active channel is relatively constant up to a valley floor width

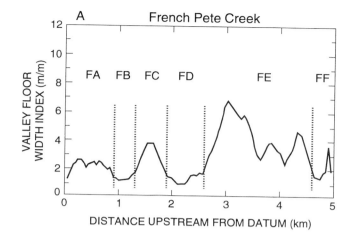

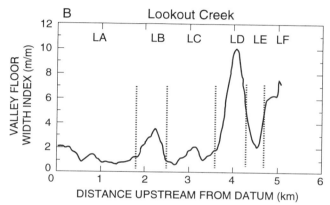

Fig. 5. Plot of valley floor width index (VFWI) as a function of distance upstream for the two study sites. VFWI was calculated as a 200-m running average of the ratio of active and floodplain widths to the average active channel width for the entire site, based on measurements of valley width taken at 50-m intervals. (a) Lookout Creek. (b) French Pete Creek.

index of 4.0, above which it increases (Figure 6). Reach-to-reach variation overwhelms any systematic downstream increase in channel width. Indeed, upstream reaches (FE, LF, LD) have the widest channels at each site. Secondary channels are most extensive in reaches with wide valley floors (VFWI > 1.8) (Figure 6). Reaches with relatively narrow valley floors (VFWI < 2.0) average only about 1600 m²/km of secondary channel, whereas reaches with wider valley floors (VFWI ≥ 2.9) average 8100 m²/km.

Floodplains 1964 and younger. These include floodplains that bear vegetation originating from fluvial disturbance in 1964 or more recently. Wide valleys tend to have wide, young floodplains, although the proportion of valley floor in this landform does not increase linearly with valley floor width. Instead, there is a marked increase at a valley floor width index of approximately 2.8; widths of post-1964 floodplains average 4.3 m below this value (n = 6) and 14.0 m above it (n = 6). The most extensive development of this landform occurs in reaches where fluvial disturbances and large debris flows from tributary streams occurred during the 1964-1965 winter (reaches LD, LE, LF, and FE) (Figure 3a, b).

Pre-1964 floodplains. Pre-1964 floodplains have vegetation older than 30 years old growing on them. Reaches vary widely in the width of this landform, from less than 1 m (FD) to over 100 m (LD). Unlike other surfaces, widths of pre-1964 floodplains increase linearly with valley floor width index (r^2 = 0.87). Little of this surface type occurs in reaches with valley floor width index less than 1.8, which include reaches constrained by earthflows (FD and LC), bedrock (LA), and terraces (FB). These reaches also have the lowest proportion of valley floor width disturbed by the 1964 flood. Development of pre-1964 floodplain surfaces is limited in reaches with low valley floor width index by two related factors which lead to either erosion of deposited material or nondeposition: limited space for deposition and very high shear stresses during flood events through the narrow canyons [*Nanson*, 1986; *Baker and Pickup*, 1987; *Kelsey*, 1988; *Miller*, 1990a, b].

The reaches showing most extensive development of pre-1964 floodplain surfaces are also those with the widest 1964 floodplains (Figure 6). A debris-flow origin for some pre-1964 floodplains is inferred, based on their slightly convex-upward morphology, and position along the margins of the valley in widening reaches downstream from debris-flow producing tributaries [*Carling*, 1989]. Wide floodplains also have developed in reaches whose downstream ends are constricted by mass movements. Reaches with extensive floodplain surfaces (LD, LE, LF, and FE) have developed in response to earthflow constriction of a downstream reach (LC in Lookout and FD in French Pete). Aggradation in upstream reaches is promoted both by constriction of the channel, which may hydraulically dam the upstream reach or physically block further downstream movement of debris flows, and by raised base level in the earthflow reach [*Vest*, 1988]. Floodplains may have once been more extensive in some reaches, such as LE, but have since become obscured by subsequent growth of alluvial fans. Large landslides may also override floodplain areas and limit their further development by narrowing the valley floor.

Terraces and alluvial fans. The proportion of valley floor width occupied by terraces and alluvial fans ranges from 18% (reach LF) to 85% (reach LA) and displays no consistent relationship with valley floor width index (Figure 6, Table 3). Factors contributing to long-term preservation

TABLE 3. Average Widths of Surfaces By Reach for French Pete and Lookout Creek Sites

| | Active Channel (m) | Secondary Channel (m) | Floodplains | | Terrace (m) | Fan (m) | Total Width 1964 and Younger Surfaces (m) | Total Width Valley Floor (m) | Proportion of Valley Floor Reset By 1964 Flood |
			1964 and Younger (m)	Pre-1964 (m)					
FRENCH PETE CREEK									
FA	18.7	8.8	5.0	14.2	36.5	13.1	32.4	46.6	70
FB	18.2	1.5	3.6	1.5	25.3	22.5	23.4	24.9	94
FC	15.3	6.0	5.1	21.5	3.5	16.5	26.4	47.9	55
FD	24.3	2.5	4.4	0.7	3.6	13.5	31.3	32.0	98
FE	20.7	5.6	12.4	30.6	16.9	21.7	38.7	69.3	56
FF	20.3	3.2	3.0	8.7	16.9	13.3	26.5	35.2	75
LOOKOUT CREEK									
LA	14.3	0.3	5.2	4.6	23.6	114.5	19.7	24.3	81
LB	20.5	3.4	6.4	30.6	29.4	78.4	30.3	60.9	50
LC	15.8	0.3	4.2	5.1	2.8	8.1	20.3	25.4	80
LD	36.4	9.1	15.2	105.4	30.1	43.4	60.6	166.0	37
LE	14.9	0.0	17.3	26.7	6.1	178.6	32.3	58.9	55
LF	32.1	11.7	27.8	40.4	9.1	14.8	71.6	112.0	64

Widths calculated by dividing total area in each surface type by reach length.

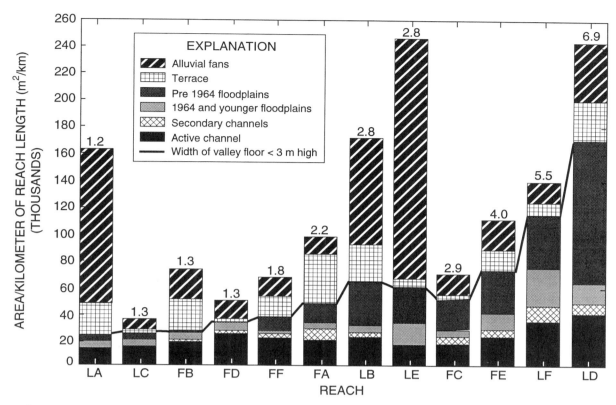

Fig. 6. Distribution of valley floor surfaces by reach for both sites ordered by valley floor width index (shown above bars). Reaches were identified by a two-letter code, where the first letter denoted the site (F = French Pete Creek, L = Lookout Creek) and the second denoted the order of the reach (A = furthest downstream, F = furthest upstream). Vertical axis is the total area by reach in each surface type divided by the reach length.

of terrace and fan deposits are complex, including location and orientation of these features in relation to the active channel, erodibility of deposits, and in the case of fans, whether they are actively growing by debris-flow deposition. More extensive fans in Lookout Creek reflect the wider valley floor (outside the zone measured by the VFWI and including terraces and fans), which allowed fans to develop and survive erosion by the mainstem channel.

Overall trends in landform distribution. Two points stand out from this analysis. First, while it is intuitively obvious that wider landforms are found on wider valley floors, not all surface types contribute equally to increasing valley width in montane valley floors (Figure 6). In general, increasing valley floor width is accommodated primarily by increasing width of floodplains. Taking both young and older floodplains together, this increase appears to be nonlinear for the limited sample size reported here. This suggests that as valleys widen, the proportion of valley floor in floodplains may increase sharply where the ratio of valley to channel width exceeds four. A second point is that the width of valleys and associated landforms in mountain streams is not controlled by the discharge alone, which is relatively constant over the short reach lengths examined here, but by the distribution of hillslope processes and resistant channel margins impinging on the valley floor.

5. VARIATION IN CHANNEL GRADIENT AND BOULDER DENSITIES BY REACHES

We examined the longitudinal profiles of the two creeks to determine whether the pattern of reaches defined by valley floor width also were reflected in the channel gradient. The longitudinal profiles of the two creeks through the two study sites are relatively straight (Figure 7). There is a weak ($r^2 = 0.15$, n = 12) negative correlation between valley floor width index and average channel gradient by reach; so narrow reaches tend to be steeper than wide reaches. The convexity in channel gradient in the French Pete profile corresponds to reach

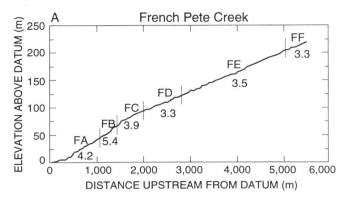

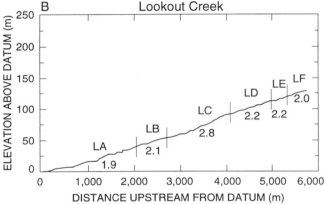

Fig. 7. Longitudinal profiles of the two study sites showing reach boundaries. (a) Lookout Creek. (b) French Pete Creek. Numbers show average channel gradient (in percent) by reach.

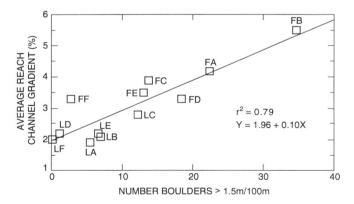

Fig. 8. Relation between average channel gradient and average boulder densities by reach.

FC, and the concavity in the Lookout Creek profile corresponds closely with reach LC.

The influence of reach-forming processes on channel gradient is expressed primarily in differing potentials for delivering boulders to and storing them in the channel. A strong correlation ($r^2 = 0.79$, n = 12) exists between average reach channel gradient and the number of large ($\geq$ 1.5 m) boulders measured in the active channel within the reach (Figure 8). Boulders of this size are moved rarely (recurrence intervals of 20 - 50 years [*Grant et al.*, 1990]) and are usually deposited in the creek bed by non-fluvial processes, such as landslides and debris flows.

Boulder densities vary by reach due to several factors. First, the potential for boulder delivery varies by the distribution of active hillslope processes or landforms capable of contributing boulders in each reach. Second, valley floors of different widths vary in their efficiency in intercepting boulders delivered by hillslope processes before they reach the channel. Channel boulder densities should be lowest where the valley floor is widest. Third, shear stresses and stream competence during floods are typically much greater in narrow than in wide reaches [*O'Connor et al.*, 1986; *Baker and Pickup*, 1987; *Kelsey*, 1988; *Miller*, 1990b]. Consequently, few subboulder-sized particles are stored in narrow reaches, so residual boulders are fully exposed. In wide reaches, on the other hand, boulders may be buried by finer sediments.

Boulder densities (number of boulders $\geq$ 1.5 m in diameter per 100 m along the channel axis) were analyzed for each study stream, because the overall boulder density in French Pete Creek (16.6) was significantly higher than in Lookout Creek (6.2). Highest boulder densities occur in both sites where the channel is bordered by deeply incised alluvial fans and terraces underlain by old debris-flow deposits (FB, FA) and along earthflow-dominated reaches (FD, LC) (Table 2). Boulder densities are lowest in wide, fan-constrained reaches (LD, LF) where the wide valley floor permits development of large, active fans. Within-fan sorting of material supplied from tributaries leads to boulder deposition near fan apexes and finer sediment deposited near distal margins of fans proximal to channels. Hence, few boulders are visible in channels adjacent to fans developed on wide alluvial floors. Extensive deposits of finer gravels and cobbles in reach LD may also have buried boulders. In both sites, boulder densities decrease with valley floor width index, approaching zero where the valley floor width index exceeds 6 (Table 2). Boulder densities are higher in French Pete as compared to Lookout Creek despite the greater proportion of channel length bordered by active mass movement processes (Table 2); this may reflect both longevity of boulders delivered to channels and prevalence of more competent lava flows and breccias in the French Pete basin [*Sherrod and Smith*, 1989].

6. VARIATION IN DISTURBANCE REGIMES BY REACH

The mosaic of valley floor landforms arrayed longitudinally along the stream reflects reach-to-reach differences in disturbances due to both fluvial and non-fluvial processes. The term "disturbance" is used in both its geomorphic and ecological sense as processes that either erode or create new landforms or remove riparian vegetation. To explore this, we examine the effects of the December 1964 storm in Lookout Creek, which varied by reach.

6.1 Fluvial Disturbance

Interpretation of the response of individual reaches to peak flows generated during the 1964 flood is limited by the fact that we can only observe the post-flood legacy of landforms and vegetation; we do not know what their distributions were prior to that event, since dense tree canopies precludes use of air photos. We can, however, infer what the magnitude of effects were by comparing the average total widths of landforms aged 1964 or younger, including the active channel and secondary channels, to the total width of the Holocene valley floor. The absolute width of valley floor reworked by the 1964 event is relatively constant up to a valley floor width index of 4.0 where it increases dramatically (Figure 6). However, the relative proportion of valley floor reworked by the 1964 storm decreases linearly with valley floor width index (Table 3). In some reaches (i.e., FB, FD, LA, LC), over 80% of the landforms on the valley floor are less than 30 years old; while in other reaches (i.e., LD), young valley floor represents only 37% of the total.

Reach-to-reach differences in hydraulic forces generated by the 1964 storm are probably responsible for these trends (Table 4). Shear stresses exerted on the bed by flood flows are much higher in narrow, constrained reaches than in wide, unconstrained reaches. The reach-level planform geometry of the valley floor strongly controls the pattern of erosion, deposition, and disturbance to riparian vegetation during major flood events. Where streams are confined by steep valley walls, high terraces, fans, landslide colluvium, or bedrock (LA, LB, LC, LE), flows are deep and channel gradients are higher than average; so shear stresses are high, and erosion of banks and surfaces and uprooting of riparian vegetation are dominant processes. Landforms created or reworked in these reaches are likely to be narrow (Figure 6). On the other hand, where the valley is wide relative to channel width (LD, LF), shear stresses are lower and deposition and lateral channel changes are dominant processes. Deposits from previous floods are less likely to be disturbed; so wide floodplains with diverse age classes of vegetation and sediment are preserved (Figure 6). The distribution of floodplain surfaces older than 1964 also follows the trend of increasing width with decreasing shear stress (Table 4, Figure 6), suggesting that control of erosion and deposition by valley floor geometry persists over long timescales.

Similar patterns have been reported elsewhere [*Wolman and Eiler,* 1958; *Nanson,* 1986; *O'Connor et al.,* 1986; *Baker and Pickup,* 1987; *Kelsey,* 1988; *Jacobson et al.,* 1989; *Miller,* 1990a, b; *Wohl,* 1992]. In particular, our data support Wolman and Eiler's observation [1958, p. 12] that scour of valley floors during floods is likely to occur where the channel occupies more than approximately one-third of the valley bottom. We plotted our reach data on their graph of the ratio of channel to valley width versus slope; the abscissa in this graph is the inverse of the valley floor width index to conform to their convention (Figure 9). All of the reaches showing significant deposition occur where the valley is greater than three channel widths wide. Although we examined only a limited range of slopes, there was no discernible trend with gradient, as suggested by the Wolman and Eiler curve. We suggest that the ratio of channel to valley width alone accounts for most of the difference in whether floods scour or deposit, by determining the height of the floodwaters, hence shear stress. As *Wolman and Eiler* [1958] point out, however, narrow valleys and steep slopes tend to be associated in nature; so distinguishing the relative importance of slope versus width may be difficult.

6.2 Non-fluvial Disturbance

Reaches also differ in their susceptibilities to disruption by non-fluvial disturbances, such as debris flows and landslides. In the case of landslides, susceptibility is related to the location of the reach with respect to sites of mass movement. Where earthflows impinge directly on the valley floor, valley floor surfaces and stream channels may experience episodic delivery of sediment and wood as toeslope areas oversteepen and fail [*Swanson et al.,* 1985]. These effects tend to be local, due to the large size and relative immobility of material delivered.

Four factors contribute to high susceptibility of reaches to debris flows: (1) presence of tributary source areas for debris flow within or immediately above a reach; (2) absence of impediments to debris-flow travel between tributary mouths and the channel, such as active fans or wide terraces or floodplains; (3) low junction angles that promote continued movement of material down the main stem [*Benda,* 1985]; and (4) abrupt decrease in gradient or

TABLE 4. Calculated Hydraulic Variables By Reach for the December 1964 Flood in Lookout Creek

Reach	Q*	n**	Slope	Width of 1964 Inundated Surface***	Calculated hydraulic radius (R)†	Shear Stress (τ)††
	(m³/s)		(m/m)	(m)	(m)	(N/m²)
LA	179	0.051	0.019	19.7	2.1	380
LB	157	0.052	0.021	30.3	1.5	300
LC	150	0.055	0.028	20.3	1.7	470
LD	136	0.045	0.022	60.6	0.8	170
LE	128	0.052	0.021	32.3	1.2	250
LF	124	0.040	0.020	71.7	0.7	130

*Discharges (Q) were calculated by reach by assuming a linear relationship between drainage areas and the discharge measured at the U.S.G.S. gaging station at 63 km² (187 m³ s) and determining the drainage area at the midpoint of each reach.

**Manning's roughness coefficients (n) were determined from *Barnes* [1967] for the roughest and least rough reaches (LC and LF, respectively), based on boulder density and other observations and the remainder of the reaches assigned a roughness coefficient from a linear regression of the log-transformed relation between boulder density and roughness for these two reaches.

***Sum of the widths of post-1964 age surfaces, including active channel, secondary channel, and 1964 and younger floodplains (Table 3). This assumes that this entire set of surfaces were occupied by flow at the same time.

†Calculated from the continuity equation Q = W*D*V and the Manning's equation V = $(R^{0.67}S^{0.5})/n$, where Q is discharge (m³ s), W is width (m), D is depth (m), and V is velocity (m/s), R = hydraulic radius, S = friction slope, and n = Manning's roughness coefficient. Assuming that S = bed slope, substituting R for D and solving for R gives R = $[(Qn/WS^{0.5})]^{0.60}$.

††Total boundary shear stress τ calculated as $\tau = \gamma RS$, where γ is the specific weight of water.

widening of the valley that promotes debris-flow stoppage. Actual transport of material down the mainstem may not be as a true rheological debris flow but as a debris-laden flood pulse, particularly if full or partial damming of the mainstem triggered a dam outburst flood [*Costa and Schuster,* 1988; *Costa,* 1988].

From these considerations, we would predict that debris flows entering narrow reaches at the head of wide reaches may result in the greatest riparian disturbances (Figure 10a). Debris flows entering directly into wide reaches may rapidly lose energy and mass and may not have sufficient volume to block the channel (Figure 10c). Debris flows into long, narrow reaches can only disturb the narrow floodplains found there (Figure 10b). This is borne out in comparing the extensive 1964-age floodplains, including debris-flow deposits, in wide reaches FE and LD with the very limited floodplains and debris-flow deposits of the same age in reaches LA, LB, and LE despite the latter reaches experiencing runout from nine separate debris flows during the 1964 storm (Figure 3a, b).

7. CONTRASTS BETWEEN VALLEY FLOORS IN LOW- AND HIGH-GRADIENT SYSTEMS

Comparing surficial morphology of valley floors in mountain and low-gradient streams is useful to highlight differences in dynamics between these two systems. Few published data, however, provide a basis for such a comparison [*Swanson and Sparks,* 1990]. In a study section of the Little Missouri River [*Everitt,* 1968], for example, the river has reworked a valley floor area of approximately 5.9 channel widths in lateral extent (0.54 km² of valley floor per km of valley length) in the century before Everitt's analysis. This meandering river section has low slope (0.00085) and an extensive history of channel change recorded in cottonwood forest up to 300 years old.

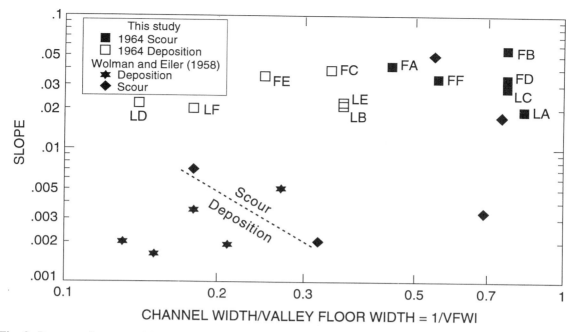

Fig. 9. Patterns of scour and deposition during floods in relation to channel gradient and the ratio of channel width to valley floor width, after *Wolman and Eiler* [1958]. Shown are Lookout and French Pete Creek reaches and data points for the Connecticut River following the 1955 flood. The dotted line is Wolman and Eiler's [1958] envelope curve (fitted by eye) distinguishing domains of scour and deposition for their data.

In French Pete and Lookout Creeks, by comparison, the area that has been fluvially reworked in the past century ranges by reach from 1.3 to 2.1 channel widths (0.023 to 0.039 km² of valley floor per km of valley length) in French Pete and 1.0 to 3.3 channel widths (0.020 to 0.072 km² of valley floor per km) in Lookout Creek. These values are calculated using the widths of active channel plus 1964 and later floodplains; this assumes that virtually all channel changes in the past 100 years occurred during the 1964 storm. Recently published analyses of long-term cross-section data from Lookout Creek confirms that very little change in cross-section area occurs between major storms [*Nakamura and Swanson*, 1993]. Constraint by bedrock, landslides, steep hillslopes, high gradients, and coarse channel beds limit the extent of valley floor reworking.

The rate and geometry, as well as the lateral extent, of channel change are different in the two systems. *Everitt's* [1968] dendrochronologic analysis of cottonwood forest indicates that channel change has been progressive, not episodic: arcuate bands of forest of increasing age defining old point bars generally trend away from the channel towards the edge of the valley floor. In French Pete Creek and Lookout Creek, on the other hand, the only conspicuous forest age class colonizing fluvial surfaces is of 1964 origin and parallels the channel. Older forest along valley floors generally exceeds 100 years in age and originated following wildfire [*Teensma*, 1987].

From this analysis, we infer that valley floor change in lowland, meandering streams is dominated by higher frequency, lower magnitude events, and the areal extent of change at the century time scale is greater than in mountain streams. The implication of this is twofold. First, the pattern of vegetation emphasizes that rare events are needed to sculpt the bottomland morphology in mountain streams. Second, due to the limited opportunity for change in steepland valley bottoms, the same reaches of the stream network are repeatedly and episodically disturbed. Consequently, the flood record, as preserved in stratigraphy and vegetation, is much less complete than in lowland streams.

What emerges from this comparison with *Everitt's* [1968] work is that a key difference between mountain and lowland streams is the relative importance of fluvial versus non-fluvial processes in determining channel and valley floor morphology. In lowland streams, channel and valley floor features are created and maintained by the interaction among flow hydraulics, sediment transport, and the channel boundaries; bankfull and higher flows that occur relatively frequently are the dominant geomorphic agents shaping

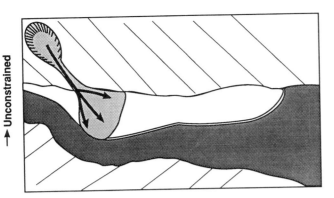
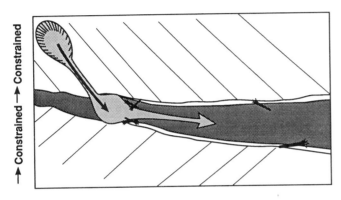
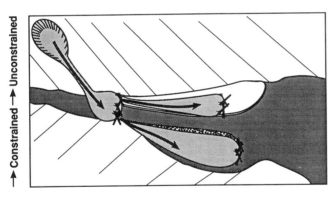

Fig. 10. Hypothesized differences in downstream disturbance potential for debris flows entering mainstem channels, depending on degree of constraint in receiving channel and downstream reach morphology. (a) Debris flow enters constrained reach situated immediately upstream of unconstrained reach. Potential for debris flow to temporarily dam creek leading to dam-break flood and major deposition and reworking of downstream surfaces is high. (b) High shear stresses and limited opportunity for sediment storage in constrained reach below debris-flow entry point limits potential for valley floor disturbance. (c) Wide valley floor surfaces in unconstrained reaches can trap debris-flow material before it reaches channel.

bottomlands. In contrast in mountain streams, fluvial processes alone are much less effective in shaping channel and valley floors. Drainage areas are small, and except for the occasional catastrophic flood, flows are generally incompetent to move the coarse material supplied by hillslope mass wasting. Under these conditions, the pattern of channel and valley floor features is much more intimately connected to the distribution and dynamics of hillslope processes.

8. CONCLUSIONS

In contrast to low-gradient, alluvial streams where most of the character of the valley bottom is the result of fluvial processes, valley floor morphology of mountain streams is strongly controlled by bedrock, hillslope, and tributary stream processes. In these environments, interactions among fluvial and non-fluvial transport processes and the degree and kind of marginal constraints determine valley floor width. Differences in valley floor width are expressed in a reach-to-reach variation in distribution of geomorphic surfaces of different types, modes of origin, ages, and geometries arrayed along the channel. Hillslope processes dominate channel morphology and disturbance regime of the valley floor. Processes creating surfaces in narrow montane valleys episodically restructure valley bottoms, preserving only a fraction of the stream's history.

Acknowledgments. This research was supported by the National Science Foundation under both the Riparian grant (BSR85-08356) and the Long-Term Ecological Research grant (BSR85-14325). We appreciate the field assistance provided by Jack Kleinman, Todd Bohle, Sallie Vest, John Moreau, Richard Harris, and Art Mckee. We wish to thank Harvey Kelsey, Robert Jacobson, Andrew Miller, Steve Wondzell, Futoshi Nakamura, John Costa, and Ed Keller, whose reviews of earlier drafts of this manuscript considerably improved its content.

REFERENCES

Ashley, G. M., W. H. Renwick, and G. H. Haug, Channel form and processes in bedrock and alluvial reaches of the Raritan River, New Jersey, *Geology, 16,* 436-439, 1988.

Baker, V. R., Flood sedimentation in bedrock fluvial systems, in Sedimentology of Gravels and Conglomerates, edited by E. H. Koster and R. J. Steel, *Canadian Soc. of Petrol. Geol. Mem. 10,* 87-98, 1984.

Baker, V. R., and R. C. Kochel, Flood sedimentation in bedrock fluvial systems, in *Flood Geomorphology,* edited by V. R. Baker, R. C. Kochel, and P. C. Patton, pp 123-137, John Wiley and Sons, N. Y., 1988.

Baker, V. R., and G. Pickup, Flood geomorphology of the Katherine Gorge, Northern Territory, Australia, *Geol. Soc. of Am. Bull., 98,* 635-646, 1987.

Barnes, H. H., Roughness characteristics of natural channels, 213 pp., *U.S. Geol. Surv. Water Supply Pap.* 1849, 1967.

Benda, L., Delineation of channels susceptible to debris flows and debris floods, *Proc. of the Internat. Symp. on Erosion, Debris Flow, and Disaster Prevention,* pp. 195-201, September 1985, Tsukuba, Japan, (publisher unknown), 1985.

Bhowmik, N. G., Hydraulic geometry of floodplains, *J. of Hydrol., 63,* 369-401, 1984.

Brakenridge, G. R., Alluvial stratigraphy and radiocarbon dating along the Duck River, Tennessee: implications regarding floodplain origin, *Geol. Soc. of Am. Bull., 95,* 9-25, 1984.

Brakenridge, G. R., River flood regime and floodplain stratigraphy, in *Flood Geomorphology,* edited by V. R. Baker, R. C. Kochel, and P. C. Patton, pp. 139-156, John Wiley and Sons, N. Y., 1988.

Carling, P. A., Hydrodynamic models of boulder berm deposition, *Geomorphology, 2,* 319-340, 1989.

Costa, J. E., Floods from dam failures, in *Flood Geomorphology,* edited by V. R. Baker, R. C. Kochel, and P. C. Patton, pp. 439-463, John Wiley and Sons, N. Y., 1988.

Costa, J. E., and R. D. Jarrett, Debris flows in small mountain stream channels of Colorado and their hydrologic implications, *Bull. of the Assoc. of Eng. Geol., 18,* 309-321, 1981.

Costa, J. E., and R. L. Schuster, The formation and failure of natural dams, *Geol. Soc. Am. Bull., 100,* 1054-1068, 1988.

Dyrness, C. T., Mass soil movements in the H.J. Andrews Experimental Forest, *Research Paper PNW-42,* 19 pp., U.S. Dept. of Ag., Forest Service, Pacific Northwest Forest and Range Experiment Station, Portland, Ore., 1967.

Everitt, B. L., Use of the cottonwood in an investigation of the recent history of a flood plain, *Am. J. of Sci., 266,* 417-439, 1968.

Fredriksen, R. L., A case history of mud and rock slide on an experimental watershed, *Research Note PNW-1,* 4 pp., U.S. Dept. of Ag., Forest Service, Pacific Northwest Forest and Range Experiment Station, Portland, Ore., 1963.

Fredriksen, R. L., Erosion and sedimentation following road construction and timber harvest on unstable soils in three small western Oregon watersheds, *Research Paper PNW-104,* 15 pp., U.S. Dept. of Ag., Forest Service, Pacific Northwest Forest and Range Experiment Station, Portland, Ore., 1970.

Frissell, C. A., W. J. Liss, C. E. Warren, and M. D. Hurley, A hierarchical framework for stream habitat classification: viewing streams in a watershed context, *Environ. Manag., 10,* 199-214, 1986.

Gottesfeld, A. S., F. J. Swanson, and L. M. Gottesfeld, A Pleistocene low-elevation subalpine forest in the Western Cascades, Oregon, *Northwest Sci., 55,* 157-167, 1981.

Grant, G. E., Downstream effects of timber harvest activities on the channel and valley floor morphology of western Cascade streams, Ph.D. dissertation, 363 pp., Johns Hopkins University, Baltimore, Md., 1986.

Grant, G. E., F. J. Swanson, and M. G. Wolman, Pattern and origin of stepped-bed morphology in high-gradient streams, Western Cascades, Oregon, *Geol. Soc. of Am. Bull., 102,* 340-352, 1990.

Gregory, S. V., F. J. Swanson, W. A. McKee, and K. W. Cummins, An ecosystem perspective of riparian zones, *BioSci., 41,* 540-551, 1991.

Hack, J. T., and J. C. Goodlett, Geomorphology and forest ecology of a mountain region in the central Appalachians, *Professional Paper 347,* 66 pp., U.S. Geol. Surv., Reston, Va., 1960.

Harr, R. D., Some characteristics and consequences of snowmelt during rainfall in western Oregon, *J. of Hydrol., 53,* 277-304, 1981.

Harris, D. D., L. L. Hubbard, and L. E. Hubbard, Magnitude and frequency of floods in western Oregon, *Open-file Report 79-553,* 15 p., U.S. Geol. Surv., 1979.

Hawkins, C. P., J. L. Kershner, P. A. Bisson, M. D. Bryant, L. M. Decker, S. V. Gregory, D. A. McCullough, C. K. Overton, G. H. Reeves, R. J. Steedman, and M. K. Young, A hierarchical approach to classifying stream habitat features, *Fisheries, 18,* 3-12, 1993.

Hupp, C. R., Stream-grade variation and riparian-forest ecology along Passage Creek, Virginia, *Bull. of the Torrey Botanical Club, 109,* 488-499, 1982.

Jacobson, R. B., A. J. Miller, and J. A. Smith, The role of catastrophic geomorphic events in central Appalachian landscape evolution, *Geomorphology, 2,* 257-284, 1989.

Jarrett, R. D., and J. E. Costa, Hydrology, geomorphology, and dam-break modeling of the July 15, 1982 Lawn Lake dam and Cascade Lake dam failures, Larimer County, Colorado, *Professional Paper 1369,* 78 pp., U.S. Geol. Surv., 1986.

Kani, T., Ecology of torrent-inhabiting insects (in Japanese), in Insects, I: Kenkyu-sha, Tokyo, edited by H. Furukuwa, pp. 171-317, 1944.

Kani, T., Stream classification in "Ecology of torrent inhabiting insects (1944)": an abridged translation," *Physiological Ecol. of Japan, 18,* 113-118, 1981.

Keller, E. A., and W. N. Melhorn, Rhythmic spacing and origin of pools and riffles, *Geol. Soc. of Am. Bull., 89,* 723-30, 1978.

Kelsey, H. M., Formation of inner gorges, *Catena, 15,* 433-458, 1988.

Kieffer, S. W., The 1983 hydraulic jump in Crystal Rapids: implications for river-running and geomorphic evolution in the Grand Canyon, *J. of Geol., 93,* 385-406, 1985.

Kishi, T., A. Mori, K. Hasegawa, and M. Kuroki, Bed configurations and sediment transports in mountainous rivers, in Comparative Hydrology of Rivers of Japan: Final Report, pp. 165-176, Japanese Research Group of Comparative Hydrology, Hokkaido University, Sapporo, Japan, 1987.

Kochel, R. C., Geomorphic impact of large floods: review and new perspectives on magnitude and frequency, in *Flood Geomorphology,* edited by V. R. Baker, R. C. Kochel, and P. C. Patton, pp. 169-187, John Wiley and Sons, N. Y., 1988.

Kochel, R. C., D. F. Ritter, and J. Miller, Role of tree dams in the construction of pseudo-terraces and variable geomorphic response to floods in Little River Valley, Virginia, *Geology, 15*, 718-721, 1987.

Leopold, L. B., and T. Maddock, The hydraulic geometry of stream channels and some physiographic implications, *Professional Paper 252*, 57 pp., U.S. Geol. Surv., 1953.

Leopold, L. B., M. G. Wolman, and J. P. Miller, *Fluvial Processes in Geomorphology*, 522 pp., W.H. Freeman, San Francisco, Calif., 1964.

Lewin, J., Changes of channel patterns and floodplains, in *Background to Paleohydrology*, edited by K. J. Gregory, pp. 303-319, John Wiley and Sons, N. Y., 1983.

Lisle, T. E. Stabilization of a gravel channel by large streamside obstructions and bedrock bends, Jacoby Creek, northwestern California, *Geol. Soc. of Am. Bull., 97*, 999-1011, 1986.

Miller, A. J., Fluvial response to debris associated with mass wasting during extreme floods, *Geology, 18*, 599-602, 1990a.

Miller, A. J., Flood hydrology and geomorphic effectiveness in the central Appalachians, *Earth Surf. Proc. and Landforms, 15*, 119-134, 1990b.

Mizuno, N., and H. Kawanabe, A topographical classification of streams, with an introduction of the system widely used in Japan: I. Reach type, stream zone, and stream type. *Verhandlungen Internationale Vereinigung Limnologie*, Bd. 21, 913, 1981.

Montgomery, D. R., and J. M. Buffington, Channel classification, prediction of channel response, and assessment of channel condition, Report TFW-SH10-93-002, 107 pp., Washington Dept. of Nat. Resour., Olympia Wash., 1993.

Nakamura, F., and F. J. Swanson, Effects of coarse woody debris on morphology and sediment storage of a mountain stream system in western Oregon, *Earth Surf. Proc. Landforms, 18*, 43-61, 1993.

Nanson, G. C., Episodes of vertical accretion and catastrophic stripping: a model of disequilibrium flood-plain development, *Geol. Soc. of Am. Bull., 97*, 1467-1475, 1986.

Nolan, K. M., and D. C. Marron, Contrast in stream-channel response to major storms in two mountainous areas of California, *Geology, 13*, 135-138, 1985.

O'Connor, J. E., R. H. Webb, and V. R. Baker, Paleohydrology of pool-and-riffle pattern development, Boulder Creek, Utah, *Geol. Soc. of Am. Bull., 97*, 410-420, 1986.

Osterkamp, W. R., and C. R. Hupp, Geomorphic and vegetative characteristics along three northern Virginia streams, *Geol. Soc. of Am. Bull., 95*, 1093-1101, 1984.

Peck, D. L., A. B. Griggs, H. G. Schlicker, F. G. Wells, and H. M. Dole, Geology of the central and northern part of the western Cascades, *Professional Paper 449*, 56 pp., U.S. Geol. Surv., 1964.

Pickup, G., and R. F. Warner, Effects of hydrologic regime on magnitude and frequency of dominant discharge, *J. of Hydrol., 29*, 51-75, 1976.

Priest, G. R., N. M. Woller, G. L. Black, and S. H. Evans, Overview of the geology of the central Oregon Cascade Range, in Special Paper 15, *Geology and Geothermal Resources of the Central Oregon Cascade Range*, edited by G. R. Priest and B. F. Vogt, pp. 3-28, Oregon Dept. of Geol. and Mineral Indust., 1983.

Richards, K. S., The morphology of riffle-pool sequences, *Earth Surf. Processes, 1*, 71-88, 1976.

Rosgen, D. L., A stream classification system, in riparian ecosystems and their management: reconciling conflicting uses, *General Technical Report RM-120*, U.S. Dept. of Ag., Forest Service, Rocky Mountain Forest and Range Experiment Station, 1985.

Rosgen, D. L., A classification of natural rivers, *Catena, 22*, 169-199, 1994.

Scott, K. M., and G. C. Gravlee, Flood surge on the Rubicon River, California- hydrology, hydraulics, and boulder transport, *Professional Paper 422-M*, 40 pp., U.S. Geol. Surv., 1968.

Sherrod, D. R., and J. G. Smith, Preliminary map of upper Eocene to Holocene volcanic and related rocks of the Cascade Range, Oregon, *Open-file Report 89-14*, 19 pp., U.S. Geol. Surv., 1989.

Stewart, J. H., and V. C. LaMarche, Erosion and deposition produced by the floods of December, 1964 on Coffee Creek, Trinity County, California, *Professional Paper 422K*, 22 pp., U.S. Geol. Surv., 1967.

Swanson, F. J., R. L. Graham, and G. E. Grant, Some effects of slope movements on river channels, Proceedings of the International Symposium on Erosion, Debris Flow, and Disaster Prevention, pp. 273-278, September 3-5, 1985, Tsukuba, Japan, (publisher unknown), Tokyo, 1985.

Swanson, F. J., and M. E. James, Geology and geomorphology of the H.J. Andrews Experimental Forest, western Cascades, Oregon: experimental forest, *Research Paper PNW-188*, 14 pp., U.S. Dept. of Ag., Forest Service, Pacific Northwest Forest and Range Experiment Station, Portland, OR., 1975a.

Swanson, F. J., and M. E. James, Geomorphic history of the lower Blue River-Lookout Creek area, western Cascades, Oregon, *Northwest Sci., 49*, 1-11, 1975b.

Swanson, F. J., and R. E. Sparks, Long-term ecological research and the invisible place, *Biosci., 40*, 502-508, 1990.

Takahashi, G., A study on the riffle-pool concept, *Transactions of the Japanese Geomorphological Union, 11(4)*, 319-336, 1990.

Teensma, P., Forest fire history of the H.J. Andrews Experimental Forest and vicinity, western Cascades, Oregon, Ph.D. dissertation, 188 pp., University of Oregon, Eugene, OR., 1987.

Tricart, J., Mechanismes normaux et phenomenes catastrophiques dan l'evolution des versants du bassin du Guil (Htes-Alpes, France) (in French), *Zeitschrift für Geomorphologie*, Band 5, Heft 4, 277-301, 1962.

Vannote, R. L., G. W. Minshall, K. W. Cummins, J. R. Sedell, and C. E. Cushing, The river continuum concept, *Canadian J. of Fisheries and Aquatic Sci., 37*, 130-137, 1980.

Vest, S., The effects of earthflows on valley floor and channel morphology, M.S. thesis, Oregon State University, Corvallis, OR., 1988.

Waananen, A. O., D. D. Harris, and R. C. Williams, Floods of December 1964 and January 1965 in the far western states, 265 pp., U.S. Geol. Surv. *Water Supply Paper 1866-A,* 1971.

Williams, G. P., Bankfull discharge of rivers, *Water Resour. Res., 14,* 1141-1154, 1978.

Wohl, E. E., Bedrock benches and boulder bars: floods in the Burdekin Gorge of Australia, *Geol. Soc. of Am. Bull., 104,* 770-778, 1992.

Wolman, M. G., and J. P. Eiler, Reconnaissance study of erosion and deposition produced by the flood of August 1955 in Connecticut, *Trans. Am. Geophys. Union, 39,* 1-14, 1958.

Wolman, M. G., and L. B. Leopold, River flood plains: some observations on their formation, *Professional Paper 282C,* pp. 87-107, U.S. Geol. Surv., 1957.

Gordon E. Grant and Frederick J. Swanson, Forestry Sciences Laboratory, 3200 SW Jefferson Way, Corvallis, OR 97331

Relative Scales of Time and Effectiveness of Watershed Processes in a Tropical Montane Rain Forest of Puerto Rico

F. N. Scatena

USDA Forest Service, International Institute of Tropical Forestry, Rio Piedras, Puerto Rico

Turnover periods and recovery times were used to quantify time scales that are relevant to watersheds processes in the lower elevation forests of the Luquillo Mountains of Puerto Rico. One hundred sixty turnover periods and 55 recovery periods of biotic and abiotic reservoirs were estimated. The transport of water and nutrients through the system was faster than the turnover of biological reservoirs by internal nutrient cycling. The turnover of biological reservoirs by nutrient cycling was faster than the turnover of biological reservoirs by natural disturbances. Landforms and physiographic features are the most stable features of the landscape and turnover on time periods that range between 1000's to millions of years. The empirical evidence suggests this landscape is a dynamic complex of subsystems that respond to a range of processes and indicates that the recovery period of a subsystem increases as the turnover period of the subsystem increases. Because different components of the system respond and recover from perturbations in different ways and at different rates, strict steady state conditions may never be achieved between all biotic and abiotic components of the landscape. Comparison of the relative timing of different watershed processes can provide the temporal scales needed to define the importance and independence of geomorphic and ecologic variables that act on the landscape.

1. INTRODUCTION

Quantifying the magnitudes and frequencies of discrete disturbance events and relative scales of time and effectiveness have been a unifying concepts in watershed geomorphology for several decades [*Wolman and Miller*, 1960, *Wolman and Gerson*, 1978]. Quantifying the role of disturbances in organizing ecosystems has also been a dominant theme in ecology [*Connell*, 1978, *Pickett and White*, 1985] and has led to the paradigm that landscapes are structured by the magnitudes and frequencies of the processes that act upon them [*O'Neill et al.*, 1986, *Urban et al.*, 1987, *Swanson et al.*, 1988]. It is now generally recognized that the ability of a disturbance to affect the morphology of a landscape or the structure of an ecosystem depends on the type of disturbance (such as flood, fire, landslide, pathogen, human activity), the force exerted, the area affected, the return period of the event, and the magnitude of the constructive or restorative processes that occur between perturbations [*Wolman and Gerson*, 1978, *Waide and Lugo*, 1992].

While the importance of determining the relevant time scales involved in sculpturing landscapes and structuring ecological communities is recognized, empirical measures of the time scales associated with different processes are scant. Nevertheless, understanding the spatial and temporal scales of watershed processes is essential to evaluating and differentiating short and long-term environmental change [*Wolman*, 1977]. This paper evaluates the time scales that are relevant to different watersheds processes that act on the lower elevation forests of the Luquillo Mountains of Puerto Rico. Because the importance and independence of geomorphic and ecologic variables can vary over the spatial or temporal scales considered [*Chorley et al.*, 1984, *Pickett and White*, 1985], this study limits comparisons to an area that has a relatively uniform climate, geology, forest cover, and environmental history. Process are considered over geographic spatial scales (*sensu Webb et al.*, 1972) and steady-time frames (*sensu Schumm and Lichty*, 1965).

Natural and Anthropogenic Influences in Fluvial Geomorphology
Geophysical Monograph 89
This paper is not subject to U. S. copyright. Published in 1995 by the American Geophysical Union

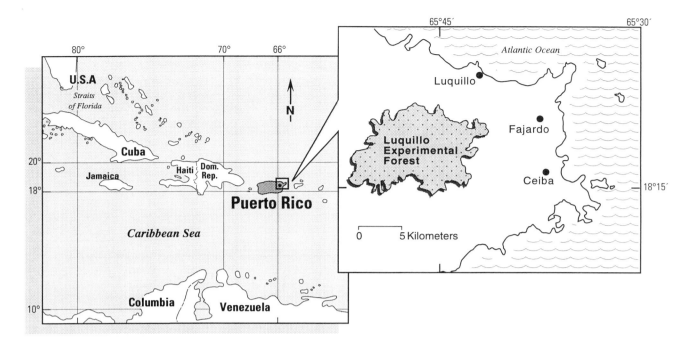

Fig. 1. Location map of the Luquillo Experimental Forest in Puerto Rico.

2. METHODOLOGY

A conceptual model of the interactions between the main inputs and subsystems of the Luquillo landscape provides a focus for this discussion (Fig. 1). The principal inputs to the landscape are climatic, tectonic, and human activities. The main subsystems of the landscape are the geologic framework of the Luquillo Mountains (i.e. geologic structure, lithology, elevation), landforms (i.e. ridges, slopes, valleys, terraces), soils, and biotic communities. In this simplified model, the geologic framework provides the landforms upon which soil and biotic communities develop. Climatic inputs to the system are modified by interactions with the different subsystems as they affect biotic components of the landscape. These interactions include the modifications that slope, aspect, and elevation can imposed on sunlight, rainfall, and other climatic inputs. In turn, the subsystems of the landscape influence the development, form, and structure of each other through cycling, mineralization and mass transport.

To compare relevant time scales associated with different watershed processes acting within and between the subsystems of Figure 1, turnover periods, recovery periods, and an index of event effectiveness were determined (Fig. 2 and 3). Average turnover periods were calculated as the quantity of storage in a reservoir within a mature forest, divided by the average annual flux of a given pathway. For example, the turnover of above-ground forest biomass by landslides was estimated at 3400 years and was calculated as the average above-ground biomass (220 Mg/ha from *Scatena et al.*, 1993) divided by the annual loss of biomass by landslides (0.065 Mg/ha/yr, from *Scatena and Lugo*, 1995). Comparisons of these average turnover times does not imply that the compartments are perfectly mixed or that the average age and average transit time of elements within a reservoir are equal [*Bolin and Rodhe*, 1973].

Recovery periods were considered as the time required for a feature or process that was produced or modified by a discrete event, to be obliterated or returned to its original form or rate. For example, the biotic recovery of forest vegetation following a landslide is the time required for a landslide scar to have the same basal area or biomass as the adjacent undisturbed forest [*Guariguata*, 1990, *Zarin*, 1993]. The geomorphic recovery of a landslide is the time required for the scar to refill with colluvium and attain the morphology of the adjacent hillslope. These recovery periods reflect the time span during which restorative processes operate between disturbances and are considered to be direct measures of the ability of a subsystem to return to a steady-state condition following a perturbation [*DeAngelis*, 1980].

Turnover and recovery periods were determined from existing studies and graphed with frequency boxes (Fig. 3).

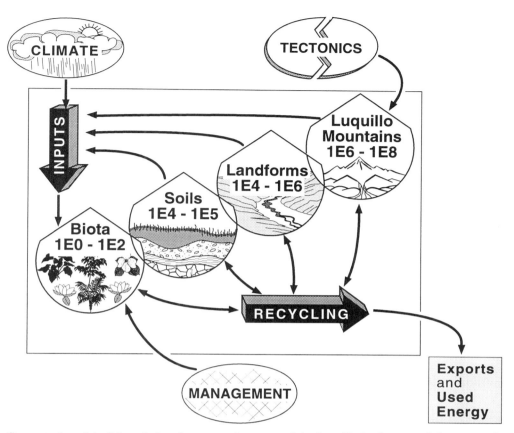

Fig. 2. Conceptual model of the relations between subsystems of the Luquillo landscape and the range of turnover periods associated with each subsystem. See text for details.

Each box encompasses the 25th through 75th percentiles, and has horizontal lines at the 10th, 50th, and 90th percentiles. Circles are data outside the 10th and 90th percentiles. In this figure, recovery periods are denoted by black boxes while turnover periods are denoted by unshaded boxes. The box labeled "mountain" refers to the residence time of the Luquillo Mountains as determined from denudation rates (see discussion). "Landforms" and "soil" refers to turnover of saprolite and soil pedons by surface erosion, respectively. "Stems/dist.", "biomass/dist.", and "crown/dist.", refer to turnovers of forest stems, biomass, and crown area on ridges, slopes, and valleys by hurricanes, treefall gaps, and landslides [*Scatena and Lugo*, 1995]. "Succession" refers to length of time required for a forest to attain pre-disturbance structure following both natural and anthropogenic disturbances [*Crow*, 1980; *Weaver*, 1986, *Weaver and Murphy*, 1990]. "Landslide" refers to recovery periods of above-ground biomass and nutrient pools on landslides [*Guariguata*, 1990; *Zarin*, 1993]. "Biomass/input" and "biomass/litter" refers to the turnover of N, P, Ca, Mg, K nutrient pools in biomass by atmospheric and weathering inputs, and litterfall respectively [*Scatena et al.*, 1993, *Lodge et al.*, 1991]. "Element" refers to the turnover of 15 common elements in the forest [*Odum*, 1970]. "Litter recovery" refers to the recovery of litter mass and nutrient fluxes following defoliation by Hurricane Hugo, "Crown/litter" refers to the turnover of nutrient pools in the forest crown by litterfall, "Floor/litter" refers to the turnover of forest floor biomass and nutrient stocks by litterfall, and "Seedling" refers to the turnover of selected seedling populations [*Smith*, 1970]. "H20" refers to the turnover of water in various ecosystem components [*Odum*, 1970] while "Soil pore" refers to recovery of soil pore pressures following intense rainfall events [*Simon et al.*, 1990; *Larsen et al.*, 1991], and "Sed. Conc." refers to the recovery of suspended sediment concentrations following Hurricane Hugo in September 1989 [*Gellis*, 1993].

The relative effectiveness of an event to affect a component of the landscape was scaled as the event flux divided by the average daily flux (Fig. 4). For example, the relative effectiveness index of Hurricane Hugo on

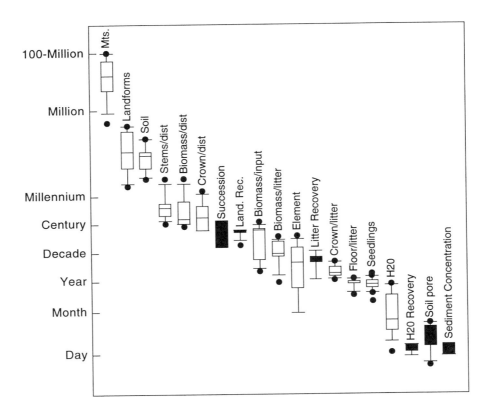

Fig. 3. Box plots of turnover and recovery periods for various components of the lower elevation forests of the Luquillo Experimental Forest of Puerto Rico. Each box encompasses the 25th through the 75th percentiles and has horizontal lines at the 10th, 50th and 90th percentiles. Circles are data outside the 10th and 90th percentiles. Black boxes are recovery periods, open boxes denote turnover periods. See text for details.

canopy litterfall was determined by dividing the litter fall that fell during the hurricane (1083 g/m^2/day from *Lodge et al.*, 1991) by the average daily litterfall (2.38 g/m^2/day). In Figure 4, "Q" refers to the relative effectiveness of stream discharge (hurricane daily discharge/average daily discharge) for the major rivers draining the Luquillo Mountains [*Scatena and Larsen*, 1991]. "Leaves", "litter", and "wood" refer to the ratio of hurricane flux to the average daily flux for both mass and nutrients in three different areas of the Luquillo Mountains [*Lodge et al.*, 1991]. "Landslides" refers to the relative effectiveness of the hurricane with regards to generating landslides (landslide area created by hurricane/average daily production of landslides) in four different parts of the Luquillo Mountains [*Scatena and Larsen*, 1991; *Scatena and Lugo*, 1995].

3. RESULTS

3.1 *Physiographic setting and local landforms*

The Luquillo Mountains rise abruptly from a narrow coastal plain and are the dominant physiographic feature of northeastern Puerto Rico. The asymmetrical, fault-block mountains have been tectonically stable since the end of Tertiary time, and under the influence of the present regime of plate motion for the past 2.5 to 4 million years [*Monroe*, 1980; *Schwab et al.*, 1991; *Ahmad et al.*, 1993]. Climatic conditions, soil morphology, forest structure, ecosystem productivity, and hurricane damage are all strongly correlated with the elevation, aspect, and lithology of the Luquillo Mountains [*Brown et al.*, 1983; *Weaver and Murphy*, 1990; *Boose et al.*, 1994]. Because of the

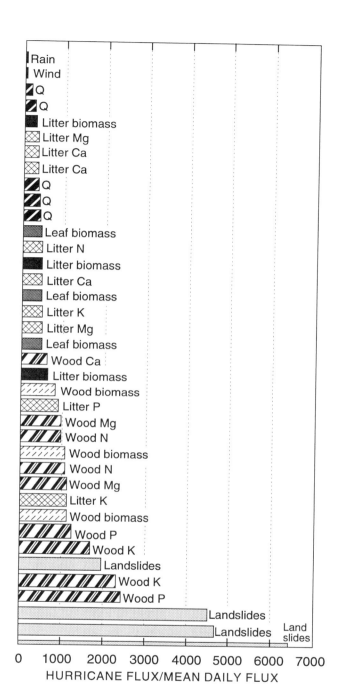

Fig. 4. The relative effectiveness of Hurricane Hugo, September 1989, for various ecosystem components of the lower elevation forests of the Luquillo Experimental Forest of Puerto Rico. Relative effectiveness of Hurricane Hugo on different components is defined as hurricane flux divided by average daily flux of that component. See text for details.

affect that elevation, aspect, and size of landmass have on climate and environmental gradients, physiography has traditionally been considered a dominant factor in determining the distribution of Caribbean and other tropical montane forests [*Beard*, 1949; *Jordan*, 1985].

The watersheds that are within the physiographic setting of the Luquillo Mountains are highly dissected assemblages of narrow ridges, steep slopes, and incised, structurally-controlled valleys. Catenas and toposequences in vegetation structure and composition, the size of soil nutrient pools, and the magnitude and frequency of slope failures, treefall gaps and hurricane damage have been identified with these landforms [*Basnet*, 1992; *Scatena and Lugo*, 1995]. Compared to other landscape positions, ridges are well drained, have fewer landslides and treefall gaps, more above-ground biomass, older forests, greater species richness, and soils with relatively low quantities of exchangeable bases but high quantities of soil-organic matter, acidity, and exchangeable iron. Valley settings have higher frequencies of treefalls and slope failures, less biomass, younger-aged stands, lower species richness, and in the frequency of treefall gaps, landslides and the impacts of hurricanes, the turnover of forest biomass by these disturbances can vary between 250 years for ridges to 40 years for valleys [*Scatena and Lugo*, 1995].

Unlike the turnover periods of many ecosystem components, the turnover of the entire Luquillo Mountains, or of individual landforms within the mountain range, are unknown and can only be approximated. Recent estimates of Pleistocene uplift along the northern coast of Puerto Rico average 40 mm/1000 years [*Taggart*, 1992] but may be locally as high as 600 mm/1000 years [*Seiders et al.*, 1972; *Briggs and Aquilar-Cortes*, 1980]. Estimates of denudation of the Luquillo Mountains average 120 mm/1000 years but range between 40 to 230 mm/1000 years [*R. Stallard*, personnel communication, *Larsen and Torres-Sanchez*, 1992; *McDowell and Asbury*, 1994]. Given these average rates of uplift and denudation, the turnover period of the 1000 m Luquillo Mountains is approximately 12 million years, but could range between 1 to 100 million years depending on local rates of uplift and denudation. At these denudation rates the average turnover time for the 5 to 8 m of saprolite that overlies Luquillo bedrock [*Simon and Larsen*, 1990] is 50,000 years but could vary locally between 2000 and 200,000 years. While these estimates are admittedly crude, they encompass the Cretaceous to Holocene-age deposits that comprise the Luquillo Mountains and indicate that although the geologic framework and local landforms are dynamic, they are the most stable components of the Luquillo landscape (Fig. 3).

Broad time constraints can also be placed on the development of soil pedons within the forest (Fig. 3). Given that creep rates for these soils have been measured at 0.2 cm/yr. [*Lewis*, 1974], and that the average length of slopes from ridges to perennial stream channels is 51 m (N = 88, SE = 4.5), it would take over 27,000 years to remove the forest's top soil by creep alone. Considering the total rates of denudation, this soil could be removed in 4000 to 25,000 years. Finally, estimates of the turnover of soil profiles by landslides range between a few thousand to 25,000 years and vary with elevation, lithology, and local topography [*Guariguata*, 1990; *Larsen and Torres Sanchez*, 1992; *Scatena and Lugo*, 1995].

While these estimates put some constraints on the time periods associated with the development of soil and saprolite, the relevant time periods for observing internal soil processes are considerable faster. The recovery of soil trace gas fluxes following a hurricane and a clear-cut have been measured on time scales ranging from 4 to 14 months and vary with the type of disturbance and the trace gas considered [*Steudler et al.*, 1991]. Time series measurements of soil oxygen indicate that soil water and gasses are mixed on the order of hours to days [*W. Silver*, personnel communication] while several days may be needed to observe the recovery of soil pore pressures following large rain storms [*Simon et al.*, 1990;, *Larsen and Torres-Sanchez*, 1990]. Weeks to months are needed to observe changes in extractable soil nutrients and root biomass following clear cutting and hurricanes [*Silver and Vogt*, 1993] and computer simulations indicate that treefalls and defoliation by hurricanes cause large increases in soil organic matter that are observable for several years and changes in phosphorus and nitrogen mineralization rates for up to 10 years [*Sanford et al.*,1991].

3.2 *Climatic inputs*

Because of its maritime trade-wind location, total monthly rainfall in the Luquillo Mountains is distributed relatively evenly throughout the year. Precipitation typically occurs in either short-duration rainfalls throughout the day, or in relatively high-intensity events associated with frontal systems that occur periodically during the year. The net result of this combination of low-intensity showers and high-intensity storms is a relatively high level of canopy interception compared to other montane tropical forests [*Scatena*, 1990]. Because of the fast-growing nature of these forests, throughfall in treefall gaps and large canopy openings typically recovers to levels of the adjacent forest within 1 to 2 years [*Scatena*, 1990].

While high frequency, short duration rainfall may account for up to 80% of the total annual rainfall, rainfalls exceeding 50 to 100 mm per day are the dominant process in sculpturing the geomorphic and ecologic landscapes of the Luquillo Mountains [*Scatena*, 1989; *Lugo and Scatena*, 1995]. Prolonged droughts occur somewhere on the island of Puerto Rico every 10 years, and 15 consecutive days without rainfall occurs in the lower Luquillo Mountains about once every 15 years. Even when these droughts occur, drought stress in canopy trees is not readily apparent. In contrast, storms that produce multiple landslides occur somewhere on the island once every 10 months [*Larsen and Simon*, 1993] and once every 1 to 2 years in the lower Luquillo Mountains. Once every 21 years a hurricane passes directly over some part of Puerto Rico. Once every 50 to 60 years a hurricane will pass directly over the Luquillo Mountains and do extensive modification to the landscape [*Scatena and Larsen*, 1991]. Maximum storm rainfall for these hurricanes ranges between 135 to 725 mm/storm (Avg. = 401, N = 6, SE=90) or between 5 to 40% of mean annual rainfall at the location of maximum storm rainfall (Avg. = 18.5, N = 6, SE = 4.95). The recovery of streamflow, and suspended sediment concentrations to background levels following these storms occurs within hours to days [*Gellis*, 1993].

3.3 *Vegetation and Nutrient Cycles*

While the composition of the tabonuco forest has been relatively stable during the last century of scientific observation, species turnover is extremely dynamic at the scale of hectares and averages about 1 species per hectare per year [*Lugo and Scatena*, 1995]. During the 500 years since European settlement of Puerto Rico, introduced dicotyledones and monocotyledones species have naturalized in the lower elevation forests a rate of approximately 1 per 23 years [*Chinea et al.*, 1993]. At these rates, it would take approximately 40 years to turnover species within a hectare, and over 7500 years to replace all the species in the forest with introduced exotica.

The average residence time of seedlings within the forest ranges from 0.17 to 1.6 years (Avg. = 0.68 yr., N=37, SE = 0.06; *Smith*, 1970). The average age of trees in the mature forest is about 30 years, and is considerably younger than the oldest individuals, which may live for 100's of years [*Scatena and Lugo*, 1995]). Non-catastrophic mortality associated with competition and stand thinning is the predominant type of tree mortality and ranges from 0.5 to 2.6% per year [*Lugo and Scatena*, 1995]. This non-catastrophic mortality is comparable to other humid tropical forests and can turn over the biomass and stems in a mature forest in 30 to 80 years compared to the 95 to 3400-year range needed by hurricanes, landslides, or treefalls [*Scatena and Lugo*, 1995].

Given that above-ground biomass accumulates in this forest at a average rate of 2.5 t/ha/yr, an average of 90 years or more are needed to accumulate the above-ground biomass in a mature forest stand (Weaver and Murphy 1990). After 50 years of unhindered growth following selective cutting of these forests, stand structure attained levels comparable to adjacent forests, but species composition and aboveground nutrient pools did not (Garcia-Montiel and Scatena 1994). Studies of regeneration on Luquillo landslides indicate that 50 to 60 years are needed for landslide scars to attain the composition, structure, and aboveground nutrient pools of adjacent forested stands (Guaruguata 1990, Zarin 1993). Considerably longer periods are needed for a landslide scar to be filled with colluvium and attain the form of adjacent hillslopes. When the recovery of above-ground biomass and nutrient capital for Luquillo landslides is compared with other tropical, temperate and boreal forests, most variation between sites can be explained by growing-season length, growing-season temperature, and site age (Zarin 1993). The limited evidence available indicates that the aseasonal tropical climate of Luquillo provides the most favorable conditions for rapid recovery.

The turnover of water stored in vegetation, animals, soil, and litter range from days to months (Avg. = 0.16 yr., N=8, SE = 0.091, Odum et al. 1970). In less than a year, the input of litter to the forest floor is sufficient to replace all the nutrients and necromass on the forest floor (Avg. = 0.7 yr., N = 6, SE = 0.095) and in 1 to 4 years litterfall can turnover a mass of organic matter and nutrients greater than that stored in the closed forest canopy (Avg. = 2.1 yr., N=6, SE = 0.49 calculated from Lodge et al. 1991 and Scatena et al. 1993). The turnover of most major nutrients reservoirs within the forests is less than a decade (Avg. = 8.4 yr., N=14, SE = 2.9, Odum 1970), while the time period needed for rainfall or weathering to provide the same quantity of nutrients into the system that is stored in the biomass of a mature forest is several decades (Avg. 47.5 yr., N=6, SE = 17.8, calculated from Scatena et al. 1993 and McDowell and Asbury 1994).

4. DISCUSSION

Over 160 turnover periods and 55 recovery periods were calculated for various components of the Luquillo landscape (Fig. 4). These time periods span 12 orders of magnitude and indicate the complexity that can exist within one, well defined landscape. In general, water and nutrients are cycled through the ecosystem in days to years, the turnover of forest biomass occurs on the order of decades, and changes in geomorphic form occur over centuries to millennium. The turnover and recovery periods associated with internal nutrient cycling is faster than the turnover induced by treefall gaps, landslides, or hurricanes. However, these disturbances turn over biological reservoirs faster than geomorphic processes turn over soil, saprolite, or landforms. The length of time needed for ecosystem processes to recover to pre-disturbance levels indicates system components recover faster than the recurrence of perturbations and that the recovery time of a subsystem increases as the average turnover period of the subsystem increases, as has been suggested by theoretical studies (DeAngelis 1980). Specifically, the recovery of forest structure following treefalls, landslides, and hurricanes is faster than forest turnover by these disturbances. The recovery of stream discharge, soil pore pressures, and other hydrologic variables following storm events are also faster than the average turnover of water in the system.

Hurricanes, landslides, and treefall gaps are the dominant disturbances in this landscape. When they occur, they initiate massive regeneration by opening the forest canopy and releasing nutrients stored in relatively immobile compartments like wood. This regeneration then dominates the structure, composition, and age of the disturbed area for decades or centuries (Crow 1980, Weaver 1986). Between these catastrophic events, nutrients and energy cycle through the system to support forest metabolism, recovery, and growth. Because such catastrophic events are relatively rare while nutrient cycling is continuous, a greater amount of "ecological work", as measured by the amount of nutrients cycled, is done by non-catastrophic processes rather than catastrophic disturbances. Like the observations that the movement of sediment from hillslopes or the transport of dissolved load by streams is not necessarily synonymous with landscape sculpturing (Wolman and Gerson 1978), the ecological work of nutrient cycling is associated with the maintenance of biological processes and not necessarily the structuring of ecosystems.

The relative time scales compiled here provide empirical support for the view of landscapes as a complex of dynamic, interacting subsystems. For example, within the soil subsystem the relevant time periods for observing different processes varies from hours to thousands of years. Over more discrete time scales, a hurricane can affect different components of the landscape in different ways (Fig. 4). When hurricanes occur, they can have winds with 50 to 60-year recurrence intervals, rainfalls with recurrence intervals of 10 years or less, and produce stream discharges with recurrence intervals of 10 to 30 years (Scatena and Larsen 1991). Because of differences in the magnitude of these forces, the relative effectiveness of an individual hurricane on different parts of the landscapes can range over 3 orders of magnitudes (Fig. 4). Moreover, where forest vegetation interfaces directly with high-magnitude winds, widespread defoliation and treefalls prevail (Scatena et al. 1992). In contrast, ecosystem components like stream channels or aquatic populations that interface with moderate-magnitude stream discharges may only experience minor modifications (Covich et al. 1991). Because different components of the landscape can respond to perturbations in different ways, time lags between the recovery of subsystems can develop such that observations of steady state conditions become scale dependent and may never be observed between all biotic and abiotic components. Comparisons of turnovers and recovery times like was done here (Fig. 3) can provide a logical construct for defining the temporal scales associated with the effectiveness of different geomorphic and ecologic variables and a famework for organizing investigations into the interactions of biotic and abiotic landscapes.

The concepts of magnitude and frequency, and relative scales of time and effectiveness have proven to be powerful organizing concepts for unraveling complex relationships between geomorphic and ecological processes in the Luquillo Mountains of Puerto Rico. It is hoped that this contribution will stimulate the compilation of other inventories and facilitate a quantitative approach to determining the relative importance of the work performed by the "dwarf, man, and huge giant" in sculpturing the landscape (Wolman and Miller 1960).

REFERENCES

Ahmad, R., F.N. Scatena, A. Gupta, Morphology and sedimentation in Caribbean montane streams: examples from Jamaica and Puerto Rico, *Sed. Geol.*, 85, 157-169, 1933.

Basnet, K., Effect of topography on the pattern of trees in tabonuco (Darcyodes excelsa)-dominated rain forests of Puerto Rico, *Biotropica*, 24, 31-42, 1992.

Beard, J.S., *The Natural Vegetation of the Windward and Leeward Islands*, Oxford Forestry Memoirs 21, 192 pp., 1949.

Brown, S., A.E. Lugo, S. Silander, L. Liegal, Research histories and opportunities in the Luquillo Experimental Forest, *Gen. Tech. Rep. SO-44*, 128 pp., Southern Experimental Forest, U.S. Dept. Agr. For. Ser., 1993.

Briggs, R.P., E. Aquilar-Cortes, Geologic map of Fajardo and Cayo Icacos Quadrangles, Puerto Rico, scale 1:20,000, *Misc. Geol. Invest. Map I-1153*, U.S. Geol. Surv., Reston, Va., 1980.

Bolin, B., H. Rodhe H., A note on the concepts of age distribution and transit time in natural reservoirs, *Tellus*, 25, 58-62, 1973.

Boose, E.R., D.R. Foster, M. Fluet, Hurricane impacts to tropical and temperate forest landscapes, *Ecol. Monographs*, 64, 4, 369-400, 1994.

Chinea, J.D., R.J. Beymer, C. Rivera, I. Sastre de Jesus, F.N. Scatena, An annotated list of the flora of the Bisley Area, Luquillo Experimental Forest, Puerto Rico 1987 to 1992, 12 pp., *Gen. Tech. Rept. SO-94*, August 1993.

Chorley, R.J., S.A. Schumm, D.E. Sugden, *Geomorphology*, 605 pp., Cambridge University Press, Cambridge U.K., 1984.

Connell, J.H., Diversity in tropical rain forests and coral reefs, *Sci.*, 199, 1302-1309, 1978.

Covich, A.P., T.A. Crowl, S.L. Johnson, D. Varza, D.L. Certain, Post-Hurricane Hugo increases in Atyid shrimp abundances in a Puerto Rican montane stream, *Biotropica*, 23, 4a, 448-454, 1991.

Crow, T.R., A rainforest chronicle: a 30-year record of change in structure and composition at El Verde, Puerto Rico, *Biotropica*, 12, 42-55, 1980.

DeAngelis, D.L., Energy flow, nutrient cycling, and ecosystem resilience, *Ecol.* 61, 4, 764-771, 1980.

Garcia-Montiel, D., and F.N. Scatena, The effect of human activity on the structure and composition of a tropical forest in Puerto Rico, *Forest Ecol. and Management*, 63, 57-78, 1994.

Gellis, A., The effects of Hurricane Hugo on suspended-sediment loads, Lago Loiza basin, Puerto Rico, *Earth Surf. Proc.*, 18, 6, 505-504, 1993.

Guariguata, M.R., Landslide disturbance and forest regeneration in the upper Luquillo Mountains of Puerto Rico, *J. of Ecol.*, 78, 814-832, 1990.

Jordan, C.F., *Nutrient cycling in tropical forest ecosystems*, 199 pp., John Wiley & Sons New York, 1985.

Lewis, L.A., Slow movement of earth under tropical rain forest conditions, *Geol.*, 2, 9-10, 1974.

Lodge, D.J., F.N. Scatena, C.E. Asbury, M.J. Sanchez, Fine litterfall and related nutrient inputs resulting from Hurricane Hugo in subtropical wet and lower montane rain forests of Puerto Rico, *Biotropica*, 23, 4a, 336-343, 1991.

Larsen, M. C., and A. Simon, A rainfall intensity-duration threshold for landslides in a humid-tropical environment, Puerto Rico, *Geografiska Annaler*, 75(a), 13-23, 1993.

Larsen, M. C., and A. J. Torres-Sanchez, Rainfall-soil moisture relations in landslide-prone areas of a tropical rain forest, Puerto Rico, *Proc. of the Intern. Symp. on Tropical Hydrology*, edited by J. Hari Krishna, V. Quinones-Aponte, F. Gomez-Gomez, G.L. Morris, pp. 121-131, 1990.

Larsen, M. C., A. J. Torres-Sanchez, Landslides triggered by Hurricane Hugo in Eastern Puerto Rico, September 1989, *Caribbean J. of Sci.*, 28, 3-4, 113-125, 1992.

McDowell, W. H., C. E. Asbury, Export of carbon, nitrogen, and major ions from three tropical montane watersheds, *Limnol. Oceanogr*, 39(1), 111-125, 1994.

Monroe, W.H., Some tropical landforms of Puerto Rico,

Prof. Paper 1159 U.S. Geological Survey, 39 pp., 1980.

Odum, H. T., Summary, An Emerging View of the Ecological System at El Verde, in a tropical Rain Forest, edited by H.T Odum and R.F. Pigeon, *Div. of Tech. Info.,* U.S. Atomic Energy Commission, pp I191-I289, Washington D.C., 1970.

Odum, H. T., A. M. Moore, L. A. Burns, Hydrogen budget and compartments in the rain forest, in A tropical Rain Forest, edited by H.T. Odum and R.F. Pigeon, *Div. of Tech. Info.,* U.S. Atomic Energy Commission, pp H105-H122, Washington, D.C., 1970.

O'Neill, R. V., D. L. DeAngelis, J. B. Waide, T. F. H. Allen, A hierarchical concept of ecosystems, 253 pp., *Monographs in Population Biology,* 23, Princeton University Press, Princeton, New Jersey, 1986.

Pickett, S. T. A., and P. S. White (Eds.), *The ecology of natural disturbance and patch dynamics*, 384 pp., Academic Press, New York, 1985.

Sanford, R. L., W. J. Parton, D. S. Oijma, D. J. Lodge, Hurricane effects on soil organic matter dynamics and forest production in the Luquillo Experimental Forest, Puerto Rico: Results of simulation modeling, *Biotropica,* 23, 4a, 364-372, 1991.

Scatena, F. N., Watershed scale rainfall interception on two forested watersheds in the Luquillo Mountains of Puerto Rico, *J. of Hyd.*, 113, 89-102, 1989.

Scatena, F.N., An introduction to the physiography and history of the Bisley Experimental Watersheds in the Luquillo Mountains of Puerto Rico, *Gen. Tech. Rep. SO-72,* Southern Forest Experiment Station, U. S. Dept. Agr. Forest Service, 22 pp., 1989.

Scatena, F. N., L. C. Larsen, Physical aspects of Hurricane Hugo in Puerto Rico, *Biotropica,* 23, 4a, 317-323, 1991.

Scatena, F. N., A. E. Lugo, Geomorphology, disturbance, and the soils and vegetation of two subtropical wet steepland watersheds of Puerto Rico, *Geomorph.*, in press, 1995.

Scatena, F. N., W. L. Silver, T. Siccama, A. Johnson, M.J. Sanchez, Biomass and nutrient content of the Bisley Experimental Watersheds, Luquillo Experimental Forest, Puerto Rico, before and after Hurricane Hugo, 1989, *Biotropica,* 25, 1, 15-27, 1993.

Schumm, S. A., and R. W. Lichty, Time, space, and causality in geomorphology, *Amer. J. Sci.*, 263, 110-119, 1965.

Schwab, W. C., W. W. Dabforth, K. M. Scalon, D. G. Masson, A giant submarine slope failure on the northern insular slope of Puerto Rico, *Marine Geol.* ,96, 237-246, 1991.

Sieders, V. M., R. P. Briggs, L. Glover III, Geology of Isla Desecho, Puerto Rico, with notes on the great southern Puerto Rico Fault Zone and Quaternary stillstands of the sea, *Prof. Paper* 739, U.S. Geological Survey, U.S. Printing Office, Washington D.C., 22 pp., 1972.

Silver, W. L., and K. A. Vogt, Fine root dynamics following single and multiple disturbances in a subtropical wet forest ecosystem, *J. of Ecol.*, 81, 729-738, 1993.

Simon, A., M. C. Larsen, C. R. Hupp, The role of soil processes in determining mechanism of slope failure and hillslope development in a humid-tropical forest, in Eastern Puerto Rico, *Geomorph.*, 3, 263-286, 1990.

Steudler, P. A., J. M. Melillo, R. D. Bowden, M. S. Castro, The effects of natural and human disturbance on soil nitrogen dynamics and trace gas flux in a Puerto Rican wet forest, *Biotropica,* 23, 4a, 356-363, 1991.

Smith, R. F., The vegetation structure of a Puerto Rican rain forest before and after short-term gamma irradiation, in *A Tropical Rain Forest,* edited by H.T. Odum and R.F. Pigeon, Division of Technical Information, U.S. Atomic Energy Commission, 1970.

Swanson, F. J., T. K. Kratz, N. Caine, R.G. Woodmansee, Landform effects on ecosystem patterns and processes, *BioScience,* 38, 2, 92-98, 1988.

Taggart, B. E., Tectonic and eustatic correlations of radiometrically dates marine terraces in northwestern Puerto Rico and Isla de Mona, Puerto Rico, PhD dissertation, 252 pp., Univ. of Puerto Rico, Mayaguez, 1992.

Urban, D. L., R. V. O'Neill, H. H. Shugart, Landscape ecology, *BioScience,* 37, 2, 119-127, 1987.

Waide, R. B., A. E. Lugo, A research perspective on disturbance and recovery of a tropical montane forest, in *Tropical Forests in Transition*, edited by J. G. Goldammer, Birkhauser Verlag, pp. 173-190, Basel, Switzerland, 1992.

Weaver, P. L., Hurricane damage and recovery in the montane forest of the Luquillo Mountains of Puerto Rico, *Caribbean J. Sci.,* 22, 53-70, 1986.

Weaver, P. L., P. G. Murphy, Forest structure and productivity in Puerto Rico's Luquillo Mountains, *Biotropica, 22,* 1, 69-82, 1990.

Webb, L. J., J. G. Tracey, W. T. Williams, Regeneration and pattern in the subtropical rainforest, *J. of Ecol.,* 60, 675-695, 1972.

Wolman, M. G., J. P. Miller, Magnitude and frequency of forces in geomorphic processes, *J. of Geol.,* 68, 54-74, 1960.

Wolman, M. G., Changing needs and opportunities in the sediment field, *Water Resour. Res.,* 13, 50-54, 1977.

Wolman, M. G., R. Gerson, Relative scales of time and effectiveness of climate in watershed geomorphology, *Earth Surf. Proc.,* 3, 189-208, 1978.

Zarin, D. J., Nutrient accumulation during succession in subtropical lower montane wet forests, Puerto Rico, Ph.D. Dissertation, Univ. of Penn., 168 pp., 1993.

Zimmerman, J. K., E. M. Everham, R. B. Waide, D. J. Lodge, C. M. Taylor, N. V. L. Brokaw, Response of tree species to hurricane winds in subtropical wet forest in Puerto Rico: implications for tropical tree life histories, *J. of Ecol.,* in press. 1994.

F. N. Scatena, International Institute of Tropical Forestry, USDA Forest Service, Call Box 25000, Rio Piedras, Puerto Rico.

Contribution of Heavy Rainfall to Rainfall Erosivity, Runoff, and Sediment Transport in the Wet Tropics of Australia

Bofu Yu

Faculty of Environmental Sciences, Griffith University, Brisbane, Australia

In the wet tropics of Australia where annual rainfall normally exceeds 2500 mm, watersheds are essentially saturated in the wet season (November - April) with high runoff coefficients. Even in this type of environment, heavy rainfall is still important in relation to rainfall erosivity, runoff and subsequent sediment transport in streams. Simple models relating daily rainfall amounts to rainfall erosivity, runoff and sediment transport were used to demonstrate that 1% of total rainfall with the highest intensity contributes 1.5% of runoff, 4% of rainfall erosivity and up to 9% of transported sediment. For one-third of total rainfall with the highest intensity, the relative contribution to runoff, rainfall erosivity, and sediment transport is 46%, 59% and 87%, respectively. These values probably represent the lower limits to the percent contribution of heavy rainfall in comparison to other climate regions in Australia. The relative contribution of heavy rainfall does not vary significantly in the study area with respect to rainfall erosivity and runoff, although a greater percentage contribution of heavy rainfall is always associated with a lower mean annual rainfall.

1. INTRODUCTION

In their seminal paper, *Wolman and Miller* [1960] demonstrate that both magnitude and frequency of occurrence are important aspects of a geomorphological event and large events are not necessarily the most significant in terms of the amount of material transported in streams. An effective discharge can be defined as one that transports the most sediment over a period of time. Other investigations have since shown that such an effective discharge can be related to, and is almost identical with, the channel-forming or dominant discharge [*Pickup and Warner*, 1976; *Andrews*, 1980; *Carling*, 1988]. However, little has been said about the role high flows above a certain discharge threshold play in carrying sediment relative to their contribution to the total streamflow. Also of interest is the question of how important rainfall above a certain intensity threshold is to soil erosion and runoff, relative to its contribution to the total precipitation.

From a different perspective, recent studies have shown that the temporal variation of high-intensity rainfall does not necessarily follow the same pattern of variation as the rainfall totals. In Australia, high-intensity rainfall in the Southern Tablelands region can occur in periods with above average rainfall as well as in periods of below average rainfall [*Yu and Neil*, 1991]. In southwest Western Australia where rainfall has significantly decreased in the last 70 to 80 years, high-intensity rainfall has not decreased concurrently [*Yu and Neil*, 1993]. *Nicholls and Kariko* [1993] found that the average rainfall intensity has not changed appreciably in areas where a significant increase in rainfall totals since the late 1940s has long been recognized [*Kraus*, 1954; *Pittock*, 1975, 1983; *Cornish*, 1977]. *Lough* [1993] found that no significant change to heavy rainfall (greater than 50 mm/day) has occurred in Queensland. In sub-Saharan West Africa, where an abrupt decrease of rainfall since the late 1960s has been well documented [*Nicholson*, 1980; *Lamb*, 1982; *Druyan*, 1989], change to the high-intensity rainfall and average rainfall intensity during the same period appears to be insignificant [*Albergel*, 1986; *Olaniran and Sumner*, 1989]. The observation that occurrence of heavy rainfall is, to

Natural and Anthropogenic Influences in Fluvial Geomorphology
Geophysical Monograph 89
Copyright 1995 by the American Geophysical Union

some extent, independent of the rainfall total is particularly relevant in the context of climate change in relation to the enhanced greenhouse effect. Temporal variation of heavy rainfall amounts warrants separate examination and analysis, and any scenarios about possible change to the magnitude and frequency of heavy rainfall occurrences must not be based on possible change to rainfall total alone.

Although the hydrological and geomorphological importance of heavy rainfall is well known, its contribution to such processes as soil erosion, runoff and sediment transport has not been quantified in a systematic way. The objective of this paper is to evaluate the relative importance of heavy rainfall in the wet tropics in terms of the percent contribution of heavy rainfall to these processes.

2. DEFINITIONS OF HEAVY RAINFALL

Heavy rainfall can be defined in different ways. A threshold intensity has often been used to define heavy rainfall. This approach is particularly attractive when a physical threshold clearly exists. With respect to soil erosion and runoff, however, physically-based thresholds are highly variable on a basin scale both spatially and temporally, depending on such factors as vegetation cover, soil property and the antecedent soil moisture condition. As a result, a threshold is usually arbitrarily chosen [*Lough*, 1993] or based on limited field observations [*Yu and Neil*, 1991; *Neil and Yu*, 1994]. Alternatively, heavy rainfall may be defined using a given recurrence interval, since the larger the recurrence interval the greater the rainfall depth for given duration. The problem with the second definition is that the contribution of heavy rainfall to climatology decreases as the recurrence interval increases and therefore it is difficult to quantify the importance of heavy rainfall in the context of climate change. In this paper, heavy rainfall refers to a specified fraction of total rainfall above a certain intensity threshold. In this definition, heavy rainfall is, first of all, associated with the highest intensity. In addition, since the contribution of heavy rainfall to the mean annual precipitation is specified, the likely change to heavy rainfall can be precisely determined for any given climate change scenarios in terms of precipitation changes. To evaluate the contribution of heavy rainfall to rainfall erosivity, runoff, and sediment transport, a series of fractional rainfall with the highest intensity is considered, ranging from 1% of total rainfall to 50% of total rainfall. To assess the variation of the contribution of heavy rainfall within the study area, a particular intensity threshold was selected so that rainfall with intensity above this threshold amounts to exactly one-third of total rainfall. Any fixed fraction of total rainfall would do for this purpose. One-third was chosen here only as a bench mark against which the relative importance of heavy rainfall to rainfall erosivity, runoff, and sediment transport can be compared on a regional basis.

3. THE STUDY AREA

Wet tropics is that region where the mean temperature of the coldest month exceeds 18°C and the duration of the wet season exceeds 9.5 months [*Chang and Lau*, 1993]. A wet month is defined as one that has more than 100 mm of rainfall, and half a wet month as one that has between 60 and 100 mm of rainfall. According to this definition, the wet tropics in Australia is spatially restricted to a narrow strip of coastal lowlands and adjacent mountains and tablelands of northeastern Queensland. Although relatively small in size (approximately 4400 km^2), the wet tropics has a significant conservation value with the most extensive tropical rainforest tracts in Australia [*Laurence*, 1993]. Any significant increase of the sediment and nutrient loadings in the streams of the region could potentially have adverse effects on the Greater Barrier Reef [*Kinsey*, 1991; *Wolanski*, 1994].

Two of the five rainfall stations in the wet tropics are located on the tablelands (Nos 2 and 4 in Figure 1) and the remaining three rainfall stations and all of the five stream gauging stations are located on the coastal lowlands. Mean annual rainfall and the intensity threshold for one-third of the total rainfall are given in Table 1. In the study area, the mean annual rainfall exceeds 2500 mm and can be as high as 7600 mm near the top of the mountain range. The fact that the rainfall is well in excess of the potential evaporation of about 1800 mm per year, especially during the wet season from November to April, suggests an essentially saturated environment with high runoff coefficients. When the ground is fully saturated, all of the rainfall becomes runoff irrespective of its intensity. Heavy rainfall, therefore, is relatively less important in comparison to temperate, semi-arid, and arid regions where significant runoff occurs only during events of high intensity and/or long duration. The relative contribution of heavy rainfall to runoff is, therefore, expected to be the lowest in the wet tropics.

4. RAINFALL EROSIVITY

4.1 Data and Methods

Soil erosion is a very complex process in the wet tropics as elsewhere [*Bruijnzeel*, 1990; *Rose*, 1993] and rainfall erosivity is one of several factors that determine the rate of soil loss at a given location. One of the most widely used rainfall erosivity indices is the R-factor in the Universal Soil Loss Equation (USLE) [*Wischmeier and Smith*, 1978], which is the long-term average of the product of the total rainfall energy and the maximum 30-minute rainfall intensity for an event, i.e. EI_{30}, accumulated for a year. The rainfall erosivity concept is used in USLE as a numerical description of the potential of rainfall to erode soil [*Wischmeier*, 1959]. Since soil erosion is proportional to rainfall erosivity *ceteris paribus*, EI_{30} as a rainfall erosivity index may be used to indicate the relative contribution of heavy rainfall to soil erosion. To determine EI_{30}, rainfall intensity data at 30 minute interval are needed. In the study area, however, there are no rainfall stations with sufficient amount of rainfall intensity data. Therefore, daily rainfall amounts must be used to derive this rainfall erosivity index.

Partly due to the lack of adequate spatial coverage of rainfall stations with continuous record of rainfall intensity at short time intervals in the United States, *Richardson et al.* [1983] developed a model relating EI_{30} and daily or event rainfall amount, P, in the form of a simple power function:

$$EI_{30} = aP^b.$$

Using EI_{30} and P data from 11 locations in the United States east of the Rocky Mountains, the exponent b showed little spatial variation and thus an average value of 1.81 was recommended [*Richardson et al.*, 1983]. This model has been tested and found to be applicable in the United States [*Haith and Merrill*, 1987; *Selker et al.*, 1990], Canada [*Bullock et al.*, 1989], Finland [*Posch and Rekolainen*, 1993] and Western Amazonia [*Elsenbeer et al.*, 1993]. To validate the relationship between this rainfall erosivity index and daily rainfall amount for the study area, annual EI_{30} at Cairns, which lies on the coast about 70 km north of South Johnstone, for the period 1957 to 1972 was used to estimate the parameters most appropriate for the study area. The Cairns EI_{30} data have previously been used in a study to determine the distribution of rainfall erosivity in Queensland [*Rosenthal and White*, 1980]. In that study, all events with rainfall of less than 12 mm were discarded [*Rosenthal and White*, 1980]. The same threshold of 12 mm/day was also used here when estimating annual EI_{30}. The least-square estimates of the parameters a and b are

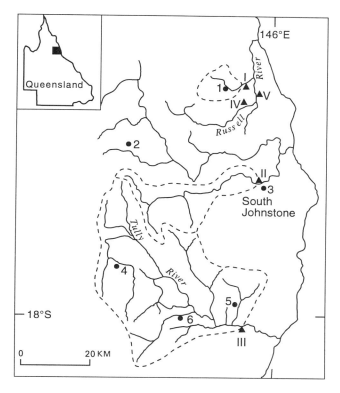

Fig. 1. Location map showing rainfall and stream gauging stations.

TABLE 1. Location, mean annual rainfall (MAR), threshold rainfall intensity (Io) and percent contribution of one-third of total rainfall with the highest intensity to rainfall erosivity (Pe).

No. on map	Station Number	Location	MAR (mm)	Io (mm/day)	Pe (%)
1	31144	Babinda (Happy Valley)	4,575	108.4	58
2	31042	Millaa Millaa	2,698	48.5	62
3	32037	South Johnstone Exp. Sta.	3,348	82.0	58
4	31083	Koombooloomba	2,725	51.0	61
5	32042	Tully Sugar Mill	3,962	90.9	58

0.059 and 1.6, respectively. The actual versus estimated annual EI_{30} for Cairns is shown in Figure 2, and this simple model using daily rainfall amounts explains 87% of the variance of the annual EI_{30}. In accordance with *Rosenthal and White* [1980], annual EI_{30} is given in metric units. One metric unit of EI_{30} equals 9.8 MJ ha^{-1} mm hr^{-1}. In this paper, the exponent, b, equal to 1.6 was used for all of the stations in the study area, and the coefficient, a, is not critical for this investigation because the focus of the paper is on the relative contribution of heavy rainfall to the rainfall erosivity.

4.2 Results

Daily rainfall amounts for a 30-year period (October 1962 - September 1992) were used for estimating the rainfall erosivity index, EI_{30}. An exception is that of Millaa Millaa, for which the 30-year period from October 1953 to September 1983 was used because of the missing daily rainfall records since 1983 at that site. For a number of rainfall intensity thresholds, EI_{30} using only those daily rainfall amounts above the given threshold was also computed. The contribution of heavy rainfall to the EI_{30} was taken to indicate the relative importance of heavy rainfall in relation to rainfall erosivity. The rainfall intensity thresholds were made dimensionless using the mean rainfall intensity for the 30-year period. The latter was defined as the amount of rainfall per day, which differs from the usual definition of the amount rainfall per rainday. In spite of a difference in the mean annual rainfall by a factor of 1.7 among the five locations (Table 1), use of the dimensionless intensity threshold has resulted in a tight cluster of data points. The pattern in Figure 3 characterizes the relative importance of heavy rainfall in relation to rainfall erosivity for the study area. Percent contribution of one-third of the total rainfall with the highest intensity to rainfall erosivity varies from 58% for the lowland stations to 61-62% for the tableland stations. This slight difference in the relative importance of heavy rainfall can also be seen in Figure 3. Data points for the tableland stations, i.e. Millaa Millaa (No. 2, Figure 1) and Koombooloomba (No. 4, Figure 1), consistently lie above those associated with the lowland stations when the dimensionless intensity threshold exceeds about 20. The heavy rainfall appears to be relatively more important when the mean annual rainfall is lower. Table 3 summarizes the average contribution of heavy rainfall to rainfall erosivity in the study area. One percent of the total rainfall with the highest intensity contributes 4% of the total rainfall erosivity, while 10% of the total rainfall contributes a quarter of the total rainfall erosivity.

5. RUNOFF

5.1 Data and Methods

Three basins in the wet tropics were considered, with the basin area ranging from 91 to 1475 km^2. Daily streamflow data for these basins were available for at least 20 years and the data quality has been rated as 'good' for all three gauging stations [*AWRC*, 1984]. A number of stations that record daily rainfall are located in and around these basins. Most of these stations are either on the coastal lowlands or on the tablelands, with few stations located where the

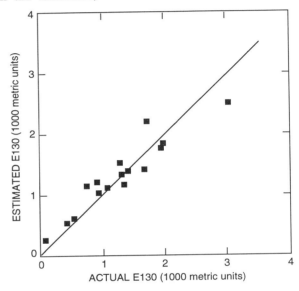

Fig. 2. Actual and estimated annual rainfall erosivity index at Cairns for the period 1957-1972.

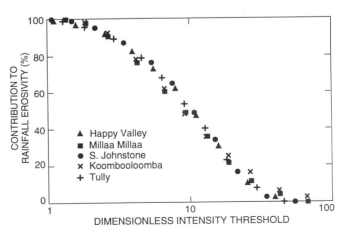

Fig. 3. Percent contribution to rainfall erosivity as a function of dimensionless rainfall intensity threshold at five locations in the wet tropics of Australia.

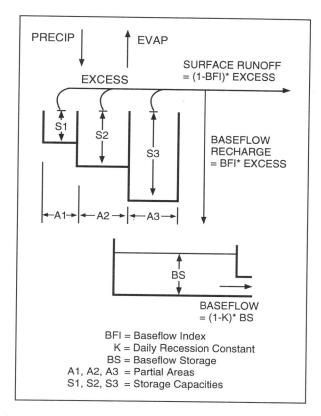

Fig. 4. Structure of the AWBM model.

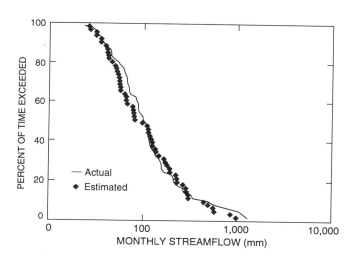

Fig. 5. Actual and estimated monthly flow duration curves for the South Johnstone River basin for the validation period from October 1978 to September 1983.

change in relief is greatest. Areal rainfall for the three basins were estimated using Thiessen-weighted average. Three rainfall stations are located inside the Tully River basin. There is only one rainfall station inside the Babinda Creek basin. For the South Johnstone River basin, the three rainfall stations (Nos 2, 3, and 4 in Figure 1) used for estimating areal rainfall are all located outside the basin boundary. Mean daily pan evaporation data were available for each month of the period from 1973 to 1994 at South Johnstone Experiment Station (Figure 1) and these evaporation data were applied as the potential rate of evaporation for all three basins. For the study area, the period for which there is concurrent evaporation, rainfall and streamflow data is short and only ten to eleven years data could be assembled for modeling purposes.

The model used to evaluate the saturation excess runoff is a modified version of AWBM [*Boughton*, 1993], in which a number of sub-areas were used to partition the basin. The model structure is shown in Figure 4 and essentially represents a piece-wise linear rainfall-runoff relationship. Other conceptual hydrological models have been developed with varying degrees of complexity. For example, the Stanford Watershed Model [*Johanson et al.*, 1980] and the Sacramento Model [*Peck*, 1976] are among the better known ones. The AWBM model was selected partly because one of its predecessors [*Boughton*, 1968] was successfully applied to two small basins (area less than 1 km^2) in the same area for a 4-year period [*Gilmour and Boughton*, 1980]. A comparison of model performance using different numbers of sub-areas showed that a 2-sub-area model with five parameters in total can adequately describe the basin water balance and a further increase of the number of sub-areas does not significantly improve model performance. Parameters were estimated by minimizing the sum of squared percentage errors using the downhill simplex method [*Press et al.*, 1989]. To a great extent, model performance depends on the accuracy of estimated areal rainfall. A lack of adequate spatial coverage of rainfall stations not only influences the model performance but also affects the estimated parameters.

5.2 Results

For each basin, monthly streamflow for the first five years was used for parameter estimation and the remaining five to six years data were used for model validation. Table 2 summarizes some of the basic basin characteristics and model performances and Figure 5 shows, as an example, the actual and estimated flow duration curves using monthly validation results for the period from October 1978 to September 1983 for the South Johnstone River

TABLE 2. Location, basin area, mean annual rainfall (MAR) for the basin, runoff coefficient (Rc), modeling period squared correlation coefficient between actual and estimated monthly streamflow for calibration r^2(C) and validation r^2 (V) periods and the percent contribution of one-third of total rainfall with the highest intensity to runoff (Pr)

No. on map	Station Number	Location	Area (km²)	MAR (mm)	Rc	Period	r^2(C)	r^2(V)	Pr (%)
I	111102	Babinda Creek at Babinda	91	4,375	0.79	1978-1988	0.88	0.96	44
II	112101	South Johnstone River at Central Mill	390	3,236	0.74	1974-1983	0.92	0.94	47
III	113006	Tully at Euramo	1,475	2,767	0.67	1980-1990	0.97	0.93	47

TABLE 3. Fractional contribution of rainfall with the highest intensity to runoff, rainfall erosivity and sediment transport, and its frequency of occurrence and dimensionless intensity threshold for the study area.

Rainfall (%)	Runoff (%)	Rainfall erosivity (%)	Sediment transport (%)	Time (%)	$I_o/\bar{I}$
1	1.5	4	9	0.02	37
2	3	7	17	0.09	31
5	7.5	14	34	0.16	24
10	15	25	53	0.41	18
20	29	42	74	1.1	12
33.3	46	59	87	2.5	7.9
50	65	77	94	5.3	4.9

basin. It can be seen that the fit is good and the percent of the variance in explained monthly streamflow is consistently in excess of 90% for the validation period (Table 2). Figure 6 shows the percent contribution to runoff above a certain intensity threshold. The intensity threshold was made dimensionless using the mean rainfall intensity for each basin. For a given dimensionless intensity threshold, the percent contribution of heavy rainfall to runoff is about the same for all three basins. Table 3 shows that 1% of the total rainfall with the highest intensity contributes 1.5% of the total runoff and 10% of the total rainfall contributes 15% of the total runoff. Contribution of heavy rainfall does not vary greatly in the study area. For one-third of the total rainfall, its contribution varies from 44% for the Babinda basin with the highest rainfall of 4375 mm/yr, to 47% for the other two basins with lower rainfall. Table 3 also shows that the relative contribution of heavy rainfall to runoff is not as high as that to rainfall erosivity.

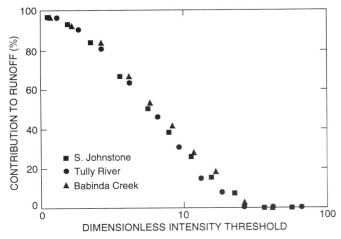

Fig. 6. Percent contribution to runoff as a function of dimensionless rainfall intensity threshold for three basins in the wet tropics of Australia.

6. SEDIMENT TRANSPORT

6.1 Data and methods

An early study by *Douglas* [1967] showed a low natural denudation rate in the region. The rate of denudation on the order of 10 t/km²/yr was estimated for the two year period of 1963-1964. More recently, *Neil* [1994, per. comm.] estimated the long-term sediment yield for the Tully River basin in undisturbed conditions to be 18 t/km²/yr. With catchment clearing and land use intensification, sediment yield could increase dramatically. Soil losses of 400 t/ha in a single intensive storm from a cultivated sloping land near South Johnstone were cited by *Rose* [1993]. *Gilmour* [1977] reported that sediment concentration for given discharge could increase by a factor of two to three following logging and clearing of a small lowland rainforest area. Given the confounding factor of land use change in the study area, sediment rating curves and measured daily discharge data were used to estimate sediment yield for a 15-year period. Such estimates are best interpreted as variations of sediment yield over time given the land use conditions at the time when the rating curves were developed. In spite of the various shortcomings of sediment rating curves [*Walling*, 1977; *Walling and Webb*, 1988] due to the complicated relationship between sediment concentration and discharge [*Olive and Rieger*, 1985; *Williams*, 1989], they are still useful for sediment yield estimation, especially in the context of examining the temporal variation of sediment yield in relation to the natural climatic and hydrologic variability of a given region.

Sediment rating curves are available for two of the three basins considered for runoff modeling purposes. Since the number of sediment samples of the Babinda Creek was insufficient to derive a reliable sediment rating curve, additional sediment concentration data at two gauging stations on the Russell River, to which the Babinda Creek is a tributary, were also used to develop a basin-wide rating curve. These two gauging stations shown in Figure 1 (Nos IV and V) have basin areas of 231 and 325 km², respectively.

A scatter plot of the instantaneous sediment load against discharge is shown in Figure 7. It can be seen that the relationship between sediment load and discharge essentially follows a similar pattern for all three sites. A power function was fitted using the combined sediment load and discharge data, and the parameters for the combined rating curve were estimated, resulting in the following relationship:

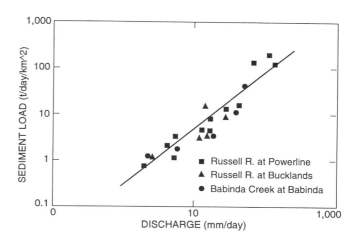

Fig. 7. Combined sediment rating curve for the Babinda-Russell basin.

$$q_s = 0.000062 \, q^{2.46}$$

where q_s is the sediment load in t/day/km² and q discharge in mm/day. The coefficient 0.000062 includes a correction factor to remove the bias due to log-transformation [*Dunn*, 1983; *Koch and Smillie*, 1986]. The correlation coefficient squared and the standard error of the estimate are 0.86 and 0.26, respectively.

For the Tullt River at Euramo, *Neil* [1994] developed a sediment-rating curve:

$$C = 0.255 \, Q^{0.858}$$

where C is the sediment concentration in mg/L and Q is discharge in m³/s.

6.2 Results

Daily flow for the period from October 1972 to September 1987 was used to calculate the sediment yield of the Babinda Creek and Tully River. The mean annual streamflow for the period is 3854 mm and 2113 mm, respectively. The mean annual sediment yield for the same period is 78.4 and 63.4 t/km², respectively. This represents a three- to four- folds increase for the Tully River basin by comparison with the sediment yield from the undisturbed areas. Threshold discharge was made dimensionless using the mean daily discharge and the percent contribution to the total amount of trans-ported sediment is shown in Figure 8 as a function of the dimensionless discharge threshold. If one-third of the total rainfall with the highest intensity for the Babinda Creek contributes 44% of streamflow (see Section 5.2 and Table 1), a dimensionless threshold

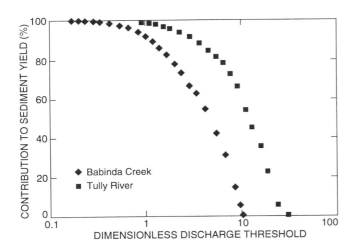

Fig. 8. Percent contribution to sediment transport as a function of dimensionless discharge threshold for Babinda Creek and Tully River.

discharge of 2.18, equivalent to 23.0 mm/day, would result. Almost 94% of sediment is transported when the discharge exceeds the 23.0 mm/day threshold. Even if it is conservatively assumed that one-third of the total rainfall contributes no more than one-third of the streamflow, the percent of sediment transported above the corresponding threshold of 34.9 mm/day is still as high as 90%. For the Tully River basin, a similar analysis showed that one-third of the total rainfall contributes between 67% and 79% of the total transported sediment.

Whereas the contribution-threshold relationship for rainfall erosivity and runoff remains essentially the same for all sites in the region, the contribution to transported sediment above a given dimensionless discharge threshold for the Babinda Creek is much higher than that for the Tully River. At least two factors can be identified that determine the contribution-threshold relationship and cause the disparity between the two basins. The first is the size of the basin. The smaller the basin area, the flashier are flood events and the more variable is the streamflow in general (Kalinin, 1971; *McMahon et al.*, 1987]. A larger proportion of total transported sediment above given dimensionless discharge threshold is therefore expected for the Babinda Creek. Secondly, the exponent of the sediment rating curve for the Babinda Creek is greater than that for the Tully River. As a result, high flows of the Babinda Creek would carry relatively a greater amount of sediment than the Tully River. In any event it seems safe to conclude that one-third of the total rainfall with the highest intensity in the study area contributes around 80% of the total sediment transported. Contribution of heavy rainfall to sediment transport is by far the greatest by comparison

with its contribution to runoff and rainfall erosivity as shown in Table 3. For example, 10% of the total rainfall with the highest intensity contributes to more than half the transported sediment, while its contributions to rainfall erosivity and runoff are only 15% and 25%, respectively. The relative contribution of heavy rainfall to runoff and sediment transport is largely determined by the variability of heavy rainfall over time, the threshold above which runoff and sediment transport occur, and the nonlinearity of the relationship between rainfall and runoff, and runoff and sediment transport as measured by, for example, the exponent of the sediment rating curve. The combined effects of the nonlinear rainfall-runoff and runoff-sediment transport relationships result in a highly nonlinear relationship between rainfall and sediment yield, and consequently a relatively large contribution of heavy rainfall to transported sediment by comparison with its contribution to runoff.

7. DISCUSSION

It is worthwhile to review and discuss some of the critical assumptions made and the limitations of this study because of these assumptions. It is also instructive to examine how rainfall erosivity, runoff and sediment transport can be related to one another given heavy rainfall as a common driving force in the wet tropics.

The most crucial assumption in relation to rainfall erosivity calculations is that rainfall erosivity is assumed to be proportional to soil erosion, other things being equal. In other words, relative contribution of heavy rainfall to rainfall erosivity was taken to mean the relative contribution to soil erosion. Validity of this assumption depends on how appropriate EI_{30} as a rainfall erosivity index is and the extent to which rainfall erosivity is related to actual soil erosion. In the absence of long-term monitoring of soil erosion for a variety of soils, topographic settings and land uses in the study area, it would be useful to illustrate the link between EI_{30} estimated using daily rainfall amounts greater than 12 mm (Section 4.1) and sediment yield estimated using sediment rating curves and daily flows (Section 6.1). Annual EI_{30} for the Babinda Creek and Tully River basins were calculated in the same way as areal rainfall was determined (Section 5.1). Annual sediment yield for the 15-year period from October 1972 to September 1987 is plotted against annual EI_{30} for these two basins (Figure 9). The percent of variance of annual sediment yield that can be explained by basin EI_{30} using a simple linear model is 72% and 96% for the Babinda Creek and Tully River, respectively. Considering that the estimated rainfall erosivity depends on daily rainfall

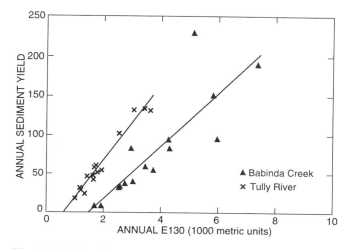

Fig. 9. Relationship between annual sediment yield estimated using the rating curve method and annual EI_{30} for Babinda Creek and Tully River.

amounts alone, the results are indeed most encouraging. In fact, the annual EI_{30} is a better indicator of the annual sediment yield than the annual streamflow for these two basins. It follows that the temporal variation of rainfall erosivity in terms of EI_{30} to a great extent determines the temporal variation of sediment yield in the region.

Rainfall erosivity is an intrinsic aspect of rainfall and it depends only on rainfall characteristics. Runoff and total transported sediment, on the other hand, depend on such factors as soil properties, vegetation cover, and topography in addition to rainfall. For undisturbed basins in the wet tropics, only rainfall effectively varies over time on a time scale of decades. With land use change, other factors, especially vegetation cover, begin to vary on a similar time scale. Moreover, land use change can be an intensive and comparatively localized phenomenon, resulting in an increased spatial heterogeneity within the watershed. The effect of land use change on the relative importance of heavy rainfall in relation to runoff and sediment transport remains to be resolved. This study only addresses the contribution of heavy rainfall given the current land use conditions for the purpose of highlighting the importance of heavy rainfall to rainfall erosivity, runoff and sediment transport.

8. CONCLUDING REMARKS

In the wet tropics, the high annual rainfall total is associated with a low natural variability [*Conrad*, 1941; *Nicholls and Wong*, 1990] and a high runoff coefficient [*Budyko*, 1974; *Shiklomanov*, 1993]. As a result, a relatively low percent contribution of heavy rainfall to runoff, rainfall erosivity and subsequent sediment transport is expected. In drier climates, where rainfall variability is higher and runoff coefficient is lower, the relative contribution of heavy rainfall could be much higher than reported here for the wet tropics. This inverse relationship between the mean annual precipitation and relative importance of heavy rainfall with respect to the hydrological and geomorphological processes can even be established in the study area. Table 1 and 2 show that the contribution of heavy rainfall is greater when the mean annual rainfall is lower, although further investigation into the relative importance of heavy rainfall for a range of climatic environments is clearly needed.

Assuming that rainfall is the driving force, and runoff, water erosion, and sediment transport are some of its hydrological and geomorphological effects, the relative effectiveness of heavy rainfall can be inferred as the proportion of these processes that can be attributed to that heavy rainfall. The percent contribution of heavy rainfall depends on the natural variability of rainfall itself and on the relationship between rainfall and these hydrological and geomorphological processes. Using simple models that relate rainfall erosivity and runoff to daily rainfall amounts, and using sediment rating curves to relate sediment load to discharge, the percent contribution of heavy rainfall to these processes can be evaluated. One-third of total rainfall with the highest intensity contributes about 45% of total runoff, 60% of total rainfall erosivity and 87% of total transported sediment. One percent of total rainfall with the highest intensity contributes about 1.5% of total runoff, 4% of total rainfall erosivity and up to 9% of total transported sediment. These values probably represent the lower limits to the relative importance of heavy rainfall in comparison to other climatic regions in Australia.

By comparison with average precipitation, heavy rainfall has higher temporal variability and greater hydrological and geomorphological effects. Because of its temporal variability, significant change to the occurrence of heavy rainfall is particularly difficult to detect. Greater impacts with respect to surficial processes implies that the occurrence of heavy rainfall is of greater importance and relevance to geomorphology and hydrology than climatology alone. Given the importance of heavy rainfall in the hydrological and geomorphological processes, the temporal variation of heavy rainfall, in contrast to that of total rainfall, warrants more attention than it has received so far.

Acknowledgments. The author would like to thank D. Neil, A. Miller, A. Gupta, and E. Wohl for their constructive comments. The project concerning the temporal variation of high-intensity rainfall in the humid tropics of Australia was funded by the Commonwealth of Australia through Griffith University.

REFERENCES

Albergel, J., Evolution de la pluviométrie en Afrique Soudano-Sahlienne: Exemple du Burkina-Faso, in Colloque International sur la Révision des Normes Hydrologiques Suite aux Incidences de la Sécheresse, CIEH, Ouagadougou, pp. 17, 1986.

Andrews, E. D., Effective and bankfull discharge of streams in the Yampa River Basin, Colorado and Wyoming, *J. Hydrol.*, 46, 311-330, 1980.

AWRC, *Stream Gauging Information, Australia*, 6th edition, 204 pp., Australian Water Resources Council, Canberra, 1984.

Boughton, W. C., A mathematical catchment model for estimating runoff, *J. Hydrol. (New Zealand)*, 7, 75-100, 1968.

Boughton, W. C., A hydrograph-based model for estimating the water yield of ungauged catchments, *Proc. Hydrol. and Water Resour. Symp.*, Newcastle, IEAust., Natl. Conf. Publ. No. 93/14, 317-324, 1993.

Bruijnzeel, L. A., *Hydrology of Moist Tropical Forests and Effects of Conversion: A State of Knowledge Review*, UNESCO IHP, Humid Tropics Programme, Paris, 1990.

Budyko, M. I., *Climate and Life*, 508 pp., Academic, New York, 1974.

Bullock, P. R., E. de Jong, and J. J. Kiss, An assessment of rainfall erosion potential in southern Saskatchewan from daily rainfall records, *Canadian Agricultural Engineering*, 32, 17-24, 1989.

Carling, P., The concept of dominant discharge applied to two gravel-bed streams in relation to channel stability thresholds, *Earth Surface Processes and Landforms*, 13, 355-367, 1988.

Chang, J-H., and L. S. Lau, Definition of the humid tropics, in *Hydrology and Water Resources Management in the Humid Tropics*, edited by M. Bonell, M. M. Hufschmidt, and J. S. Gladwell, pp. 571-574, Cambridge University Press, Cambridge, UK, 1993.

Conrad, V., The variability of precipitation, *Mon. Wea. Rev.*, 69, 5-11, 1941.

Cornish, P. M., Changes in seasonal and annual rainfall in New South Wales, *Search*, 8, 38-40, 1977.

Douglas, I., Natural and man-made erosion in the humid tropics of Australia, Malaysia and Singapore, in *River Morphology*, pp. 17-29, IAHS Publ. No. 75, 1967.

Druyan, L. M., Advances in the study of sub-Saharan drought, *Int. J. Clim.*, 9, 77-90, 1989.

Dunn, N., Smearing estimate: A nonparametric retransformation method, *J. Am. Stat. Assoc.*, 78(383), 605-610, 1983.

Elsenbeer, H., D. K. Cassel, and W. Tinner, A daily rainfall erosivity model for Western Amazonia, *J. Soil and Water Conservation*, 48, 439-444, 1993.

Gilmour, D. A., Effects of rainforest logging and clearing on water yield and quality in a high rainfall zone of north-east Queensland, *Proc. Hydrol. Symp.*, Brisbane, IEAust, Natl Conf. Publ. No. 77/5, 156-160, 1977.

Gilmour, D. A., and W. C. Boughton, Estimation of runoff from small disturbed tropical rainforest catchments using the Boughton model, Research Paper, 13 pp., Department of Forestry, Queensland, Australia, 1980.

Haith, D. A., and D. E. Merrill, Evaluation of a daily rainfall erosivity model, *Transactions, American Society of Agricultural Engineers*, 30(1), 90-93, 1987.

Johanson, R. C., J. C. Imhiff, and H. H. Davis, User manual for the Hydrological Simulation Program - Fortran (HSPF), Environmental Protection Agency Report FPA-600/9-80, 1980.

Kalinin, G. P., *Global Hydrology*, 311 pp., Jerusalem, Israel Program for Scientific Translations, 1971.

Kinsey, D. W., Can we resolve the nutrient issue for the Reef, *Search*, 22, 119-121, 1991.

Koch, R. W., and G. M. Smillie, Comments on 'River loads underestimated by rating curves' by R.I. Ferguson, *Water Resour. Res.*, 22(13), 2121-2122, 1986.

Kraus, E. B., Secular change in the rainfall regime of SE Australia, *Quarterly Journal of Royal Meteorological Society*, 80, 591-601, 1954.

Lamb, P. J., Persistence of sub-Saharan drought, *Nature*, 299, 46-48, 1982.

Laurence, W. F., Research challenges and opportunities in the wet tropics of Queensland World Heritage Area, *Pacific Conservation Biology*, 11, 3-6, 1993.

Lough, J. M., Variations of some seasonal rainfall characteristics in Queensland, Australia: 1921-1987, *Int. J. Clim.*, 13, 391-409, 1993.

McMahon, T. A., B. L. Finlayson, A. Haines, and R. Srikanthan, Runoff variability: A global perspective, in *The Influence of Climate Change and Climatic Variability on the Hydrologic Regime and Water Resources*, pp. 3-11, IAHS Publ. No. 168, 1987.

Neil, D. T., The sediment yield response to land use intensification in a humid, tropical catchment: The Tully River catchment, northeast Queensland, Australia, unpubl. Ph.D. Dissertation, The Australian National University, Canberra, 1994.

Neil, D. T., and B. Yu, Variation in estimated rainfall erosivity on the Southern Tablelands, NSW, over the last one hundred years, *Australian Journal of Soil and Water Conservation*, 7(1), 23-28, 1994.

Nicholls, N., and K. K. Wong, Dependence of rainfall variability on mean rainfall, altitude, and the Southern Oscillation, *J. Clim.*, 3, 163-170, 1990.

Nicholls, N., and A. Kariko, Eastern Australian rainfall events: interannual variations, trends and relationships with the Southern Oscillation, *J. Clim.*, 6, 1141-1152, 1993.

Nicholson, S. E., The nature of rainfall fluctuations in subtropical West Africa, *Mon. Wea. Rev.*, 108, 473-487, 1980.

Olaniran, O. J., and G. N. Sumner, Climatic change in Nigeria: Variations in rainfall per rain-day, *Weather*, 44(6), 242-248, 1989.

Olive, L.J., and W.A. Rieger, Variation in suspended sediment concentration during storms in five small catchments in southeast New South Wales, *Australian Geographical Studies*, 23, 38-51, 1985.

Peck, E. L., Catchment modeling and initial parameter estimation for the National Weather Service River Forecast System, NOAA Technical Memorandum NWS HYDRO-31, U.S. Department of Commerce, Silver Spring, Maryland, 1976.

Pickup, G., and R. F. Warner, Effects of hydrologic regime on magnitude and frequency of dominant discharge, *J. Hydrol.*, 29, 51-75, 1976.

Pittock, A. B., Climatic change and the pattern of variation in Australian rainfall, *Search*, 6, 498-503, 1975.

Pittock, A. B., Recent climatic change in Australia: Implications for a CO_2 warmed earth, *Climatic Change*, 5, 321-340, 1983.

Posch, M., and S. Rekolainen, Erosivity factor in the Universal Soil Loss Equation estimated from Finnish rainfall data, *Agricultural Science in Finland*, 2(4), 271-279, 1993.

Press, W. H., B. P. Flannery, S. A. Teukolsky, and W. T. Vetterling, *Numerical Recipes: the Art of Scientific Computing*, 702 pp., Cambridge University Press, Cambridge, UK, 1989.

Richardson, C. W., G. R. Foster, and D. A. Wright, Estimation of erosion index from daily rainfall amount, *Transactions, American Society of Agricultural Engineers*, 26, 153-157, 160, 1983.

Rose, C. W., Erosion and sedimentation, in *Hydrology and Water Resources Management in the Humid Tropics*, edited by M. Bonell, M. M. Hufschmidt, and J. S. Gladwell, pp. 301-343, Cambridge University Press, Cambridge, UK, 1993.

Rosenthal, K. M., and B. J. White, Distribution of a rainfall erosion index in Queensland, Division of Land Utilisation, Report 80/8, 29 pp., Queensland Department of Primary Industries, Brisbane, Australia, 1980.

Selker, J. S., D. A. Haith, and J. E. Reynolds, Calibration and testing of a daily rainfall erosivity model, *Transactions, American Society of Agricultural Engineers*, 33(5), 1612-1618, 1990.

Shiklomanov, I. A., World freshwater resources, in *Water in Crisis: A Guide to the World's Fresh Water Resources*, edited by P.H. Gleick, pp. 13-24, Oxford University Press, New York, 1993.

Walling, D. E., Limitations of the rating curve technique for estimating suspended sediment loads with particular reference to British rivers, in *Erosion and Solid Matter Transport in Inland Waters*, pp. 34-78, IAHS Publ. No. 122, 1977.

Walling, D. E., and Webb, B. W., The reliability of rating curve estimates of suspended sediment yield: some further comments, in *Sediment Budgets*, pp. 337-350, IAHS Publ. No. 174, 1988.

Williams, G. P., Sediment concentration versus discharge during single hydrologic events in rivers, *J. Hydrol.*, 111, 89-106, 1989.

Wischmeier, W. H., A rainfall erosion index for a universal soil-loss equation, *Soil Sci. Soc. Am. Proc.*, 23, 246-249, 1959.

Wischmeier, W. H., and D. D. Smith, Predicting rainfall erosion losses - A guide to conservation planning, Agricultural Handbook No. 537, USDA, Washington, 1978.

Wolanski, E., *Physical Oceanographic Processes of the Great Barrier Reef*, 194 pp., CRC Press, Boca Raton, Florida, 1994.

Wolman, M. G., and J. P. Miller, Magnitude and frequency of forces in geomorphological processes, *J. Geol.*, 68, 54-74, 1960.

Yu, B., and D. T. Neil, Global warming and regional rainfall: the difference between average and high intensity rainfalls, *Int. J. Clim.*, 11, 653-661, 1991.

Yu, B., and D. T. Neil, Long-term variations in regional rainfall in the south-west of Western Australia and the difference between average and high intensity rainfalls, *Int. J. Clim.*, 13, 77-88, 1993.

Bofu Yu, Faculty of Environmental Sciences, Griffith University, Brisbane, Australia

Magnitude, Frequency, and Special Factors Affecting Channel Form and Processes in the Seasonal Tropics

Avijit Gupta

Department of Geography, National University of Singapore, Singapore

The rivers of the seasonal tropics are controlled by three variables: (1) the difference between the discharges of the wet and dry seasons; (2) the occurrence of high-magnitude floods on a recurrence interval of decades; and (3) the volume and texture of the available sediment. The general channel form is box-shaped with steep banks and a channel-in-channel morphology. The entire channel is maintained by high-magnitude floods whereas a smaller inner channel carries high flows of the wet season. Variations from this model occur when one of the three variables, or a particular combination thereof, is dominant. A modified version of the magnitude-frequency concept explains channel forms and sediment in the seasonal tropics.

1. INTRODUCTION

A number of studies in the United States and Britain have demonstrated a relationship between channel size and bankfull discharge. A combined factor of magnitude and frequency, often measured by bankfull discharge, controls channel size and to some extent channel forms. The effects of large floods are eliminated in a short time period (months to years) from streams in the humid temperate climate zone [*Costa*, 1974; *Gupta and Fox*, 1974; *Wolman and Gerson*, 1978]. In contrast, in arid areas, flood disturbances are not modified and the channels remain enlarged. A step-wise increment of channel width for streams draining small basins has been proposed by *Wolman and Gerson* [1978].

The rivers in the seasonal tropics fall somewhere in between these two examples for two reasons. First, they undergo a seasonal fluctuation of discharge that significantly alters the width-depth ratio and stream power. Second, a large part of the seasonal tropics is visited at irregular intervals by tropical storms that may at times reach hurricane force. The effect of such storms on stream channels is usually conspicuous. The channels display a nested pattern: a large channel for the storm and a small channel for the high discharge of interstorm periods [*Gupta*, 1988]. Such nested appearance of landforms, suggesting activity at widely different and discrete scales, was also reported for Central Australian floodplains [*Pickup*, 1991]. Earlier *Pickup and Rieger* [1979] proposed that channel forms result from a series of discharges rather than a single dominant one. The dimensions of the channel can be explained by a single discharge of a given frequency only when either the temporal fluctuation in discharge is low or the particular channel dimension is insensitive to flow variations.

The discharge which transports most sediment has also been perceived as a channel-controlling discharge [*Benson and Thomas*, 1966; *Nash*, 1994], and termed the effective discharge. As the capacity of sediment transport is related to multiple factors such as stream velocity, discharge, bed and bank materials and the amount of sediment in storage within the channel, it is perhaps not the best criterion for identifying the channel-forming discharge. Effectiveness has been defined as the ability of an event or a combination of events to affect the form of the landscape [*Wolman and Gerson*, 1978]. The control of channel size and form is therefore best studied by relating channel attributes to a particular discharge or a set of discharges.

This paper discusses the variety of forms associated with channels of the seasonal tropics and explains why a nested set of channel-forming discharges is necessary in order to account for their size, shape and sedimentary characteristics. The magnitude and frequency concept is applied towards understanding the size and forms of the rivers of the seasonal tropics. The explanation is not clear-cut as variations are also imposed by other factors such as the drainage basin geology, location near plate margin activities, and the Quaternary inheritance of the river system. The explanation is hampered by the minimal number of measurements of both channel forms and discharges from the seasonal tropics, but, based on what is available, certain semi-quantitative conclusions are presented.

2. RIVER HYDROLOGY IN THE SEASONAL TROPICS

The tropics are located between the two anticyclonic belts at about 30° north and south latitudes. The seasonal tropics are characterized by a marked concentration of rainfall within a few months. The annual rainfall shows a wide range across the seasonal tropics. About 6000 mm may fall in the wet months where the geographic conditions are especially favorable. Usually, however, seasonality is associated with the drier version of the humid tropics, although the annual rainfall seldom drops below 1000 mm. The seasonality is transferred to the streamflow, and the hydraulic geometry of the rivers changes dramatically between the wet and dry seasons. This is obvious even to a casual observer.

Rivers remain at a high stage throughout the wet season. Hydrographs illustrating this are found in several studies on rivers of the seasonal tropics: the Brahmaputra at Bahadurabad and the Ganga at Hardinge Bridge [*Coleman*, 1969]; the Brahmaputra at Ranaghat, Pandu and Jogighopa [*Goswami*, 1985]; the rivers Godavari and Krishna at unspecified stations [*Rao*, 1979]; and the Yallahs at Mahogany Vale, Jamaica (Figure 1). As Table 1 indicates, more than 80 percent of the flow may occur in 4-5 months.

Areas of the seasonal tropics are also affected by episodic high-magnitude rainfall caused by synoptic events ranging in force from tropical storms to tropical cyclones. As tropical storms require a warm ocean temperature for their formation, these storms arrive in summer, which in many places coincides with the wet monsoon period when the soil is saturated and rivers are quite full. In India, between 7 and 10 tropical storms arrive annually in an average season [*Rao*, 1981; *Sikka*, 1977]. The storms usually last 2-5 days inland, although monsoonal storms up to 9 days long have been recorded. The storms are large, 1000-1300 km across, with a 7-9 km deep cyclonic circulation below an anticyclonic outflow [*Sikka*, 1977]. Very heavy episodic rainfall, totalling over a thousand mm in several days, may occur [*Gupta*, 1988].

Rivers of the seasonal tropics have to adjust to distinctly separate periods of high and low flows. Depending on their geographic locations, some of the rivers also have to adjust to high-magnitude floods arriving in the wet season. The magnitude and recurrence of these floods have been described by *Gupta* [1975, 1988], *Ahmad et al.* [1993], and *Kale et al.* [1994]. The effect of such floods could be preserved given suitable geologic and physiographic environment [*Baker*, 1977; *Gupta*, 1983]. Equilibrium of river forms requires adjustments to multiscale discharges, and at times, river forms may appear to be in disequilibrium with the current discharge conditions [*Stevens et al.*, 1975]. The paper explores this complexity.

3. CHANNEL FORM AND PROCESSES

In the seasonal tropics, it is possible to construct three starting hypotheses. 1. Almost all the work of the rivers in the seasonal tropics will be carried out during the wet season. During the dry season river discharges will be too low to be effective. 2. Rivers in certain geographical areas within the seasonal tropics will also carry high-magnitude floods on an irregular basis [*Gupta*, 1988]. The frequency of such floods will be high enough for one large flood to be expected within a specific time period (usually decades) resulting in typical channel morphology. 3. Rivers may also display morphology and sediment, modified from what is expected under hypotheses 1 and 2, in areas affected by externalities such as active plate margin locations which may result in frequent mass movements within the drainage basin and the arrival of large quantities of sediment (part of which could be very coarse) in the channel.

It is difficult to test such hypotheses quantitatively as good hydrological and morphological data for these rivers are seldom available. I am therefore reviewing three examples I am familiar with, to illustrate each of these three cases. The accounts are mainly based on detailed fieldwork, plus whatever amount of hydrological data are available for these rivers. The three factors that are proposed as controls on channel forms and processes in the seasonal tropics are usually present in a varying degree for most of the rivers. Each of the rivers chosen is dominated by one of the three factors. Detailed accounts for these rivers were published earlier [*Gupta and Dutt*, 1989; *Ahmad et al.*, 1993; *Rajaguru et al.*, 1995]. The rivers are the Auranga, eastern India; the Narmada, central India; and a set of rivers draining the southern slopes of the Blue Mountains, Jamaica (Figure 2). A discussion regarding the expected channel forms follows such presentations.

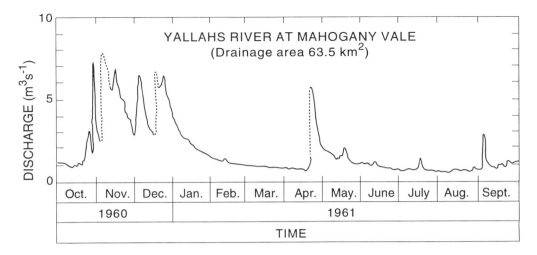

Fig. 1. Yallahs River at Mahogany Vale; annual pattern of flow variation.

TABLE 1. Seasonality in Discharge, the Ganga Basin

River and station	Mean Annual flow (m^3s^{-1})	Wet monsoon flow[a] ($10^6 m^3$)	Percent[b]
Ganga at Rishikesh (1971-81)	856	21,631	80.1
Ganga at Kanpur (1960-81)	1184	30,763	82.4
Ganga at Allahabad (1970-81)	4126	112,278	86.3
Ganga at Varanasi (1960-81)	4105	112,206	86.0
Ganga at Patna (1968-81)	7626	192,625	80.1
Ganga at Azamabad (1960-81)	10,159	235,357	73.5
Yamuna at Delhi (N.A.)	251	6,500	82.3
Son at Chopra (1960-81)	748	14,867	72.8
Tons at Meja Road (1960-81)	156	4,159	84.4

[a] West monsoon flow refers to discharge of 4 months between June and September.
[b] Percentage of annual rainfall [Dasgupta, 1984].

3.1 *The Auranga, primarily a seasonal stream*

The 85 km Auranga River drains a 1664 km² basin located in the Palamau district, Bihar near the northeastern corner of Peninsular India. The river basin includes escarpment-bounded Precambrian granite-gneiss plateaus and low hills and plains of Gondwana (late Paleozoic-early Mesozoic) sedimentary rocks. A paleo-alluvium occurs in the valleys in which the major rivers have excavated their courses. Given such a geology, the sediment load of the Auranga is primarily sand with a small percentage of fine pebbles. This sediment is mainly brought in by a number of small tributaries. The main channels are in a relatively undisturbed state.

About 80 percent of the 1200 - 1500 mm annual rainfall arrives between June and September during the southwestern monsoon. Up to nearly 300 mm of rain has fallen in 24 hours on the basin. Floods therefore may be expected to superimpose on the high discharge of the wet monsoon. No stream gauges operate on this river and no hydrologic data are available beyond rainfall measurements.

Figure 3 is a section across the Auranga, the most striking feature of which is the complex bar formation. In the dry season, the lower part of the channel is braided, point bars exist at meander bends, and a high bar appears at some of the bends. During the wet season, the braid bars are under water and presumably destroyed as the velocity is high enough to move sand as bed load. The Auranga behaves as a normal meandering river with point bars at bends. It is suggested that large floods on the Auranga create the higher bars at some of the bends where the flood velocity slackens and deposition takes place. The boundary of the flood bars is very sharp and a distinct 1 m step separates them from the point bars below. Both banks of the channel are steeply marked.

The sediment of the Auranga is almost entirely medium to coarse sand with some fine pebbles. Scattered large cobbles and small boulders, however, are found on the

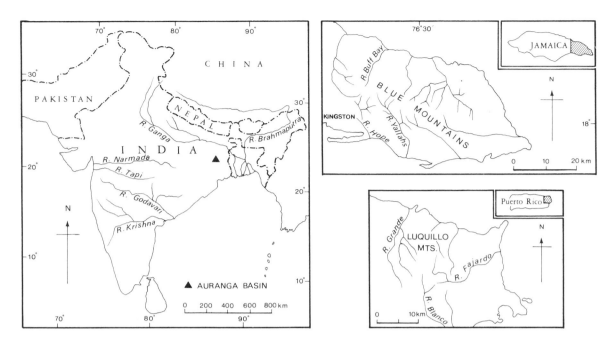

Fig. 2. Location map for rivers mentioned in the text.

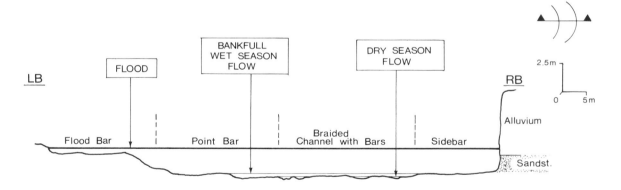

Fig. 3. Cross-section of the Auranga River at Rajdaha Nala.

flood bar. An association between channel physiography and sediment characteristics exists. Generally, the channel, braid bars and point bars consist of coarse, moderately sorted sand with a near-symmetrical distribution. The flood bar is made of medium sand, also moderately sorted, but with a strong coarsely skewed distribution. The flood bar sediment is characterized by upper regime plane beds, with a minor component in dunes or bars. In contrast, the point bars show evidence of low regime dune movement on top of a basal flood-deposited alluvium. Locally, the dunes are topped by sand deposited by upper regime flows.

During floods, the large volume of sediment stored in the tributaries, mostly sand, moves down the channels to the Auranga to build the flood bar and the channel floor alluvium. The point bars are built during lower flow regimes (as shown by their trough and planar cross-bedded structure) during the high discharge of the wet season. Once built, the point bars are occasionally topped by floods that accounts for the local presence of upper regime flood markers (plane beds) over the cross-bedded sand. During the dry season, the discharge falls and braid bars are created in the lower part of the channel by dissection of transverse bedforms, as was described by *Smith* [1970] for the Platte and the South Platte rivers.

Thus the channel physiography of the Auranga is a 3-tier organization resulting from the seasonal pattern of flow and the large supply of non-cohesive sand that can be moved as single grains even during low flows. Sand movement, grain by grain, was seen during low flow in February. Braid bars are formed during every dry monsoon at the lowest level of the channel. These bars are submerged and removed during the following wet season. Point bars that occur at every bend are <u>annual relicts</u> of the previous wet period, and the few flood bars are also relicts, of relatively rare high-magnitude floods. It is the seasonality of high flows that primarily shapes the river.

3.2 The Narmada River, flood processes and forms

The 1300 km long Narmada river (Figure 2) drains a 99,000 km^2 basin. The forms and processes of this river are flood dominated in spite of the seasonality of its discharge pattern and active tectonic movements along its course [*Rajaguru, et al.,* 1995]. Nearly 90 percent of the rain falling on the basin arrives between June and September. During this period, tropical storms arrive from Bay of Bengal and travel downstream nearly parallel to the basin axis. The upper basin is intersected by the most expected path of these storms [*Rao,* 1981]. The storm, surface runoff and the flood peak may travel near-simultaneously down the basin and the channel of the Narmada. The geology of the basin provides the river with a sandy sediment load with coarser material near the rapids and gorges. The course of the river is tectonically controlled within a seismically active rift zone. It is postulated that tectonic movements are responsible for the rocky sections of Narmada's course.

The Narmada flows along a structural lineament, alternating between narrow rocky gorges and rapids, and meandering wide alluvial reaches. Channel forms and processes were studied along a 120 km alluvial meandering reach of the upper Narmada.

Gaging on the Narmada started in the 1950s, but it has been possible to prepare a list of very large floods on this river from various sources [*Kale et al.,* 1992; *Kale, et al.,* 1994; *Rajaguru et al.,* 1995]. All estimates or measurements of high-magnitude floods made so far on this river indicate large discharge and high velocity. Unit discharge could be around 1 m^{3}s^{-1}km^{-2}, and velocities as high as 11 m s^{-1} (with a Froude number of about 1 indicating near-critical flow) when the channel is full. Calculated bed shear stresses vary between 300 and 3000 Nm^{-2} and power per unit area between 2600 and 12800 Wm^{-2}. These figures locally can be even higher. The Narmada in flood has eroded a scabland topography across hard rocks, carried cobbles in suspension, and moved boulders as bed load [*Rajaguru et al.,* 1995].

The channel of the Narmada is conspicuously bounded by two high alluvial cliffs between 10 and 15 m in height (Figure 4). These cliffs are in the sandy silt and sand of Pleistocene and Holocene age. Between the cliffs occur the meandering channel, the point bars, and also a floodplain in fine material which is submerged during every wet monsoon.

In this century, the Narmada was nearly full or overflowed about three times in extremely large floods, which occurred in 1923, 1926 and 1991. During such floods, the entire 400 m wide channel of the study reach was utilized and the high cliffs served as channel banks on both sides (Figure 5). These floods also eroded the rocky stretches and build the point bars [*Rajaguru et al.,* 1995]. Even the high flows of the wet monsoon are incapable of filling the entire channel. Their activities are limited to building of discontinuous floodplains within the cliffs and modifying bedforms and bars. Inset floodplains and other comparable forms have been reported from other rivers as flood-built features [*Wohl,* 1992; *Miller and Parkinson,* 1993; *Bourke,* 1994]. Features similar to the dry season braid bars of the Auranga are not seen in the channel.

The Narmada therefore carries a channel-in-channel physiography. The meandering movement is limited both by the restrictions imposed by the stable rocky sections at intervals and the high cliffs that hinder lateral movement. Avulsions affecting small lengths of the channel, however, have happened during past floods. In spite of being located in a seismically active zone and a tropical monsoon setting, it is the rare high-magnitude floods that form and maintain the morphology and sediments of the Narmada River.

3.3 Forms and processes of the Caribbean montane streams

The streams draining the southern slopes of the mountains of eastern Jamaica and Puerto Rico (Figure 2) show seasonality in their annual discharge. This is also an area of very high rainfall occurring from tropical storms which range up to hurricane force. A total rainfall of about 1000 mm in 3-4 days is not uncommon [*Gupta,* 1988]. The flood-dominated rivers show a channel-in channel-physiography. A braided pattern appears during the dry season. Where there is no distinct dry season, the rivers meander. This area, however, is also part of a 200 km wide seismically active zone of Neogene left-lateral strike-slip deformation that marks the plate boundary between the Caribbean and North American Plates. This is an area of steep hillslopes scarred by mass movements, from which a large amount of coarse material arrives at intervals to the streams. The mass movements on these steep slopes may

130 CHANNEL FORM AND PROCESSES IN THE SEASONAL TROPICS

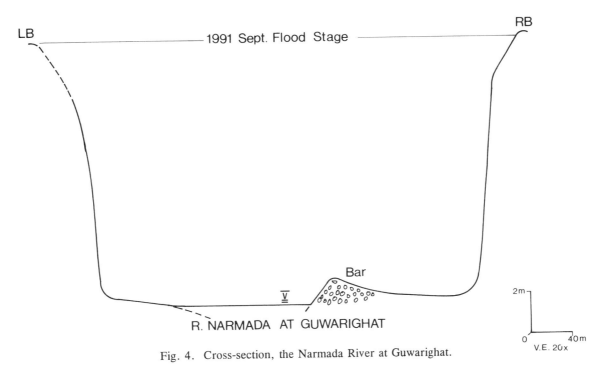

Fig. 4. Cross-section, the Narmada River at Guwarighat.

Fig. 5. The Narmada at Guwarighat, December 1991. The channel was full to the top of the alluvial banks during the flood of September 1991.

Fig. 6. The Yallahs River, Blue Mountains, eastern Jamaica.

be triggered either by intense rainfall or by seismic disturbances. Such events cause numerous small shallow slides on the hillslopes or, at times, huge ones moving tens of thousands cubic meters of soil and rock.

The local lithology governs the type of sediment reaching the streams and the subsequent channel forms and sediment transfer. In the Luquillo Mountains of eastern Puerto Rico the volcaniclastics provide boulders and clay-size material. The mountain streams therefore are boulder strewn, as the fine material is easily transported out of the system. In contrast, streams draining a quartz-diorite body carry a large amount of sand and build point bars. In the Blue Mountains of Jamaica, large boulders coming down the side slopes often block the streams. This subsequently gives rise to a series of features in the downstream direction: (1) a boulder zone across the channel is replaced by (2) scattered boulders, and further downstream by (3) bars of pebbles and cobbles with several large boulders perched on top. The bars gradually become larger and the material smaller in dimension.

The morphology of the channels consists of steep slopes, near-vertical banks, low sinuosity and an accumulation of boulders (Figure 6). The wider reaches are characterized by floodplains and flood terraces at various levels [*Gupta*, 1988]. These channels experience high-magnitude floods with a periodicity of decades. The unit discharges could be very high, in tens of $m^3s^{-1}km^{-2}$ [*Ahmad et al.*, 1993]. The seasonality on the southern slopes causes the streams to braid when the bed material consists of pebbles or sand. The channel forms and sinuosity, however, are also influenced by the active plate margin location, and the final appearance could be a trade-off between the three controlling environmental factors.

4. DISCUSSION

The case studies (Table 2) presented above contain both common features and deviations from such shared characteristics. Very few field measurements of the geometry of the streams of the seasonal tropics are available. Some generalization, however, is possible using data personally collected in the field, surrogate measurements, and a few approximations. For example, drainage area can be used as a surrogate of discharge [*Knighton*, 1984 using data from *Emmett*, 1975; *Kale*, 1990; *Miller*, 1990]. Discharge information, especially at several points

TABLE 2. Four Tropical Rivers: Summary Observations

Characteristics	Rivers			
	Auranga	Narmada	Yallahs	Buff Bay
Bed material	Sand (pbl)	sand (coarse)	pbl, cbl bld	pbl, bld
Bank material	sand	sand (silt)	pbl, sand	pbl, sand, silt
Channel pattern	M, (B)	M	B	M
Seasonal	yes	yes	yes	no
Large floods	yes	yes	yes	yes
Steep alluvial banks	yes	yes	yes	yes
Channel-in-channel morphology	yes	yes	yes	yes
Flood features	yes	yes	yes	yes
Point bars	yes	yes	no	yes
Braid bars	during dry season	no	yes	no
Features indicative of	seasonal high flow	flood	flood	flood

Note: M = meandering, B = Braided.
Features within parentheses indicate low occurrence.

along a river, is rarely available and almost never on the stream of one's choice.

Figure 7 shows the plot of channel area against drainage area for both the inner channel and the flood channel for two Jamaican streams: the Yallahs and the Buff Bay. All four cases, as expected, show an increase in channel area downstream. This diagram primarily illustrates the large difference in channel size between the flood channel and the smaller channel. Variations that exist between the slope of the regression lines and the magnitude of the scatter are reflections of local basin characteristics. The rivers have to adjust to two complete sets of discharges. The high-magnitude flood effects are visible irrespective of seasonality, as the Buff Bay is not a seasonal river [*Gupta*, 1975].

The adjustment by form is carried out by creating a box-shaped channel with high banks. Such channels have a relatively high width-depth ratio during low flows, or even during the wet season flow that fills the smaller channel. The width-depth ratio decreases strikingly with the arrival of high-magnitude floods when the entire channel is used

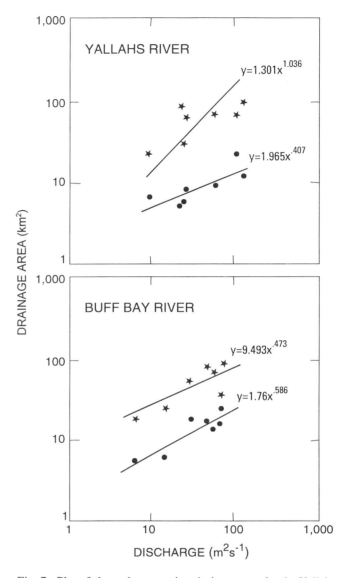

Fig. 7. Plot of channel area against drainage area for the Yallahs and the Buff Bay Rivers, Jamaica for both entire and inner channels.

(Figure 8). In Figure 8, an increase of the width-depth ratio occurs only twice in high-magnitude floods, and both times on limestone. The biggest decrease occurs in the Auranga river which carries primarily sand. Such box-shaped channels are common in parts of the seasonal tropics which also experience large floods [*Kale, et al.*, 1994]. Traditional geomorphological explanation involves incision but as such channels are filled a few times in a century [*Gupta*, 1988; *Kale et al.*, 1994; *Rajaguru et al.*, 1995], these are active flood channels of the present-day river.

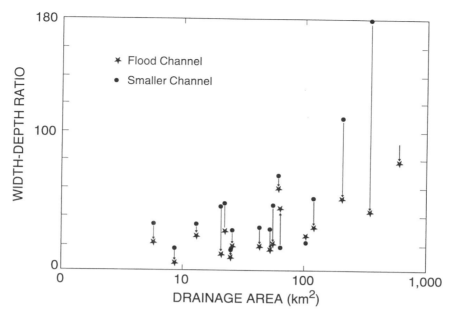

Fig. 8. Changing width-depth ratio in high-magnitude floods, selected rivers.

I calculated the bed shear stress and unit channel power [*Baker and Costa*, 1987] for a few available measurements on these rivers (Table 3) using the equations $\tau = \gamma YS$, and $\omega = \gamma QS/W$, or $\omega = \tau V$, where τ is the bed shear stress (Nm^{-2}), γ is specific weight of the fluid (9807 Nm^{-3} for clear water), Y is depth (m), S is channel slope, ω is power per unit area of the bed (Wm^{-2}), Q is discharge (m^3s^{-1}), W is channel width (m), and V is mean velocity (ms^{-1}). The numbers in Table 3 have been rounded off as some of the measurements are approximations.

Even with such restricted data three conclusions are reached.

1. There is a very large difference in bed shear stress and power per unit area of the bed between the annual seasonal high floods and the high-magnitude floods that occur at much greater intervals. This is true for all four rivers under very different basin conditions.

2. Three of the rivers (excluding the Auranga) can move quantities of very coarse sediment during the high-magnitude floods. It has been shown that a 1.5m boulder can be moved under conditions that exceed 1083 Wm^{-2} stream power, 225 Nm^{-2} bed shear stress, and 2.52 ms^{-1} mean velocity [*Williams*, 1983]. These thresholds are routinely exceeded in floods and boulder size material is transported downstream, except in the Auranga. Small boulders may also move during seasonal high flows in the headwaters where the slope is steep. This difference in shear stress and stream power determines that the flood effects will be preserved in the channels for decades until the next large flood arrives. The Auranga, in contrast, is a sand-bed river and sand is removed and the forms adjusted both during the large floods and seasonal floods. The wet and dry monsoons therefore alter Auranga every year, and flood impressions are sparse. This is an example of how local geology, through sediment texture, influences channel morphology.

3. The seasonal high flood that perhaps every year fills the inner channel of the rivers of the type being discussed, is capable of moving material certainly up to pebble size. This leads to (1) accumulation of finer material around the boulders, (2) formation and modification of bars, and (3) maintenance of a smaller channel. Material for such channel modification is available within the channels themselves as indicated by *Thomas* [1991] for the daily suspended sediment flow in the Hope River of Jamaica which includes transport in a hurricane. The daily suspended sediment load for this river for the 10 days between 9 and 18 September 1988 were 14, 14, 35, 27051, 35727, 14479, 4491, 2234, 1167, and 762 tons per day, A considerable part of this sediment may get flushed out of the system but a significant amount also stays in storage in sheltered areas in the channel and the valley flat. The relatively finer part of this sediment in storage travels each year.

Wohl [1992] working in the drier seasonal tropics of northern Australia opined that the role of large floods may be of particular importance. In the rocky Burdekin gorge, extreme floods (tens of thousands cubic meter per second)

TABLE 3. Shear Stress and Unit Stream Power, Selected Rivers

River	Location	Shear Stress (a) (Nm^{-2})	Shear Stress (b) (Nm^{-2})	Unit Stream Power (a) (Wm^{-2})	Unit Stream Power (b) (Wm^{-2})
Narmada	Guwarighat	290	1280	1500	8000
	Barmanghat	---	3000	---	12800
Auranga	Richughutu	70	80	180	350
	Gowa fording	20	100	10	400
	Deobar	20	30	20	70
	Kechki	60	200	130	1100
Yallahs	Silver Hill	400	1400	3000	20000
	Robertsfield	100	700	100	11000
	Potsdown	100	350	370	2600
	Mavis Bank	200	700	1000	7500
	Mahogany Vl	100	600	370	6200
	Ramble	130	200	750	1200
	Llandewey	70	200	170	1300
Buff Bay	Wakefield	270	650	2800	7000
	Silver Hill	550	2240	8000	35000
	Balcarres	170	500	980	1200
	Tranquility	190	480	800	4100
	Charles Town	200	350	580	1200
	Buff Bay	130	160	500	700

The figures are approximations. (a) refers to seasonal high discharge filling the inner channel, and (b) to the high-magnitude flood filling the entire channel.

may control the location of channel forms in boulders followed by wet season high flows reworking sediment across pre-existing forms. During the wet season (mean discharge 1260 m^3s^{-1}), the Burdekin River carries large volumes of sand and silt across the pre-existing channel forms in boulders. During the dry season (April-November; mean discharge 60 m^3s^{-1}), the river carries only very fine silt and clay in suspension [*Wohl*, 1992].

Severe erosion of bottomland in the Upper Potomac and Cheat River Basins in the flood of November 3-5, 1985 was associated with reach-average unit stream power of 200 - 500 Wm^{-2} [*Miller and Parkinson*, 1993]. The figure at some location may have exceeded 2500 Wm^{-2}. Such figures again support the hypothesis that it is the high-magnitude flood (Table 3) that forms and maintains river channels from the seasonal tropics discussed in this paper. The wet season high flows (except for sand channels) modify the channel forms to a limited extent.

Channel forms are usually inset within the bordering high banks associated with the high-magnitude floods. This has been reported as a post-flood feature even from rivers in the humid temperate areas [*Miller and Parkinson*, 1993] and arid regions [*Bourke*, 1994]. Bourke has proposed a model for 'mega-scale' floodplain morphodynamics. Floods remove the entire valley bottom aggradational features which are then rebuilt, mostly as insets. Such forms have been seen on many tropical rivers (Figure 9). These are rarely reported in the literature because of the small number of geomorphological studies on the rivers of the seasonal tropics.

It seems likely that the channel forms and sediment in the rivers of the seasonal tropics reflect three variables: the nature of seasonality, the frequency of high-magnitude floods, and the volume and texture of the ambient sediment. Usually a combination of these variables will operate. The Auranga is a stream with strong seasonality, sandy sediment, and a few large floods. The Narmada is a flood-dominated river in spite of a monsoon location and with both bed and suspended sediment primarily in sand. The Jamaican streams discussed have very coarse sediment

Fig. 9. The Mekong River, Lao PDR. Depositional forms inset against the high channel bank.

in their channels, and are flood-dominated irrespective of whether a seasonal variation in their discharge occurs or not. Rivers of the seasonal tropics therefore tend to display characteristic forms such as the channel-in-channel physiography, inset depositional features, and the box-shape. Certain variations of morphology from the general trend may occur depending on the strength of the seasonality and the texture of the sediment being transported. As *Wolman and Gerson* [1978] concluded, events which are described as catastrophic may in reality be common in certain environments.

5. CONCLUSIONS

The rivers of the seasonal tropics usually display a channel-in-channel physiography and a box-shaped appearance. This is due to the river being affected by both high flows of the wet season and, for a large part of this region, the occurrence of high-magnitude floods at intervals of decades. The volume and texture of sediment also determine the channel form. The channel form and processes are modified if one of these three factors is dominant. The seasonal tropics has its own style of adjusting to the magnitude-frequency concept.

Acknowledgments. I would like to thank James C. Knox, John E. Costa, and A. Fraser Gupta for their critical reading of an earlier draft of this paper.

REFERENCES

Ahmad, R., F.N. Scatena, and A. Gupta, Morphology and sedimentation in Caribbean montane streams: examples from Jamaica and Puerto Rico, Sed. Geol., 85, 157-169, 1993.

Baker, V.R., Stream-channel response to floods with examples from central Texas, Geol. Soc. Am. Bull., 88, 1057-1071.

Baker, V.R. and J.E. Costa, Flood power, in Catastrophic Flooding, edited by L. Mayer and D. Nash, pp.1-21, Allen and Unwin, London, 1987.

Benson, M.A. and D.M. Thomas, D.M., A definition of dominant discharge, Bull. Int. Assoc. Sci. Hydrol., 11, 76-80, 1966.

Bourke, M.C., Cyclical construction and destruction of flood dominated flood plains in semiarid Australia, in Variability in Stream Erosion and Sediment Transport edited by L.J. Olive,

R.J. Loughran and J.A. Kesby, pp.113-123, IAHS Publication No. 224, Wallingford, U.K., 1994.

Coleman, J.M., Brahmaputra River: channel processes and sedimentation, Sed. Geol., 8, 129-239, 1969.

Costa, J. E., Response and recovery of a Piedmont watershed from tropical storm Agnes, June 1972, Water Resour. Res., 10, 106-112, 1974.

Dasgupta, S.P., Ed., The Ganga Basin, Part II, 204 pp., Central Board for the Prevention and Control of Water Pollution, New Delhi, 1984.

Emmett, W.W., The channels and waters of the Upper Salmon River area, Idaho, U.S. Geol. Sur. Prof. Paper 870A, 116 pp., 1975.

Goswami, D.C., Brahmaputra River, Assam, India: physiography, basin denudation, and channel aggradation, Water Resour. Res., 21, 959-978, 1985.

Gupta, A., Stream characteristics in eastern Jamaica, an environment of seasonal flow and large floods, Am. J. Sci., 275, 825-847, 1975.

Gupta, A., High-magnitude floods and stream channel response, Spec. Publ. Int. Assoc. Sedimentol. 6, 219-227, 1983.

Gupta, A., Large floods as geomorphic events in the humid tropics, in Flood Geomorphology edited by V.R. Baker, R.C. Kochel and P.C. Patton, pp. 301-315, Wiley, New York, 1988.

Gupta, A. and A. Dutt, The Auranga, description of a tropical monsoon river, Z. Geomorphol., 33, 73-92, 1989.

Gupta, A. and H. Fox, Effects of high-magnitude floods on channel form: a case study in Maryland Piedmont, Water Resour. Res., 10, 499-509, 1974.

Kale, V.S., Morphological and hydrological characteristics of some allochthonous river channels, western Deccan Trap upland region, India, Geomorphology, 3, 31-43, 1990.

Kale, V.S., L.L. Ely, Y. Enzel, and V.R. Baker, Geomorphic and hydrologic aspects of monsoon floods on the Narmada and Tapi Rivers in central India, Geomorphology, 10, 157-168, 1994.

Kale, V.S., S. Mishra, Y. Enzel, L. Ely, V.R. Baker, and S.N. Rajaguru, Geomorphic investigations of floods in Central Narmada Basin, India, First Nat. Symp. on Environmental Hydraulics, C.W.P.R.S., Pune, 211-220, 1992.

Knighton, D., Fluvial Forms and Processes, 218 pp. Edward Arnold, London, 1984.

Miller, A.J., Flood hydrology and geomorphic effectiveness in the central Appalachians, Earth Surf. Process. Landforms, 15, 119-134, 1990.

Miller, A.J. and D.J. Parkinson, Flood hydrology and geomorphic effects on river channels and flood plains: the flood of November 4-5, 1985 in the South Branch Potomac River Basin of West Virginia, U.S. Geol. Sur. Bulletin 1981E, 96p, 1993.

Nash, D.B., Effective sediment-transporting discharge from magnitude-frequency analysis, Jour. Geol., 102, 79-95, 1994.

Pickup, G., Event frequency and landscape stability on the floodplain systems of arid Central Australia, Quat. Sci. Rev., 10, 463-473, 1991.

Pickup. G. and W.A. Rieger, A conceptual model of the relationship between channel characteristics and discharge, Earth Surf. Process., 4, 37-42, 1979.

Rajaguru, S.N., A. Gupta, V.S. Kale, S. Mishra, R.K. Ganjoo, V.R. Baker, L.L. Eli, and Y. Enzel, Channel form and processes of the flood-dominated Narmada River, India, Earth Surf. Process. Landforms, 20, in press, 1995.

Rao, K.L., India's Water Wealth, 267 pp., Orient Longmans, New Delhi, 1979.

Rao, K.N., Tropical cyclones of the Indian Seas in Climates of Southern and Western Asia, edited by K. Takahashi and H. Arekawa, pp. 257-324, Elsevier, Amsterdam, 1981.

Sikka, D.R., Some aspects of the life history, structure and movement of monsoon depressions, Pure Appl. Geophys. 115, 1501-1529, 1977.

Smith, N.D., Braided stream depositional environment - comparison of the Platte river with some Silurian clastic rocks, north-central Appalachians, Bull. Geol. Soc. Am., 81, 2993-3013, 1970.

Stevens, M.A., D.B. Simons, and E.V. Richardson, Non-equilibrium river form, J. Hydraul. Div. Am. Soc. Civ. Eng., 101, 557-566, 1975.

Thomas, H., Water quality analysis for the period 1988-1990: UNDP, UNEP, and Government of Jamaica Project - environmental management of the Hope River Watershed, JAM/87/008-009 UNDP, 1991

Williams, G.P., Palaeohydrological methods and some examples from Swedish fluvial environments. I. cobble and boulder deposits, Geogr. Ann., 65A, 227-243, 1983.

Wohl, E.E., Bedrock benches and boulder bars: floods in the Burdekin Gorge of Australia, Bull. Geol. Soc. Am., 104, 770-778, 1992.

Wolman, M.G. and R. Gerson, Relative scales of time and effectiveness of climate in watershed geomorphology, Earth Surf. Process., 3, 189-208, 1978.

Avijit Gupta, Department of Geography, National University of Singapore, Singapore 0511

Rivers and the Art of Stream Restoration

Robert Newbury

Newbury Hydraulics Ltd., Gibsons, British Columbia, Canada

Water flowing in rivers has a structure and pattern that are too complex to analyze completely. Antoine Chézy, a keen observer of rivers, presented a partial analysis of river flow in 1768 by assuming the flow to be steady and uniform. In the two centuries that followed, his flow equation became a tool for designing uniform river channels and canals, often to the detriment of fish and other aquatic organisms adapted to complex river flows. To restore and enhance habitats in uniformly channelized rivers, the founding observations of Chézy must be re-interpreted using natural river characteristics and the eye of an angler.

1. INTRODUCTION

This is the tale of two rivers, in reality the Seine River in France and the Pine River in western Canada, and in perception, the idealized channel of the engineer Chézy and an angler's vision of a perfect trout stream. On the Seine River in 1769, the formula for uniform open channel flow was proposed by Chézy. Since that time, a little over two centuries ago, Chézy's equation (or Manning's equation, a later derivative) has been used to predict the velocity and thus the rate of flow of water in canals and rivers.

Two hundred years later on the Pine River, the limitations of the uniform flow formula became apparent in a project to create a meandering reach of river with habitats preferred by rainbow trout. Chézy knew about these limitations in natural rivers and cautioned users of his formula. However, the cautionary notes are often disregarded, and the idealized formula is used to define the ideal river. This is not the trout angler's ideal river. The habitat for trout is formed by diverse flow conditions that occur in non-uniform natural channels. These two ideals must be reconciled in order to restore a channelized stream.

2. ANTOINE CHÉZY'S EXPERIMENTS

Antoine Chézy was born in Chalors-sur-Marne, France in 1718 (Figure 1). He was educated as a teacher and entered a new engineering school l'Ecole des Ponts et Chaussées in Paris in 1748. He was appointed the Inspecteur Général des Ponts et Chaussées for Paris and continued to teach at the school until 1790. Chézy proposed his famous equation in 1768 while undertaking a study to augment the Paris water supply [*Herschel*, 1897]. It was proposed that water from the Yvette River be diverted into Paris with a small canal. Chézy was asked to design the channel. He cast the problem of predicting the velocity of flow in his 1775 memoir as follows (translated from *Chézy* 1921):

After having designed a channel, and carefully specified its slope, it is very interesting to know if the cross-section of the channel will be sufficient to conduct the required discharge. To do this, it is necessary to know the velocity of the flow in the uniformly sloping channel . . .

To develop a theoretical method for predicting velocity, Chézy measured velocities in existing canals and rivers to see if there was a general relationship between the slope of the channel and the rate of flow. He seemed to have an inkling of what he would find. A century earlier, Sir Isaac Newton had described the laws of gravity in which the velocity of a falling object could be predicted by the

Natural and Anthropogenic Influences in Fluvial Geomorphology
Geophysical Monograph 89
Copyright 1995 by the American Geophysical Union

Fig. 1. Antoine Chézy 1718 - 1796 [*Herschel* 1897].

distance it had fallen. Chézy observed that water flowing down the bed of a uniform sloping channel, maintained a velocity that was constant regardless of the length of travel. He proposed that the accelerating forces of gravity had been counterbalanced by frictional forces that resist the flow on the bottom and sides of the channel, and that the square root of the frictional forces appeared to vary with the velocity. This could only be applied to uniform flow where the velocity, depth, slope, and width of the river are constant. To find the balance point between the velocity and slope under these uniform conditions, he conducted his first experiment in the Courpallette canal in a plank-lined reach called la Gibonnière. In his words [*Chézy*, 1921]:

Pour connaître la vitesse relative à la pente, j'ai choisi la partie de la rigole la plus droite et la plus égale, à l'endroit dit de la Gibonnière, par un temps très calme. Une boule de cire jetée dans le milieu de la rigole a parcourou 101 toises en 23 minutes . . .

Chézy observed that the ball of wax that he threw into the center of the uniform wooden canal travelled a distance of 101 toises (196.85 meters) in 23 minutes. Chézy conducted a larger scale experiment in the Seine River on October 7, 1769. This time the measured reach was 100 meters wide and 2.6 kilometers long, stretching between the Suresnes ferry and the pont de Neuilly. The average time of travel for the balls of wax in two trials was 55 minutes. One can imagine this distinguished scholar dropping the balls from the ferry as it crossed the river, landing on the other side and galloping to the pont de Neuilly to observe them passing under the bridge. He was not a casual observer. His advice to those who would study rivers was strict (translated from *Chézy*, 1921):

It is important to remember that all these experiments demand the greatest care; that it is very difficult to make them with sufficient precision, and that one ought not to rely upon any except those made by persons known to give them the most scrupulous attention.

Based on his observations, Chézy presented his famous equation in 1775:

$$v = C (RS)^{1/2}$$

where v is the mean velocity, R is the hydraulic radius of the flow (the cross-sectional area of the flow divided by the wetted perimeter), S is the slope of the channel, and C is the empirical constant that became known as Chézy's C. A century later, Chézy's equation was modified by Robert Manning, an Irish engineer. Manning replaced Chézy's C with $R^{1/6}/n$, where n is the channel resistance or roughness factor [*Manning*, 1891]. Manning's form of the equation, $v = R^{2/3}S^{1/2} / n$ is widely used today.

3. THE IDEAL HYDRAULIC CHANNEL

In designing a channel, where the slope is adjusted to the valley topography and the materials of the channel bed are known, the minimum channel size for a given discharge will have dimensions that maximize the hydraulic radius for a given cross-sectional area. This occurs when the channel cross-section is a half circle and is known as the "best hydraulic section" [*Rangaragu*, 1981, *Roberson and Crowe*, 1993].

Several factors prohibit the construction of a semicircular channel; the surveys required and excavation methods are difficult and the vertical upper banks collapse because the stable angle of repose of most river valley materials is less than 90 degrees [*Chow*, 1959]. A simple compromise exists that is almost as efficient as the semicircular channel, the trapezoidal channel. This channel form may be simply

Fig. 2. Wilson Creek, Manitoba is regularly re-excavated to maintain a trapezoidal channel form with a discharge capacity that is greater than the natural bankfull flow.

surveyed and excavated with a flat bottom and two flat banks inclined at less than the stable angle of repose of the bank materials. Uniform flow conditions that occur in the channel can be predicted with Chézy's or Manning's equation.

The near "best hydraulic section" trapezoidal channel is often recommended for water supply and drainage canals to minimize excavation. In many stream channelization projects, the natural channel is straightened, dyked and deepened to create the narrower, more efficient trapezoidal cross-section as well (Figure 2). Unfortunately, the increased depth and gradient of the channelized reach promote erosion of streambed materials that were in equilibrium with the shallower natural channel [*Newbury and Gaboury*, 1988]. As the stream bed erodes, the banks steepen, eventually leading to their collapse when the stable angle of repose is exceeded. The channel must then be re-excavated to a wider cross-section to reduce the bank slopes. Alternatively, the stream gradient may be reduced by adding drop structures and stilling basins or the stream bed may be paved with less erodible materials (Figure 3).

The widespread use of uniform channel designs in river engineering is not the legacy of Chézy. He knew the limitations of his experiments in the Courpalette canal and the Seine River and emphasized the need for more studies of different channels, including natural streams and rivers as follows [*Chézy*, 1921]:

Pour tirer tout l'avantage possible de cette théorie, il faudrait avoir un grande nombre d'observations sur les rigoles [channels] de sections différentes; on se servirait des observations faites sur celles qui ressembleraient les plus à la rigole projetée. Il serait intéressant, aussi, d'avoir de pareilles observations sur différents ruisseaux et rivières.

4. THE IDEAL TROUT STREAM

The Pine River is typical of many trout streams with a moderate gradient (2%) broken into regularly spaced rapids or shallow runs (Figure 4). In the middle reaches where the stream is 12 meters wide, rapids and runs are separated by pools every 70 meters or so along the stream length. The pools are deepest under the overhanging banks carved

Fig. 3. Black Creek, Ontario has been paved to prevent erosion of the hydraulically efficient trapezoidal channel that is deeper and narrower than the former natural channel.

on the outside curve of meander bends. Large boulders scoured from the valley bottom and tree trunks that have fallen from the stream bank form local pockets of calm water followed by turbulent currents downstream. Locally there are beaver dams and ponds that impound the rapids and pools for one or two meanders upstream. A trout angler reads the varying pattern of flow conditions in the stream by observing the drift of the line and lure rather than wax balls. The angler uses the natural geometry of the valley and channel for cover, moving along the low inside bank of the meanders to approach the upstream-facing trout from behind. Fishing from the inside bank keeps the angler's profile low and difficult for the fish to see from the pool below the higher outside bank (Figure 5). The first cast from location 1 (Figure 6) is made upstream and across the channel into the tail of the rapids [*Knowles and Nelson*, 1992]. The fly hurries downstream through the decelerating curl of whitewater to the head of the pool. Active trout are waiting just below the surface for bobbing insects that have been swept from their habitats in the rocky rapids. If the fly is not taken, it drifts with the main current to the outside of the bend, into the pool and overhanging bank. In the next few casts from location 2, the fly is set closer and closer to the high outside curve, allowing it to be captured by the rising current that is emerging from under the overhanging bank. More trout are waiting for the drifting insects as they rise slowly to the surface just below the pool towards the center of the channel.

When a pool has been tested, the angler moves upstream, crossing the river in the riffle where footsteps are lost in the noise of the rapids and the bubble-filled water is opaque, blocking the view from the next pool. Disturbed sediments from the riffle move downstream giving no warning to the upstream fish. This is the noisy part of the stream, splashing across the rocky bed is sharply in contrast with the stealthy approach to the quiet pools. On the way, back-eddies below boulders and logs are fished by casting the fly above the obstructions into the line of accelerating flow that sweeps closely by the downstream pockets of calm water. Cold water zones, indicated by groundwater seepage on the channel banks, are carefully fished for trophy trout.

As the fish hunt continues up the channel, the pattern of

Fig. 4. Rapids and an upstream pool on the Pine River, Manitoba.

ponds, pools, rapids, and rolling current is repeated with local variations in every reach. The successful angler must read the pattern of the flow and place the fly in just the right current that leads to the wary trout. These are the diverse flow conditions that are created in the wave-like form of pool and riffle reaches in a natural stream. They are not the uniform flow conditions created in channelized streams that were modelled by Chézy from his experiments in the Courpalette canal.

5. MEANDERS, POOLS, AND RIFFLES

The repeated wave-form of meanders, pools and riffles in a natural stream has been observed and measured by stream researchers as well as anglers. One of the earliest observers was Tohkichi Kani, a student of stream habitats in Japan [*Kani*, 1981]. Kani found that the preferred habitats of benthic insects and fish could be related to the pattern of pools and riffles on the stream bed. In the 1930's, he developed a classification system for a range of streams based on their pool and riffle patterns (Figure 7). The significance of pools and riffles to trout stream habitats was observed by Stuart in Scotland as well [*Stuart*, 1953]. He was one of the first stream restorers to have naturally-spaced pools and riffles built on a channelized stream bed as described in *Leopold et al.*, [1964]:

Being concerned with the effect of diversion and re-alignment of certain gravel streams in Scotland on their ability to maintain trout, Stuart noted that new stream beds dredged by a dragline were, when just constructed, of uniform depth without pools and riffles. With the aim of producing the usual pool and riffle sequence, he directed the operator of the dragline to leave piles of gravel on the stream bed at intervals appropriate to riffles .. that is, 5 to 7 [stream] widths apart. After a few flood seasons, these piles had been smoothed out and presented to the eye a picture that in all respects appeared natural for a pool and riffle sequence. Moreover, the riffles so formed have been stable over a number of years of subsequent observation.

Fig. 5. Fishing from the point bar formed on the inside bank of a meander on the Pine River, Manitoba.

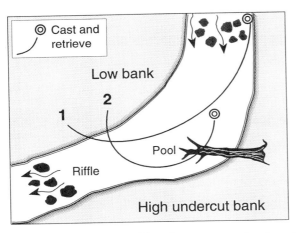

Fig. 6. Preferred casting positions for a trout angler from the inside bank of a river meander [*Knowles and Nelson*, 1992].

Many uniformly-channelized streams with erodible bed materials will return to their original pool and riffle pattern without rebuilding the riffles or rapids [*Gregory et al.*, 1994]. For example, in the older channelized reaches of the Mink River, pools and riffles have re-formed with the same frequency and length as those of the pre-excavated natural stream [Figure 8, *Newbury and Gaboury*, 1993a].

The natural dimensions of pools and riffles and their relationship to river size (Figure 9) was summarized by *Leopold et al.*, [1964]. The average spacing of this

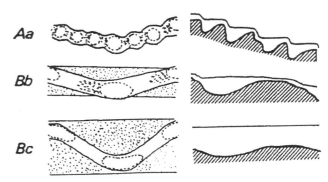

Fig. 7. The classification of pools, riffles and meanders for several Japanese streams (*Kani* [1944] 1981).

undulating wave form on the river bed profile was found to be 6 times the bankfull width of the river. For some combinations of discharge and slope, rivers were found to meander horizontally with the same wave form as well. The average meander wave length was found to be 12 times the bankfull width as it consists of two pool and riffle reaches. The average radius of curvature of the meander bends was 2.3 times the bankfull width. The natural width and corresponding flood discharge for channels that range in size from those of small streams to the Amazon River follow a surprisingly unified relationship [Figure 10, *Kellerhals and Church*, 1989]. Similar graphs of the natural geometry of a reach (the bankfull width and depth) may be prepared as a function of the drainage area in the same or similar drainage basins [*Dunne and Leopold*, 1978]. The relationships for several small basins including the Pine River are shown in Figure 11. In a natural river, the width of the channel increases more rapidly than the depth, causing the width to depth ratio to increase from 10:1 to 20:1 as the drainage area increases from 5 to 500 square kilometers. The average spacing of pools and riffles increases in proportion to the width, from 15 to 120 meters over this range of drainage areas.

For many streams, the duration of flood conditions near or above the bankfull stage is short compared to long periods of low and moderate flows. For example, in the mid-continent streams of Figure 11, bankfull and greater flood flows occur on average for only 14 per cent of the year. However, during this period the channel geometry and profile are established. The shear stress in the pools increases as the depth increases and the local slope of the water surface approaches the average gradient of the stream, scouring fine sediments and debris that have accumulated between floods (Figure 9, high flow profile).

In meandering sections, the steep overhanging banks on the outside bends are scoured and trimmed by the flood flows (Figure 9 plan). A fly cast into the center of the

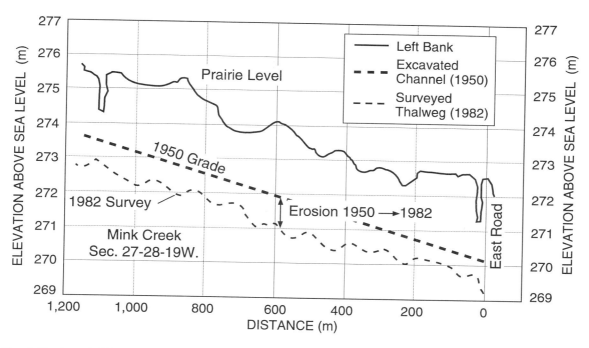

Fig. 8. Pools and riffles re-formed in the uniformly graded channel of Mink Creek, Manitoba [*Newbury and Gaboury*, 1993a].

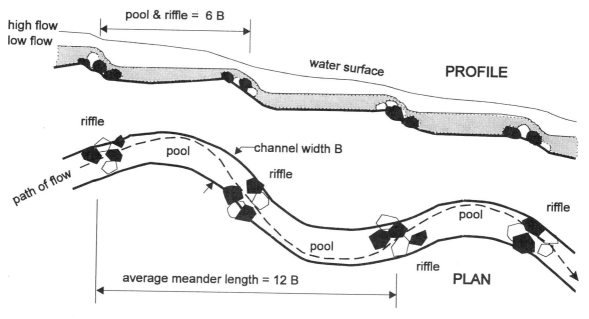

Fig. 9. Average meander, pool, and riffle dimensions as a function of the channel width B.

144 STREAM RESTORATION

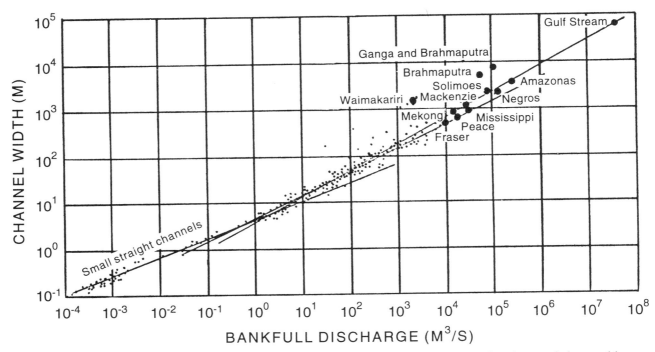

Fig. 10. The relationship between channel width and the bankfull discharge for rivers of the world [*Kellerhals and Church*, 1989].

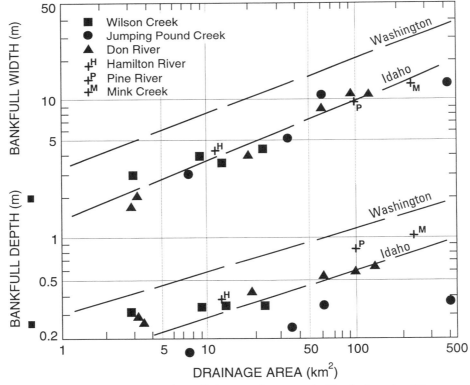

Fig. 11. The channel geometry of several small rivers and streams in North America [*Dunne and Leopold*, 1978, *Newbury and Gaboury*, 1993a].

riffle above the meander pool is carried to the outside of the bend under the overhanging bank and back across the lower edge of the pool by the helical flow that occurs as the stream changes direction. On the Pine River, the reaches preferred by trout occur in meanders with an average curvature of 2.3 river widths, suggesting that the fish may have adapted to the flow patterns and habitats created by the most frequently encountered meander curves in natural rivers.

At intermediate and lower discharges, water is stored in the pools impounded above the riffles or rapids (Figure 9, low flow profile). This maintains deeper fish habitats required for all life stages of the fish and provides water for passage up and down the river [*Bovee*, 1986]. In the riffles, the shallow flows are broken into chutes and waterfalls by cobble bars and boulders (Figure 4). Where the flow drops over an obstacle or is drawn through a narrow gap between boulders, it often reaches the critical state, a condition where the velocity is maximized for the total static and kinematic head of water that exists above the obstruction. If the water continues to accelerate over and past the obstruction, it attains super-critical velocities. Achieving this state of flow is important to maintaining the dissolved oxygen level in the river water. The super-critical flow sweeps air bubbles into the water and forms a hydraulic jump as it enters a deeper pocket or pool of sub-critical water downstream. The collapsing air bubbles rapidly re-aerate the flow. This is the source of noise in the river as the breaking bubbles on the surface make the sounds of babbling brooks or roaring rapids.

The shallow flows in riffles are efficient habitats for caddisflies, blackflies, and other benthic insects [*Statzner et al.*, 1988; *Wetmore et al.*, 1990]. By locating on the tops and sides of boulders, they are able to expand their capture nets and cephalic fans to gather detritus from the flow that converges to the narrow passages over and around the boulders. The varied structure of the flow creates protected areas that allow fish to follow a path between resting places through the rapids. Salmon may also jump over obstacles in the rapids by launching themselves from tops of local hydraulic jumps.

This is the scientist's river of gauging and measurements. Some of the phenomena observed by Chézy and a trout angler are explained by the laws of motion and by the characteristics of natural river forms. Many more still defy analysis, particularly at the local level of turbulent currents in pools and rapids.

6. "IT WAS NEVER A PROPER RIVER, BUT A LARGE TREE."

There are other ways to understand rivers. The Limbaruka River, near Cape Wilberforce in north-eastern Arnhem land, Australia is the aboriginal's river of honey [*Berndt and Berndt*, 1988]. In their myth of the origin of the river, Blanket Lizard man finds a tree laden with honey on the edge of the ocean. While gorging himself he swallows a sliver and runs up the tree, causing it to topple inward across the land and become the honey river. In the myth, the branches of the tree become branches of the river.

Rivers and trees have the same branching characteristics [*Stevens*, 1974]. In both cases there are on average 3 to 4 tributary branches to every limb or reach, spreading in a pattern that is recognized easily as a tree or river. Instead of trembling in the wind, the flow in each branch vibrates with a frequency and wave length that is set by the flood discharge. In the steep headwaters, the steps of pools and riffles are closely spaced in the narrow channel. Further downstream as the drainage area and discharge increase, the steps grow longer and the pools and riffles more pronounced. In the lower reaches, where water from all of the tributary branches has gathered, the river is widest and often meanders across the lower gradient of the valley floor. The pattern of the river's connected stairways leading into the continent is known by insects and fish, and by people living closely on the land. Knowing the branch of a river, and the characteristics of its position in the river network, is also essential for stream restoration.

7. STREAM RESTORATION

Restoration or enhancement of a river to decrease erosion, increase aeration, and promote the formation of insect and fish habitats can be accomplished by adjusting a degraded or habitat-poor channel so that the diverse hydraulic conditions of a natural stream can occur [*Chang*, 1988, *Newbury and Gaboury*, 1993a]. This was the biologist Stuart's idea when he constructed naturally spaced pools and riffles in channelized streams in Scotland. The use of average natural dimensions for designing meanders has been recognized by an engineer as well. *Henderson* [1966] advised:

Fig. 12. A plane table survey of a preferred trout habitat reach on the Pine River, Manitoba.

Values of the ratio [bankfull width to meander curvature] are found to be as small as 1.5 and as large as 10; but the median lies in the range 2 - 3, and it is within that range that the river engineer should look when planning to simulate natural meanders in river training works.

The principle of designing meanders with natural dimensions that will create hydraulic habitats for rainbow trout was applied successfully in a restoration and enhancement project on the Pine River in 1990 [*Newbury and Gaboury*, 1993b]. In the restoration reach, two factors prevented the formation of meanders and a pool and riffle profile that were typical of this branch of the river: (1) the stream bed was paved with a local glacial lag deposit of boulders that were larger than could be moved at the bankfull flood stage, and (2) the central part of the reach was straightened and fixed in place with large boulders at a highway bridge crossing.

To design the new channel, channel geometry data were gathered in stream branches throughout the basin to determine the relationships between channel width, depth, bankfull discharge and pool and riffle spacing. The location of preferred trout habitat reaches were identified on creel census forms distributed to volunteer anglers who fished the Pine River and other trout streams in the region. The most productive reaches identified by several anglers were then surveyed to determine the average dimensions of their meanders, pools, and riffles. A plane table survey [*Brinker and Wolf*, 1977] was used that allowed the plan and contours of the reach to be drawn to scale on site (Figure 12). Large boulders, overhanging banks, tree deadfalls and other local features that provided cover for the trout were located on the plan. The size and spacing of cobbles and boulders were measured in the rapids above and below the pools. The velocity and direction of the flow were plotted on a clear overlay placed on the plane table map. The pattern of current in the reach and the extent of back eddies and other hydraulic features were sketched in from the velocity vectors and by observing the progress of floats that were thrown into different locations across the upstream

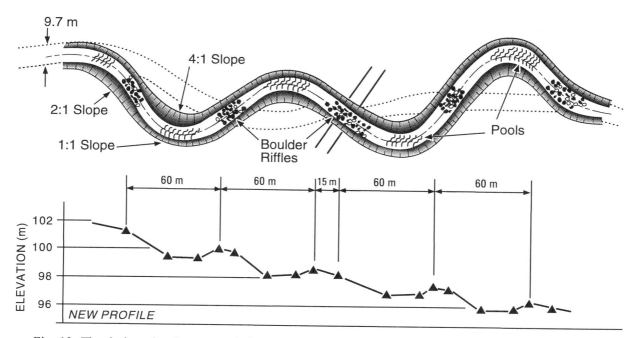

Fig. 13. The design plan for a meandering pool and riffle reach constructed on the Pine River, Manitoba to increase adult trout habitat (dotted lines are the original channel boundaries).

riffle. Thin-skinned neutrally-buoyant oranges were used as floats rather than wax balls.

Pools and riffles were designed for the restoration reach spaced 60 meters apart, approximately 6 times the bankfull width of 9.7 meters (Figure 13). Meanders with steep outside banks and shallow inside bars were excavated with a radius of curvature that was 2.3 times the bankfull width, with a few minor adjustments to straighten the alignment of the channel with the highway bridge. A pool and riffle profile was excavated along the channel that would impound 1 to 1.5 meters of water in the pools at low flow. Boulders and cobbles uncovered during construction were placed in the riffle zones at the same spacing and height as observed in natural rapids. The new channel banks and floodplains were revegetated with grasses and native trees that were moved from the highway right-of-way. Trout were found in the meander pools as the project was completed (Figure 14). In 1993, an infrequent flood passed through the reach (2.5% annual probability of exceedance, North Pine River Gauging Station No. 05LG001). The pools were deepened and overhanging banks were carved in the outside curve of the meanders. The rapids were rearranged slightly and some of the floodplain vegetation was lost. Anglers monitoring the reach reported that the trout fishing improved after the flood.

There are other examples of the use of natural river dimensions in stream habitat restoration projects that do not require the reconstruction of the entire channel. For

Fig. 14. Aerial view of the meandering pool and riffle reach constructed on the Pine River, Manitoba shown in Figure 13.

Fig. 15. Riffles added to the channelized lower reaches of Mink Creek, Manitoba have successfully restored walleye spawning and rearing habitats and made fish passage through the reach possible during low flow periods.

example, spawning and rearing habitats added to channelized streams tributary to Lake Dauphin in central Canada have restored the walleye fishery in the lake [*Newbury and Gaboury*, 1988]. The trapezoidal channels were enhanced by building spawning riffles and rearing pools spaced at intervals of six times the natural bankfull width (Figure 15). The riffle design was based on surveys of the gradients and bed materials in successful natural spawning areas. The riffle crest elevations were set to store water in the channel under low flow conditions so that drifting fry would return to the lake downstream. Similarly, boulder clusters and rock-filled rapids spaced at six times the bankfull width have been added to diverted coastal streams to enhance fish passage and accrue spawning gravels for salmon [Figure 16, *Department of Fisheries and Oceans Canada,* 1994]. In all cases, the channelized streams have been more stable in the restored pool and riffle reaches than in the adjacent uniformly graded reaches [*Newbury and Gaboury*, 1993a].

Fig. 16. Boulder rapids constructed on a steep coastal stream to create coho spawning beds and rearing pools (Twin Creeks, British Columbia).

8. CONCLUSIONS

Antoine Chézy predicted that it would be interesting to make observations of the flow in natural streams and rivers as well as in uniform channels. It was found that in most rivers, the flow is not uniform. At low and intermediate discharges, pools and riffles create eddies and locally-varied flow conditions that form and maintain the river's habitats. Natural stream geometry, the eye of an angler, and Chézy's careful approach to observing rivers, are all needed to restore complex hydraulic habitats to channelized streams. Even then, some mystery will remain in just how the water flows.

Acknowledgements. I am indebted to Rick Wowchuk, a master trout angler and teacher at the Swan Valley Regional Secondary School, for sharing his insights into trout fishing on the Pine River. Rick discovered, fished, and sketched the prime trout reaches on the river and organized the Pine River trout enhancement project with his students and the Swan Valley Sport Fishing Enhancement group. The Pine River project was undertaken as part of a series of small projects to gather background for a handbook on stream restoration with Marc Gaboury of the Manitoba Fisheries Branch. Marc had the final responsibility for negotiating and constructing the sample projects (to my relief). We learned biology and engineering from each other. The founding textbook on river morphology by *Leopold, Wolman and Miller* [1964] was our principal reference. In every stream, I am reminded of the curiosity and ideas so generously given to me at Johns Hopkins University by Dr. M. Gordon Wolman. Thanks Reds.

REFERENCES

Berndt, R.M. and C.H. Berndt, *The Speaking Land: Myth and Story in Aboriginal Australia*, Penguin Books Australia, Ringwood, 439 pp., 1988.

Bovee, K.D., Development and evaluation of habitat suitability criteria for use in the instream flow incremental methodology. *Instream Flow Info. Paper No. 21, Biol. Rpt. 86(7)*, US Fish and Wildlife, 1986.

Brinker, R.C. and P.R. Wolf, *Elementary Surveying, 6th ed*, Harper and Row, New York, 568 pp., 1977.

Chang, H.H., *Fluvial Processes in River Engineering*, 432 pp., Wiley, New York, 1988.

Chézy A., Mémoire sur la vitesse de l'eau conduite dans une rigole donnée [1775], *Annales des Ponts et Chaussées, II-LXI*, 241-251, 1921.

Chow, Ven Te, *Open Channel Hydraulics*, 680 pp., McGraw-Hill, New York, 1959.

Dept. Of Fisheries and Oceans Canada, Rebuilding Twin Creeks, *Streamtalk, 2(3)*, 1, 1994.

Dunne, T. and L.B. Leopold, *Water in Environmental Planning*, 818 pp., Freeman, San Francisco, 1978.

Gregory, K.J., Gurnell, A.M., Hill, C.T., and S. Tooth, Stability of the pool-riffle sequence in changing river channels. *Regulated Rivers Res. and Mgmt., 9*, 35-43, 1994.

Henderson, F.M., *Open Channel Flow*, 522 pp., McMillan, New York, 1966.

Herschel, C., On the origin of the Chézy formula, *J. Assoc. of Engineering Soc., 18*, 363-368, 1897.

Kani, T., Ecology of torrent-inhabiting insects [1944]. *Physiol. Ecol. Japan, 18*, 113-118, 1981.

Kellerhals, R. and M. Church, The morphology of large rivers: characterization and management, *Can. J. Fish. Aquat. Sci., Special Public. 106,* 1989.

Knowles, D. and L. Nelson, Factoring stream trout, *In-Fisherman, Spring Issue*, 92-100, 1992.

Leopold, L.B., M. G. Wolman, and J.P. Miller, *Fluvial Processes in Geomorphology*, 522 pp., Freeman, San Francisco, 1964.

Manning, R., On the flow of water in open channels and pipes, *Trans. Inst. Civil Eng. Ireland*, 20, 161-207, 1891.

Newbury R. and M. Gaboury, The use of natural stream characteristics for stream rehabilitation works below the Manitoba escarpment. *Can. J. Water Res., 13/4*, 35-51, 1988.

Newbury, R. and M. Gaboury, Stream Analysis and Fish Habitat Design. Man. Habitat Heritage Corp./Newbury Hydraulics Ltd., 256 pp., 1993a.

Newbury, R. and M. Gaboury, Exploration and rehabilitation of hydraulic habitats in streams using principles of fluvial behaviour. *Freshwater Bio., 29*, 195-210, 1993b.

Rangaraju, K.G., *Flow Through Open Channels*, 330 pp., Tata McGraw-Hill, New York, 1981.

Roberson, J.A. and C.T. Crowe, *Engineering Fluid Mechanics*, 823 pp., 5th Ed. Houghton-Mifflin, 1993.

Statzner, B., Gore, J.A. and V.H. Resh, Hydraulic stream ecology: observed patterns and potential applications, *J. N. Am. Benthol. Soc., 7*, 307-360, 1988.

Stevens, P.S., *Patterns in Nature*, 240 pp., Little-Brown, 1974.

Stuart, T.A., Spawning, migration, reproduction and young stages of loch trout, (Salmo trutta L.), Report No. 5. Scottish Home Department, 39pp., 1953.

Wetmore, S.H., R.J. Mackay and R. Newbury, Characterization of the hydraulic habitat of *Brachycentrus occidentalis*, a filter feeding caddisfly, *J. N. Am. Benthol. Soc., 9(2)*, 157-169, 1990.

Robert Newbury, School of Resource and Environmental Management, Simon Fraser University, and Newbury Hydraulics Ltd., Post Box 1173, 331 Headlands Road, Gibsons, BC, Canada VON 1VO.

Effective Discharge and the Design of Channel Maintenance Flows for Gravel-Bed Rivers

E. D. Andrews

U.S. Geological Survey, Boulder, Colorado

James M. Nankervis

U.S. Forest Service, Fort Collins, Colorado

Water resource developments which deplete the quantity or reduce the range of streamflows usually have a number of unintended effects on the channel downstream, including loss of channel capacity, loss of aquatic and riparian habitat, and channel instabilities. A method for identifying a flow regime sufficient to maintain desired stream characteristics, while permitting significant development, would have great practical value. Over the past decade, important advances have been made in our understanding of fluvial processes in gravel-bed streams. Using these advances as a basis, one can outline a method for determining channel maintenance flows for gravel-bed streams typical to the western United States. A common characteristic of gravel-bed streams is that bed particles are transported only about 5-10 percent of the time during the highest flows, and, even then, at a very low rate. Although occasional motion of bed particles begins at a discharge as small as 60 percent of the bankfull value, general motion of the bed surface is exceedingly rare. The proposed method relies on an appropriate bedload transport function and specific reach information to compute the quantity of bed material in each size fraction transported by increments of discharge in the pre- and post-regulation regimes. Evaluation of possible flow regimes indicates that bankfull channel capacity can be maintained in its pre-regulation condition where as much as 60 percent of the natural flows are diverted.

1. INTRODUCTION

Alteration of natural streamflows by regulation and diversion is common throughout the arid western United States. Indeed, there are relatively few streams with drainage areas greater than a hundred square kilometers even in mountainous regions that have not been affected substantially. The most significant change in the natural streamflow regime is a decrease in the magnitude of annual peak flows, commonly by a factor of 2 to more than 10, *Hirsch et al.* [1990]. An associated, although frequently unintended, consequence of streamflow storage and diversion is an appreciable change in quantity and size of sediment supplied to the stream channel. These changes, together with more diffuse, but often extensive, land use changes have greatly altered the quantity, seasonal distribution, and relative proportion of water and sediment supplied to stream channels throughout the western United States.

When streamflow and sediment supply are substantially altered over a period of years to decades, changes in the stream channel, floodplain, and riparian margin are common. The adjustment may include changes in channel size, morphology, pattern, rate of channel migration, bed-material particle-size distribution, floodplain morphology, and the composition and density of riparian vegetation, e.g., *Petts* [1979], *Galay* [1983], and *Williams and Wolman* [1984]. Frequently, these adjustments affect downstream aquatic and riparian resources. Increased frequency of flooding due to reduced channel size and floodplain storage, loss of ecological habitat in the stream channel and adjacent floodplain, as well as the impairment of recreational and esthetic resources commonly occur following substantial modification to the magnitude and frequency of natural streamflows.

The need and desire to minimize or avoid alteration of the stream channel and floodplain downstream from water storage and diversion facilities arise in nearly every instance. Typically, the loss of aquatic and riparian resources involve significant costs and foregone opportunities. In many instances, the stream and riparian resources are subject to state and federal statutes, e.g. National Environmental Policy Act of 1969, Clean Water Act of 1972, Endangered Species Act of 1973, Federal Land Policy and Management Act of 1976, and National Forest Management Act of 1976. State and federal courts have provided water rights for streamflows needed to achieve the objectives for which lands were reserved from the public domain, i.e. National Parks, Indian Reservations, etc. Thus, the design and operation of streamflow storage and diversion facilities invariably will include an analysis of the relation between flows and the nature of the stream channel and floodplain. That is, what quantity and distribution of streamflows throughout the year are required, together with a given contribution of sediment, to construct and maintain the channel and riparian resources of a particular reach? (See *Rosgen et al.* 1986 and *Reiser et al.* 1989 for examples of instream flow regimes formulated to maintain desired channel features.) Conversely, what portion of the natural streamflow may be stored or diverted without causing an unacceptable impairment of fluvial resources? These questions concern the fundamental issues of fluvial geomorphology and river mechanics. Clearly, the answers will have great practical significance in the management of water resources.

The concept of a dominant discharge has existed for decades in the hydrologic literature. The design of canal channels which would carry a specific discharge and a sediment load of silt, clay and fine sand without deposition or erosion was studied intensively in the 19th century, [*Ackers,* 1972]. The relation between discharge and the characteristics of a natural channel are much more complex, *Leopold and Maddock* [1953]. Stream discharge varies over a considerable range from extreme low flow to extreme flood. In addition, the range of sediment particle sizes in a naturally stream channel is typically much greater. The characteristics of naturally formed stream channels reflect this variability. Thus, one must consider a range and frequency of occurrence for the channel-forming flows rather than a single discharge.

Wolman and Miller [1960] developed a conceptual model to describe the influence of flow magnitude and frequency on the relative sediment transporting effectiveness of various discharges in natural channels. This model has become one of the fundamental paradigms of geomorphology. The Wolman-Miller model is illustrated by Figure 1. The variation in sediment transport with discharge is represented by curve A. Sediment transport begins when discharge attains a threshold value, and then increases rapidly and continuously. The threshold discharge depends to a substantial degree on the size of sediment particles composing the channel bed and banks. Relatively small sediment particles, silt and fine sand, will be transported to some extent even at very small discharge in most river channels. Conversely, relatively large sediment particles, gravel and boulders, typically will only be moved by the quite large and uncommon discharges. Because transport rate is strongly influenced by particle size, it is usually desirable to calculate separate sediment transport relations for each particle size class. The frequency of discharges in a given stream reach is described by curve B, which represents all discharges from extreme low flow to extreme flood. The product of these two relations gives a third curve, C, which is the frequency distribution of sediment transport, i.e. the percent of time a specific sediment transport rate has or will occur. Integrating the frequency of sediment transport rates over a period of time, typically one year, equals the total quantity of sediment transported.

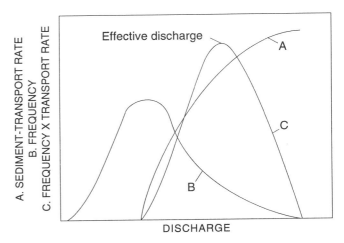

Fig. 1. Wolman-Miller model showing the relative importance of magnitude and frequency of streamflows.

The principal result of this model is that most of the sediment transported over a period of time is carried by a range of intermediate discharges. Discharges less than the threshold required for initiation of sediment motion transport no sediment. Discharges only somewhat greater than the threshold discharge may be quite common, but transport sediment at such a small rate that they carry only a small quantity of sediment over a period of years. The very largest discharges transport sediment at an exceedingly high rate, however, they occur so infrequently, perhaps only a few hours per century, that they carry only a small portion of the total quantity of sediment transported over a long period of time. The most effective sediment transporting discharges over a period of years are those which transport sediment at a moderately high rate and occur as frequently as a few days per year.

The specific shape and characteristics of the sediment transport function, A, and the frequency of discharges, B, vary greatly from river to river. Thus, the range and occurrence of the most effective sediment transporting discharges also vary greatly from one river to another. At one extreme are those rivers which carry some sediment at even the smallest discharge and do not have a particularly wide range of floods. In such streams, effective discharges may occur for several weeks to perhaps months per year. At the other extreme are those rivers where only relatively rare, extreme floods are sufficient to move the available sediment particle sizes. In such streams, the range of effective sediment transporting discharges may occur only a few days per century.

The Wolman-Miller model for magnitude and frequency of effective sediment transport discharges has been evaluated by numerous investigations, [*Pickup and Warner*, 1976, *Baker*, 1977, *Wolman and Gerson*, 1978, *Newson*, 1980, and *Miller*, 1990]. *Nash* [1994] found 138 citations to the original paper between 1960 and 1990. The vast majority of the instances where the model has been evaluated using field measurements have considered only suspended sediment transport. Measured suspended sediment concentration, in some cases collected daily, are available for several gaging stations in the United States over periods of as long as 40 years. These comparisons typically agreed quite well with the model. There is a well-defined range of relatively frequent discharges which transport most of the suspended sediment over a period of years. Furthermore, the duration of the effective discharge commonly varies from several days per year to perhaps one day in several years, [see *Wolman and Miller*, 1960, and *Pickup and Warner*, 1976].

Evaluations of streamflow magnitude and frequency based upon suspended sediment transport consider 90 percent or more of the sediment transported out of a drainage basin and, thus, are probably a good approximation of the effective discharges for watershed denudation. Such evaluations, however, are not appropriate tests for the second hypothesis of the Wolman-Miller model, namely that the effective discharges are the channel-forming discharge, because most of the suspended sediment is finer than the material composing river bed and banks. This distinction is especially significant for gravel-bed rivers where clay, silt, and fine sand commonly represent 90 percent of the total sediment load and are transported in suspension even at relatively small discharges, whereas the river bed and banks consist of much coarser sized particles that may only be transported by flows equalled or exceeded a few days or less per year.

Analysis of bed-material transport magnitude and frequency in gravel-bed rivers has been limited due to substantial uncertainty in the transport rate at a specific discharge. Extensive bedload transport measurements have been made at relatively few sites, approximately 10, in North America, and then only for a period of a few years to at most a decade or so, [*Milhous*, 1973, *Leopold and Emmett*, 1976, *Jones and Seitz*, 1980, and *Andrews*, 1994]. Thus, one must rely upon establishing a relation between flow and bedload transport rate, which can then be applied over a longer period of recorded discharges. For some of the bedload measurement sites, however, the period of record for discharges are insufficient to adequately define the frequency of moderately large to extreme flows. Given the very large commitment of funds and time needed to measure bedload transport rates in a gravel-bed river over a wide range of relatively large discharges, it is improbable that there will be a significant number of sites where the magnitude and frequency of measured bedload transport will be determined solely from measurements.

Accordingly, the analysis of bedload transport magnitude and frequency must rely, to a substantial degree, upon calculated transport rates at streamflow gaging stations with long periods of record. This is the approach taken by the investigation described herein. The approach depends upon the ability to predict these bedload transport rates in gravel-bed rivers to a reasonable degree of accuracy.

The primary objective of this study is to formulate a regime of streamflows sufficient to maintain the existing bankfull channel for a given quantity and particle size distribution of sediment supplied to the channel. The approach is to analyze the magnitude and frequency of bed-material transport in gravel-bed rivers that are typical of the western United States, and determine whether the bankfull channel is related to the range of effective transporting discharges. This approach is, in fact, an evaluation of the Wolman-Miller hypotheses as they apply to gravel-bed rivers. A second objective concerns only the bankfull discharge, i.e. the conveyance of the naturally formed channel. In many, perhaps most instances, other channel attributes, e.g. channel sinuosity or relative pool depth, also may support significant or essential resources. Maintaining a given bankfull channel by providing the regime of streamflows necessary to move the several sizes of particles present in the bed and banks may not be sufficient to maintain the aquatic and riparian resources one might desire to preserve in a given stream reach. Additional mechanical or fluid dynamical processes as well as biological activity may be essential to form and maintain particular channel resources. Thus, the streamflow regime that preserves the magnitude and frequency of transport rate is a necessary, although perhaps not a sufficient condition for maintaining all channel resources.

2. CHARACTERISTICS OF RIVER REACHES STUDIED

The basic hypothesis of the Wolman-Miller model and the principle which this study relies upon to design a regime of channel maintenance flow is that similar river channels will have similar magnitudes and frequencies of transport rates. Conversely, differences between stream channels are a consequence, wholly or in part, of different magnitudes and frequencies of transport rate. Accordingly, the appropriate test of this hypothesis must concern a group or class of identifiable, i.e. similar, rivers and streams. For this study, single-thread gravel-bed rivers with mobile beds and stable banks have been selected, because they are common throughout the mountainous areas of the western United States and frequently subject to extensive flow regulation and diversion.

Andrews [1984] determined the hydraulic geometry relation of 24 gaging stations located on gravel-bed rivers in Colorado. Most of the information required to determine the magnitude and frequency of transport had been collected at these gaging stations. Eleven streams considered in the previous study were eliminated for various reasons, including a relatively short period of record, appreciable quantities of sand in the bed-material, or a complicated stage-discharge relation affected by the backwater of a tributary. Four river reaches not included in the previous investigation have been added.

All of the stream reaches studied are in mountainous parts of Colorado and are typical of alluvial gravel-bed streams throughout the Rocky Mountain region. Several geomorphic factors, including bed and bank material, floodplain development and long-term stability of the riverbed elevation were considered when selecting the study reaches. The channel bed and banks were composed primarily of sediment transported and deposited by fluvial action. Many reaches contain a small fraction of very coarse material, including boulders, which did not appear to be moved by the stream except perhaps during the most extreme floods. A well-defined floodplain indicative of the bankfull elevation was an essential characteristic of all reaches selected for study. Although the floodplain was well-defined along all streams, floodplains frequently were discontinuous and had limited areal extent. In each study area, three to five cross sections were surveyed in relatively straight parts of the reach. Longitudinal profiles of the water surface and bankfull elevation were also surveyed through a reach of approximately 30 channel widths that included a streamflow-gaging station. The bankfull discharge of the reach was determined from the bankfull longitudinal profile and the stage-discharge relation at the gage. The bankfull hydraulic characteristics of the several cross sections were calculated using the surveyed bankfull cross sections, the bankfull discharge, and mean reach slope. The size distribution of the riverbed surface was determined by a random sampling method [*Wolman*, 1954]. The study reaches and their associated hydraulic and bed-material transport characteristics are summarized in Table 1.

Bankfull discharges range from 0.70 to 85.2 m³/sec. Median bed-material sizes vary from 24 to 91 mm, and bankfull water surface slopes vary from 0.0014 to 0.26. Computed mean annual bed-material loads are generally small relative to the contributing drainage area and vary from 0.02 to 166 ton/km²/year.

The seventeen reaches were selected for study based upon a well-defined bankfull channel in the vicinity of the gaging station, as well as an appreciable period of record uninterrupted by flow diversion and storage. The natural streamflow regime of approximately one-half of the study reaches has been altered to some degree by flow depletion. In all instances, these developments have existed for several decades during which the channels will have adjusted to the altered regime. The period of records shown in table 1 refer to a generally constant extent of streamflow alteration.

3. COMPUTATION OF BED-MATERIAL TRANSPORT RATE

Parker et al. [1982] formulated an empirical bedload transport function for poorly-sorted of gravel and cobbles. The Parker bedload function is

$$W^*_i = 0.0025 \exp[14.2(\phi_i - 1) - 9.28(\phi - 1)^2];$$
$$0.95 < \phi_i < 1.65 \quad (1a)$$

and

$$W^*_i = 11.2[1 - (0.822/\Phi_i)^{4.5}];$$
$$\Phi_i > 1.65 \quad (1b)$$

Table 1. Measured and computed bed-material transport and hydraulic characteristics at selected gravel-bed rivers.

USGS station number	Station name	Period of Record (water years)	Drainage Area (km)²	Mean Annual Discharge (m³/sec)	Bankfull Discharge (m³/sec)	Duration of Bankfull Discharge (%)	Effective Discharge (m³/sec)	Mean Annual Bedload Transport (Ton/yr)
06614800	Michigan River nr. Cameron Pass	1974-93	3.96	0.086	0.70	1.76	0.73	660
06620000	North Platte River nr. Northgate	1916-93	3706	12.3	85.2	1.21	88.4	67
06724500	Lefthand Creek nr. Boulder	1930-31, 1947-53, 1956-57, 1977-80	135	1.06	4.9	3.53	4.82	1400
06725500	Middle Boulder Creek at Nederland	1908-93	93.7	1.53	9.5	1.74	9.9	2300
06748530	Little Beaver Creek nr. Rustic	1961-73	31.9	0.23	1.6	2.10	1.56	290
06748600	South Fork Cache La Poudre River nr. Rustic	1957-79	239	1.80	9.4	3.25	10.3	770
07083000	Halfmoon Creek nr. Malta	1947-93	61.1	0.82	7.08	0.42	5.27	1380
09022000	Fraser River at upper station nr. Winter Park	1969-73, 1985-93	27.2	0.40	2.69	1.70	2.69	160
09035900	South Fork Williams Fork nr. Leal	1966-93	70.5	0.92	8.36	0.32	5.86	1080
09036000	Williams Fork nr. Leal	1940-93	231	2.72	22.6	1.1	23.5	210
09074800	Castle Creek above Aspen	1970-93	83.4	1.23	4.45	6.51	4.39	6200
09078100	North Fork Fryingpan River abv. Cunningham Creek nr. Norrie	1964-78	31.1	0.55	3.17	4.66	3.74	16
09078200	Cunningham Creek nr. Norrie	1964-78	18.4	0.29	2.52	1.73	2.48	150
09081600	Crystal River abv. Avalanche	1956-93	433	8.44	49.0	2.63	55.8	210
09124500	Lake Fork at Gateview	1938-93	865	6.68	42.0	1.77	46.4	120
09249750	Williams Fork at mouth nr. Hamilton	1985-93	1085	5.83	46.7	1.18	49.0	130
09253000	Little Snake River nr. Slater	1944-93	738	6.46	72.2	0.37	55.4	84

where $W^*_i = [q_{bi} (\gamma_{s/\gamma} - 1)] / [f_i g^{1/2} (RS)^{3/2}]$, $\phi_i = \tau^*_i / \tau^*_{ri}$, q_{bi} is the volummetric bedload transport rate of the ith particle fraction per unit width of channel, f_i is the percent of bed particles in the ith fraction, and g is the acceleration of gravity.

An essential aspect of this approach was the development of a reference dimensionless shear stress, τ^*_{ri}, such that $\tau^*_{ri} = f(d_i/d_{50})$ where d_i is the diameter of particles of the ith size fraction of bed material and d_{50} is the median particle diameter of bed material. *Parker et al.* [1982] found that the use of ϕ_i, rather than τ^*_i, resulted in a similarity collapse, so that W^*_i is approximately a single valued function of ϕ_i. The Parker bedload function for the domain $\phi_i > 1.65$ was derived by fitting ϕ_i and W^*_i to the Einstein bedload function. For the domain $0.95 < \phi_i < 1.65$, the Parker bedload function was derived from bedload transport rates measured in Oak Creek [*Milhous*, 1973].

As formulated, equation (1) should apply to any mixture of gravel-sized material, from uniform to poorly-sorted, so long as the reference shear stress correctly represents the response of the bed material to the fluid forces (i.e. when $\phi_i = 1$, then $W^*_i = 0.0025$). Therefore, equation 1 should be generally applicable. The relation $\tau^*_i = f(d_i/d_{50})$, however, may vary from stream to stream depending upon the nature of the bed material, particle size-distribution, shape and packing. The extremely steep slope of the ϕ_i vs W^*_i relation, when $\phi_i \leq 1.65$, however, means that relatively small errors in the reference shear stress, τ^*_{ri}, will lead to substantial errors in the predicted bedload transport rate. Thus, determination of the correct values of τ^*_{ri} for a given mixture of bed particles is essential, especially when calculating marginal bedload transport rates.

Parker et al. [1982] determined the dependence of τ^*_{ri} on d_i/d_{50} for Oak Creek by calculating the value of τ^*_{ri} at a dimensionless transport rate of $W^*_i = 0.0025$. The particle size distribution of subsurface bed material was used for most of the analysis of Oak Creek. Their approach, however, is not limited to the subsurface material. The size distribution of surface bed material can be used, and is equally valid [*Andrews and Parker*, 1987]. The median particle size of surface bed material is used to scale the relative particle protrusion because it represents the assemblage of bed particles from which the bedload material is derived, [*Wiberg and Smith*, 1987, *Wilcock and McArdell*, 1993, and *Andrews*, 1994]. This approach avoids the need to assume that the particle size distribution of bedload and subsurface material are similar.

Wilcock and Southard [1988], *Kuhnle* [1992], and *Andrews* [1994] have taken a slightly different approach than *Parker et al.* [1982] used to determine the function $\tau^*_{ri} = f(d_i/d_{50})$. Instead of calculating the value of τ^*_{ri} at $W^*_i = 0.0025$, they varied τ^*_{ri} to obtain the best fit of equation 1 to a wide range of measured transport rates of each ith size fraction. Bedload transport rates have been measured at three of the river reaches selected for this analysis, Middle Boulder Creek at Nederland, Lefthand Creek near Boulder, and South Fork Cache La Poudre River near Rustic. Using these measurements, the variation of τ^*_{ri} with (d_i/d_{50}) was determined for each site. The empirically determined functions, $\tau^*_{ri} = f(d_i/d_{50})$, are shown in Figure 2 together with the previously determined functions for Oak Creek, [*Parker et al.*, 1982] and Sagehen Creek [*Andrews*, 1994]. The respective equations are

$$\tau^*_{ri} = 0.033 \, (d_i/d_{50})^{-0.982} \qquad (2)$$

for Oak Creek,

$$\tau^*_{ri} = 0.0384 \, (d_i/d_{50})^{-0.887} \qquad (3)$$

for Sagehen Creek near Truckee, CA,

$$\tau^*_{ri} = 0.0354 \, (d_i/d_{50})^{-0.975} \qquad (4)$$

for Middle Boulder Creek at Nederland,

$$\tau^*_{ri} = 0.0376 \, (d_i/d_{50})^{-0.994} \qquad (5)$$

for Lefthand Creek near Boulder,

$$\tau^*_{ri} = 0.035 \, (d_i/d_{50})^{-0.942} \qquad (6)$$

for the South Fork Cache La Poudre River near Rustic.

Among the five relations, the one determined for Middle Boulder Creek at Nederland, is the nearest to an "average" relation over the range of (d_i/d_{50}) values. For a given ratio (d_i/d_{50}), values of the referenced dimensionless shear stress determined for the 5 streams vary no more than ± 10 percent from the Middle Boulder Creek relation, see Figure 2.

Bed-material transport rates for particle size fractions from 4 mm-180 mm over the range of recorded discharges were computed for the 17 sites using the Parker bedload function. The reference shear stress function, equation 5, determined for Middle Boulder Creek was applied for all sites, except Lefthand Creek and the South Fork Cache La Poudre where the site-specific functions were applied.

Sensitivity of the computed magnitude and frequency of bed-material transport to uncertainty in the measured reach hydraulic characteristics and the streamflow regime due to an insufficient period of record were evaluated using the

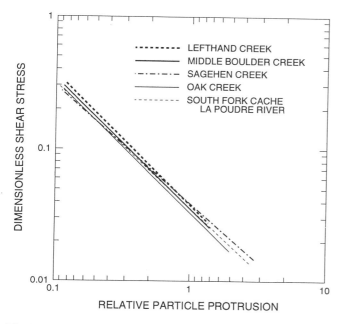

Fig. 2. Comparison of reference dimensionless shear stress, τ^*_{ri}, versus relative particle protrusion (d_i/d_{50}) determined for five gravel-bed rivers.

Middle Boulder Creek gage. This gage was selected for the sensitivity analysis because the most extensive measurements of bed-material transport and the longest period of record have been collected there.

4. EFFECTIVE BED-MATERIAL TRANSPORTING DISCHARGES

The quantity of bed-material in each particle size fraction transported by increments of discharge over the period of record at each gaging station may be determined either by calculating the quantity of bed material transported each day during the period of record or by the flow duration - sediment transport method. Both approaches will give the same result, and have been used in this study. Calculating the quantity of bed-material day-by-day preserves valuable information concerning temporal variability. In many instances, such as the evaluation of a proposed project, however, a flow duration relation will be more readily available and reliable than a time series of estimated daily values. The observed relation between dimensionless shear stress and discharge at a gage was combined with the bedload transport function for each particle size fraction. The duration of a given discharge increment over the period of record was multiplied by the quantity of bed material in a size fraction transported by the discharge per unit time. Results of these calculations at six gaging stations are summarized in Figures 3a-f. The six graphs were chosen to illustrate the range of results. The several curves in each graph represent curve C in Figure 1 for the identified particle size fractions. The area between adjacent curves is the mean annual quantity of bed material transported in the size-fraction. The mean annual quantity of bed material transported in all size fractions is the total area under the top curve.

In general, the range of significant bed-material transporting discharges is well-defined. Bed material begins to move on average at discharges approximately 60 percent of the bankfull discharge and 2-3 times the mean annual discharge. Many of the effective discharge plots are skewed to the larger discharges; however, discharges greater than twice the effective value transport a relatively small fraction of the mean annual load. For all river reaches, the middle 80 percent of the mean annual load is transported, on average, by flows between approximately 0.8-1.6 times the bankfull discharge. For the 17 streams studied, the middle 80 percent of the mean annual bed-material load is transported on average during 15.6 days per year by discharges that represent 27 percent of the mean annual runoff and range from 17 to 40 percent of the mean annual runoff. On average, discharges less than bankfull transport 39 percent of the mean annual bed-material load.

The interval of discharge that transports the largest portion of the mean annual bed-material load over a period of years, i.e. the modal value, is termed the effective discharge [*Andrews*, 1980]. It should be noted that no special significance is attached solely to a single discharge. The effective discharge is a simple and straightforward representative for the range of discharges which transport the vast majority of the mean annual bed-material load over a period of years. The effective discharge computed for each river reach is listed in Table 1, and compared in Figure 4 with the field-determined bankfull discharge. Generally, there is a good one-to-one agreement between the field-determined bankfull discharge and the computed effective discharge. The mean normalized difference between the effective and bankfull discharges is 11 percent. This result supports the Wolman-Miller hypothesis that river channel characteristics, and specifically, the capacity of the naturally formed channel, will be determined by the range of discharges that transport most of the bed-material over a period of years. The close agreement between the effective and bankfull discharge for the 17 river reach study demonstrates that the range of effective bed-material transporting discharges are the flows which construct and maintain these channels.

The effect of uncertainty in the hydraulic characteristics

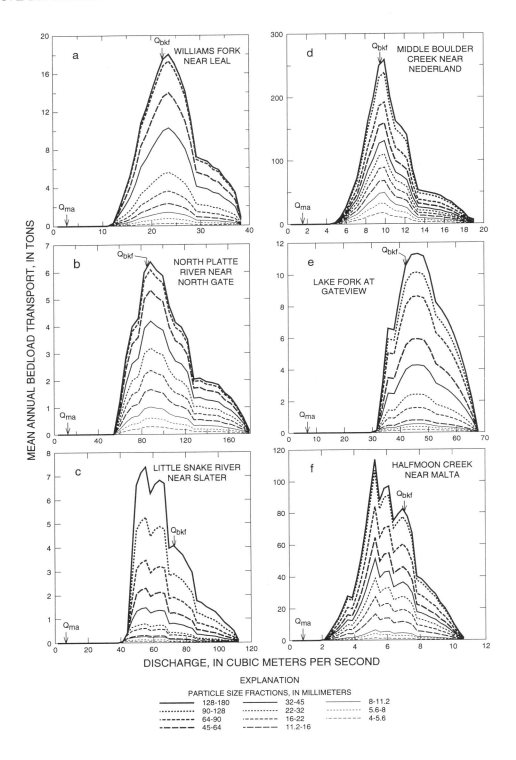

Fig. 3. Accumulative quantity of bed material in size fractions transported by increments of discharge in the (a) Williams Fork near Leal, CO; (b) North Platte River near Northgate, CO; (c) Little Snake River near Slater, CO; (d) Middle Boulder Creek at Nederland, CO; (e) Lake Fork near Gateway, CO; and (f) Halfmoon Creek near Malta, CO.

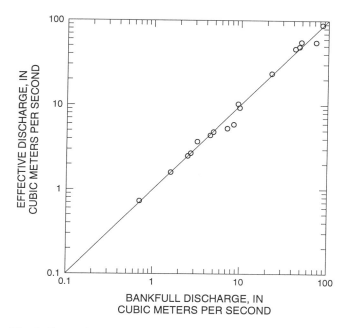

Fig. 4. Comparison of effective discharge and bankfull discharge.

of a reach was evaluated by varying the water surface slope by ± 10 percent. This test also is equivalent to a ± 10 percent change in the flow depth and median bed-material particle size. Uncertainty in the reach hydraulic characteristics has a very significant effect, as large as ± 300 percent on the computed mean annual bed-material load, but only a very small effect on the effective discharge or range of dominant bed-material transporting discharge. The computed effective discharge is 9.09 m^3/sec using a 10 percent increase in water surface slope and 9.66 m^3/sec using a 10 percent decrease in water surface slope compared to 9.94 m^3/sec as shown in Figure 3d.

As described above, the dominant bed-material transporting discharges, i.e. those discharges that transport the central 80 percent portion of the mean-annual bed-material load, occur about 15 days per year on average. There is, however, substantial year-to-year variation in the duration of bed-material transporting discharges and, correspondingly, the quantity of bed material transported. Years having relatively large runoff typically will have many more than the average number of days with discharge in the range of dominant bed-material transporting flows. Conversely, relatively small runoff years frequently will have few or no days of dominant bed-material transporting discharge. The accumulative distribution of days over the period of record when flow occurred within the range of dominant bed-material transporting discharges are shown for the six gaging stations with the longest periods of record in Figure 5a-f. The abscissa values were computed by ordering the annual runoff from smallest to largest and divided by the length of record. The ordinate values were computed by summing the number of days of dominant bed-material transporting discharge from the smallest to the largest annual runoff years and divided by the total number of days the dominant discharge occurred.

Approximately 50 percent of all days when the daily mean flow was within the range of dominant bed-material transporting discharges occur in about 20 percent of the years. Conversely, dominant bed-material transporting discharges did not occur during approximately one-third of the years. Those years when dominant bed-material transporting discharges occurred on a relatively large number of days were, with rare exception, years during which annual runoff significantly exceeded the long-term mean. For example, the 17 largest runoff years recorded at the Middle Boulder Creek at Nederland gage, 20 percent of the period of record, had 57 percent of all days with discharges within the range of dominant bed-material transporting flows. Furthermore, 51 percent of all bed material transported during the period of record, 1908-1993, occur in the 17 largest runoff years.

Finally, an additional factor complicates a simple description of the occurrence of the dominant bed-material transporting discharges. In those years with several days of dominant discharge, the period of significant bed-material transport is frequently broken into two, three, and, sometimes, four segments of a few days each and occurring over a period of 30-40 days.

5. CHANNEL MAINTENANCE FLOWS

The preceding analysis has identified the magnitude and frequency of discharges which transport most of the bed-material load over a period of years in gravel-bed rivers common throughout the western United States. These discharges have a relatively narrow range, between 0.8 and 1.6 times the bankfull discharge, and occur approximately 15 days per year on average. Furthermore, the bankfull discharge of the rivers studied lies within the range of dominant bed-material transport discharges and is very nearly equal to the interval of discharge which transports the largest portion of the long-term bed-material load. Therefore, it is concluded that the dimensions, morphology, and other physical characteristics of these gravel-bed rivers are primarily determined by a well-defined, relatively narrow range of discharges. These results establish the basis for formulating a regime of streamflows

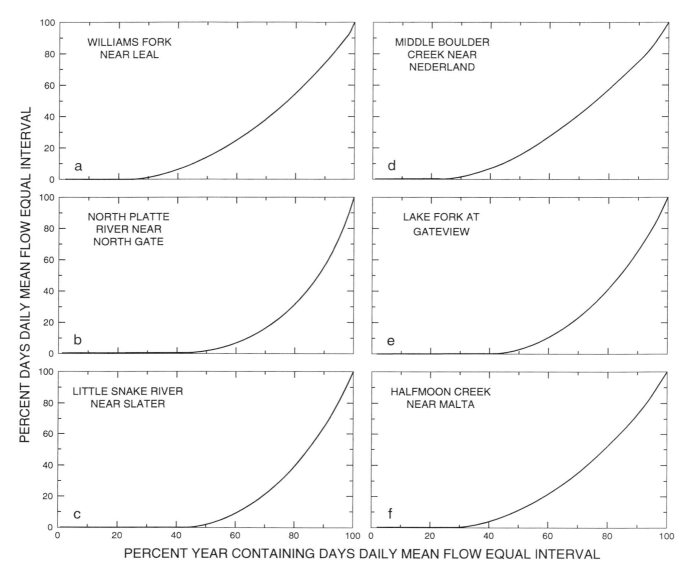

Fig. 5. Accumulative distribution of days when dominant bed-material transporting discharges occurred during the period of record, (a) Williams Fork near Leal; (b) North Platte River near Northgate; (c) Little Snake River near Slater; (d) Middle Boulder Creek at Nederland; (e) Lake Fork near Gateway; and (f) Halfmoon Creek near Malta.

which will substantially maintain the existing physical characteristics of these river channels when the natural streamflows are appreciably altered. The focus of our analysis will be the situation where natural streamflows are reduced; however, the approach is equally suitable for the situation where natural streamflows are increased. The following analysis and discussion will consider the issues and difficulties that arise in formulating a regime of channel maintenance flows.

Before proceeding, it is important to restate that the regime of channel maintenance flows is a necessary, but, perhaps, not a sufficient condition to maintain an existing channel. Additional factors frequently influence or determine specific characteristics of a channel.

The circumstances for which one would need to formulate a regime of channel maintenance flows are quite diverse. The availability of information describing the existing and future streamflows and supply of sediment to the channel will vary greatly. The ideal situation is the one in which detailed information exists concerning the natural flow regime, bed-material transport, as well as the type of structure and the manner of operation that will store or divert flow. With this information, magnitude and frequency of bed-material transporting discharges can be calculated directly for the reach of interest. The quantity of bed material supplied to the affected reach will depend upon the structure built to store or divert flow. A reservoir will usually trap all gravel sized material except where provisions for sluicing are incorporated into a dam. In contrast, flow diversion structures typically retain significantly less coarse sediment than a dam.

Formulating a practical operational regime of channel maintenance flows from an analysis of the magnitude and frequency of bed-material transport is not a simple exercise. The channel maintenance regime will be part of a development project whose primary purpose is to store and/or divert flow from the stream. Such projects inevitably involve significant economic, legal, and operational constraints on the possible instream flow regime. A proposed regime of channel maintenance flow will be evaluated and judged on whether the flows are (1) efficient and (2) predictable. An efficient regime will maintain the desired channel characteristics with the minimum quantity of water possible. The regime of channel maintenance flows also must be predictable. That is, the channel maintenance discharge on a given day must be specified. The numbers of times when mean daily flow was within the range of the dominant bed-material transporting discharge on a given date over the entire period of record are shown in Figure 6a-f, for several of the rivers considered in the study. As shown in Figure 6a-f, even during that period each year when discharges are most likely to occur within the range of the dominant bed-material transporting flows, the probability of occurrence on any given day is less that 50 percent. For most of the snowmelt period, the probability of a dominant bed-material transporting discharge occurring on a given day is significantly less than 50 percent. Consequently, any channel maintenance regime structured to retain all streamflows for a specific period of time each year, for example May 15-May 31, will be inefficient most of the time, because the actual streamflow is less than the range of dominant bed-material transporting discharges. Such a regime is predictable; however, it is also inefficient because the flows retained in the channel during most years will be insufficient to transport appreciable quantities of sediment. Alternative channel maintenance regimes that are more efficient are less predictable. In fact, efficiency and predictability are in nearly all practical circumstances mutually exclusive. For the type of stream considered in this study, gravel-bed streams with snowmelt floods, any channel maintenance regime will require a balancing of efficiency and predictability.

As water supplies become more extensively developed and competition for aquatic resources intensifies, it seems certain that greater efficiency will be demanded at the expense of predictability. Accordingly, operation of a diversion or storage structure will become much more complex than is now the common practice. The principal consequences of this shift will be a need to develop sophisticated diversion structures which can bypass the dominant bed-material transporting flows whenever they occur.

A highly efficient and effective channel maintenance flow regime would retain the historical magnitude and frequency of bed-material transport utilizing the least quantity of water. One possible channel maintenance regime for Middle Boulder Creek would be the natural flow whenever it is sufficient to move bed-material particles. In Middle Boulder Creek, bed-material transport begins at about 4.8 m^3/sec, which is approximately 3 times the mean annual discharge and one-half of the bankfull discharge.

The resulting channel maintenance flow regimes are compared with the natural flow in Figure 7 for 3 years, 1915 average runoff, 1954, below average runoff and 1957, well above average runoff. The channel maintenance flow regime has essentially the same magnitude and frequency of bed-material transporting discharge.

Over the entire period of record the proposed channel maintenance flows equal 35 percent of the long-term mean annual runoff. During 3 of the 85 years of record, natural daily mean flows did not exceed 3 times the mean annual discharge, and, therefore, under the proposed regime no channel maintenance flow would have occurred. Conversely, during several of the largest runoff years, the natural flow exceeded 3 times the mean annual flow for several weeks.

6. SUMMARY AND CONCLUSIONS

1. A method for determining a regime of channel maintenance flows that will preserve the physical characteristics of self-formed streams when an appreciable quantity of the natural flow is diverted or stored has been formulated. The method relies upon identifying the magnitude and frequency of bed-material transporting discharges in the natural channel, and then, retaining these flows following any modification of the natural flow regime.

2. To demonstrate this method, the magnitude and frequency of bed-material transporting discharges were computed for 17 gravel-bed rivers typical throughout the Rocky Mountain Region. The range of flow which transported the vast majority of bed-material over a period

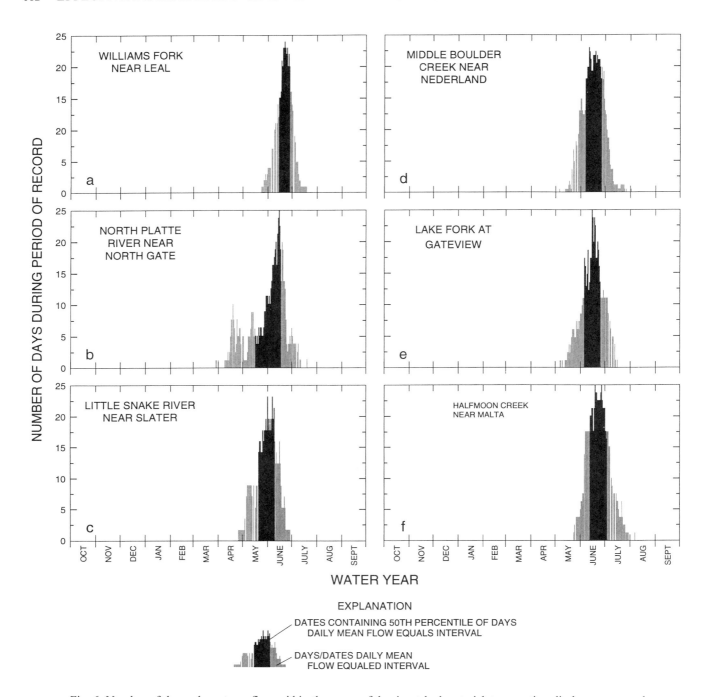

Fig. 6. Number of days when streamflow within the range of dominant bed-material transporting discharge occurred on a given date during the period of record, (a) Williams Fork near Leal; (b) North Platte River near Northgate; (c) Little Snake River near Slater; (d) Middle Boulder Creek at Nederland; (e) Lake Fork near Gateway; and (f) Halfmoon Creek near Malta.

of years was generally well-defined. On average, those flows that transported the modal 80 percent of the long-term mean bed-material load ranged from 0.8 to 1.6 times of bankfull discharge and occurred 15.6 days per year.

There are, however, substantial year-to-year variations in the duration of bed-material transporting discharge and correspondingly, the quantity of bed material transported.

3. The bankfull discharges of the 17 gravel-bed rivers

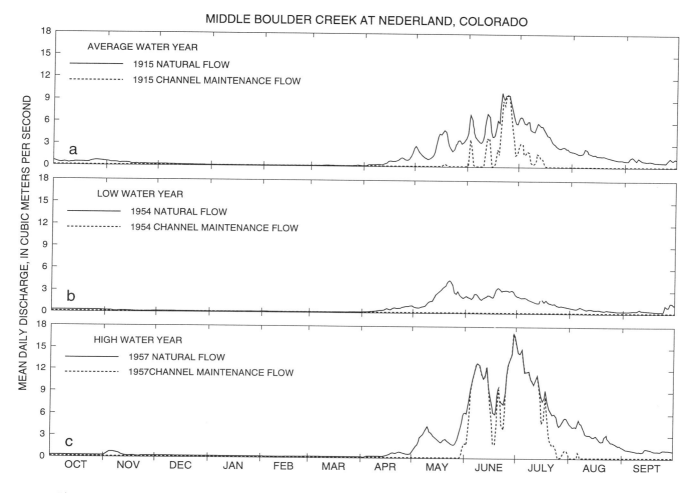

Fig. 7. Channel maintenance flows determined for Middle Boulder Creek at Nederland during 3 years: (A) 1915, (B) 1954, and (C) 1957.

are in excellent one-to-one agreement with the interval of discharge that carries the largest quantity of bed-material over the period of record. Based on this agreement it was concluded that the range of effective bed-material transporting discharges are the flows which construct and maintain these channels over time.

4. In practice, an operational regime of channel maintenance flow will involve a balancing between (1) minimizing the quantity of water required for instream flows, and (2) the predictability of these flows.

5. A highly efficient channel maintenance regime consisting of all flows sufficient to initiate bed particle motion and greater was evaluated. This regime would preserve the natural range of dominant bed-material transporting discharges with 35 percent of the mean annual flow. A substantial majority of channel maintenance flows, both number of days and volume, would occur during large runoff years. Little or no maintenance flows in excess of the baseflow would occur during years with below average runoff.

Acknowledgments. We appreciate the support and encouragement of Andrew Walch and Larry Schmidt. This investigation benefitted greatly from numerous discussions with Luna Leopold and Dave Rosgen. Their comments and suggestions are appreciated very much. Peter Wilcock and John Potyondy reviewed a draft manuscript and made several especially helpful suggestions.

REFERENCES

Ackers, P., River regime: Research and application, *J. of the Instit. of Wat. Engineers, 26,* 257-281, 1972.

Andrews, E.D., Effective and bankfull discharge of streams in the Yampa River Basin, Colorado and Wyoming, *J. of Hydr., 46,* 311-330, 1980.

Andrews, E.D., Bed-material entrainment and hydraulic geometry of gravel-bed rivers in Colorado, *Geol. Soc. of Am. Bull.*, 95, 371-378, 1984.

Andrews, E.D., Marginal bedload transport in a gravel-bed stream, Sagehen Creek, California, *Wat. Resour. Res.*, 30, 2241-2250, 1994.

Andrews, E.D. and Parker, G., Formation of a coarse surface layer as the response to gravel mobility, in Thorne, C.R., Bathurst, J.C., and Hey, R.D. eds., *Sediment Transport in Gravel-bed Rivers,* John Wiley and Sons, New York, 269-300, 1987.

Baker, V.R., Stream-channel response to floods, with examples from central Texas, *Geol. Soc. Am. Bull.*, 88, 1057-1071, 1977.

Galay, V.J., Causes of river bed degradation, *Wat. Resour. Res.*, 19, 1057-1090, 1983.

Hirsch, R.M., Walker, J.F., Day, J.C., and Kallio, Raimo, The influence of man on hydrologic systems, in Wolman, M.G., and Riggs, H.C. eds, *Surface Water Hydrology:* Boulder, Colorado, Geol. Soc. of Am., The Geology of North America, 0-1, 329-359, 1990.

Jones, M.L. and Seitz, H.R., Sediment transport in the Snake and Clearwater Rivers in the vicinity of Lewiston, Idaho: *U.S. Geol. Surv. Water-Resour. Invest. Open-File Rept. 80-690,* 179 p., 1980.

Kuhnle, R.A., Fractional transport rates of bedload on Goodwin Creek, in *Dynamics of Gravel-bed Rivers,* edited by P. Billi, R.D. Hey, C.R. Thorne, and P. Tacconi, John Wiley and Sons, Ltd., New York, NY, 141-155, 1992.

Leopold, L.B. and Maddock, Thomas, The hydraulic geometry of stream channels and some physiographic implications, *U.S. Geol. Surv. Prof. Paper 252,* 57 p., 1953.

Leopold, L.B. and Emmett, W.W., Bedload measurements in the East Fork River, Wyoming; *Proc. of the Nat. Acad. of Sci.*, 73, 1000-1004, 1976.

Milhous, R.T., Sediment transport in a gravel-bottom stream, Ph.D. Thesis, Oregon State University, 238 p., 1973.

Miller, A.J., Flood hydrology and geomorphic effectiveness in the central Appalachians, *Earth Surf. Proc. Landforms,* 15, 119-134, 1990.

Nash, D.B., Effective sediment transporting discharge from magnitude-frequency analysis, *J. of Geol.*, 102, 79-95, 1994.

Newson, M., The geomorphic effectiveness of floods--a contribution stimulated by two recent events in Mid-Wales, *Earth Surf. Proc. Landforms,* 5, 1-16, 1980.

Parker, Gary, Klingeman, P.C., and McLean, D.C., Bedload and size distribution in paved gravel-bed streams, *J. of the Hydraul. Div., Proc. Am. Soc. of Civil Eng.*, 108, 544-571, 1982.

Petts, G.E., Complex response of river channel morphology subsequent to reservoir construction, *Prog. in Phys. Geog.,* 3, 329-362, 1979.

Pickup, G. and Warner, R.F., Effects of hydrologic regime on magnitude and frequency of dominant discharge, *J. of Hydro.*, 29, 51-75, 1976.

Reiser, D.W., Ramey, M.P., and Wesche, T.A., Flushing flow needs in regulated streams, in Gore, J.A. ed., *Alternatives in Regulated Flow Management,* CRC Press, Boca Baton, FL, Chap. 4, 91-135, 1989.

Rosgen, D.L., Silvey, H.L., and Potyondy, J.P., The use of channel maintenance flow concepts in the Forest Service, American Institute of Hydrology, *Hydrol. Sci. and Tech.: Short papers,* 2, 19-26, 1986.

Wiberg, P.L. and Smith, J.D., Calculation of critical shear stress for motion of uniform and heterogeneous sediments, *Wat. Resour. Res.*, 23, 1471-1480, 1987.

Wilcock, P.R. and McArdell, B.W., Surface-based fractional transport rates: Mobilization thresholds and partial transport of a sand-gravel sediment, *Wat. Resour. Res.*, 29, 1297-1312, 1993.

Wilcock, P.R., and Southard, J.B., Experimental study of incipient motion in mixed-size sediment, *Wat. Resour. Res.*, 24, 1137-1151, 1988.

Williams, G.P. and Wolman, M.G., Effects of dams and reservoirs on surface-water hydrology; Changes in rivers downstream from dams, *U.S. Geol. Surv. Prof. Paper 1286,* 83 p., 1984.

Wolman, M.G., A method of sampling coarse river bed material, *Am. Geophys. Union Trans.*, 35, 951-956, 1954.

Wolman, M.G., and Miller, J.P., Magnitude and frequency of forces in geomorphic processes, *J. of Geol.*, 68, 54-74, 1960.

Wolman, M.G., and Gerson, R., Relative scales of time and effectiveness of climate in watershed geomorphology, *Earth Surf. Proc. Landforms,* 3, 189-208, 1978.

E. D. Andrews, U.S. Geological Survey, 3215 Marine Drive, Boulder, CO 80303

James M. Nankervis, U.S. Forest Service, 240 W. Prospect St., Fort Collins, CO 80526

Managing Bedload Sediment in Regulated Rivers: Examples from California, U.S.A.

G. Mathias Kondolf

Department of Landscape Architecture, University of California, Berkeley, California

The continuity of bedload transport has been profoundly altered in many rivers by construction of reservoirs and mining of sand and gravel for construction aggregate from the active channel. These impacts are especially well illustrated in California because of the extent of reservoir construction and strong demand for aggregate. As a result, many river channels have become starved of bedload sediment and have incised. Salmonid spawning gravel has been lost, and bridges and other structures have been undermined. State policies regulating reservoir sedimentation and sediment sluicing from reservoirs are often contradictory and are not based on an understanding of the role of bedload sediment in maintaining the river system. Regulation of gravel mining from active channels has been ineffective at preventing resource degradation, despite numerous permit requirements from various federal, state, and local agencies. The principal burden for regulating gravel mining falls on local governments, which lack resources or technical expertise to analyze geomorphic effects of gravel mining. Despite their common occurrence and inter-related nature throughout California, the problems resulting from disruption of natural bedload transport have thus far been treated on a site-specific basis only, without explicit recognition of links among processes and impacts. Environmentally sound management of bedload sediment resources requires basin-level and regional analysis of bedload sediment sources and transport rates, recognition of hydrologic alterations caused by reservoirs, and identification of suitable aggregate sources (including quarries and recycled concrete rubble). Gravel mining should generally be prohibited from channels downstream of reservoirs.

1. INTRODUCTION

Most management-related studies of sediment have focused on fine sediment, silt and clay, on which pollutants commonly adsorb [*Jobson and Carey*, 1989], and which constitute the bulk of sediment transported by most rivers into reservoirs, reflected in the widespread use of the term "siltation" to describe reservoir sedimentation [*Mahmood*, 1987; *Vanoni*, 1975]. The terms "sediment" and "sedimentation" often have negative connotations among non-geologists probably because they have connotations of fine sediment. However, "sediment" properly includes particles ranging in size from clay to large boulders, and gravel- and cobble-sized sediment are critically important for aquatic habitat and river maintenance. For example, gravel of suitable size and sufficient quantity is needed by salmonids (salmon and trout) for spawning [*Kondolf and Wolman*, 1993]. Further, removal of bedload sediment from a river typically leads to channel incision, bank erosion, undercutting of structures, and a decline in the alluvial water table.

The transport of gravel and sand occurs largely as bedload and thus is interrupted by dams (even reservoirs with low trap efficiencies for suspended sediment), resulting in loss of reservoir storage capacity and elimination of gravel supply to downstream reaches. Dams also change the flow regime downstream, typically reducing flood peaks, potentially depriving downstream reaches of high flows needed to flush fine sediment from spawning gravel. In many rivers, sand and gravel have been mined from the river bed to provide aggregate, further reducing the availability of spawning gravel and inducing incision, resulting in undermining of bridges and other structures.

These problems occur in rivers throughout the world, but are particularly well illustrated in California because of the extent to which the state's rivers have been dammed, and the large volume of sand and gravel mined from these channels. Despite the recurrence of these problems in rivers throughout the state, to date these problems have not been recognized as related parts of a statewide problem, but have been addressed, often inconsistently, on a site-specific basis only. For example, in 1986, a release of gravel and sand from a dam on the Middle Yuba River was referred to as a release of "deleterious material" by one state agency (the Regional Water Quality Control Board) [*Ebasco Environmental*, 1989]. At the same time another state agency (the Department of Fish and Game) was artificially adding gravel to the Upper Sacramento River to restore spawning habitat lost below Shasta Dam, using gravel purchased from gravel miners on tributaries to the Upper Sacramento River.

The purpose of this paper is to review the extent of problems involved in managing bedload sediment (gravel and sand) on regulated rivers. While I discuss management strategies employed in other areas, the paper focuses on interactions among physical processes, human impacts, and the existing regulatory and management structure in California, based in part on data compiled in a comprehensive planning-level study of the problem [*Kondolf and Matthews*, 1993]. A more coherent management approach to these problems in California is timely, given plans to undertake extensive gravel enhancement projects under the Central Valley Project Improvement Act of 1991, and given widespread discontent with the existing system of management and regulation of gravel mining in the state, which gave rise to four proposed legislative amendments to the state's Surface Mining and Reclamation Act in 1994.

2. DEPLETION OF BEDLOAD SEDIMENT BY RESERVOIRS AND GRAVEL MINING

2.1 Reservoirs

Reservoirs can effect a range of changes on the downstream hydrograph depending upon local runoff patterns, reservoir size (in relation to seasonal runoff volumes), and reservoir operation rules (which vary with reservoir usage for municipal supply, irrigation, hydro-power production, or flood control). However, most reservoirs reduce the magnitude of (at least frequent) flood flows. Reservoirs trap all bedload sediment delivered to them except in the rare cases in which sediment pass-through strategies are employed, and a portion of the suspended load depending the reservoir's trap efficiency, which is based upon the ratio of reservoir capacity to volume of runoff [*Brune*, 1953].

2.2 Effects of Reservoirs on Bedload Transport

2.2.1 *Trapping of bedload sediment.* Reservoirs trap all bedload, releasing sediment-starved water to downstream reaches. This *hungry water* may entrain gravel, resulting in armoring, and erode bed and banks, resulting in incision. These effects are most pronounced below smaller reservoirs that have minimal effect on flood peaks. For example, on the Bear River below Camp Far West Reservoir, what was formerly a gravel bar (F. Meyer, California Department of Fish and Game, pers. comm. 1991) has coarsened to cobble and boulders.

2.2.2 *Alteration of flood hydrographs.* Sediment-transport capacity has been substantially reduced as a result of the reduction in flood magnitudes effected by large reservoirs. In the Sacramento-San Joaquin river system of California, the post-dam discharge with a 2-year recurrence interval, Q_2, varies from 0.05 to 0.65 of pre-dam values, the 10-year flood, Q_{10}, varies from 0.12 to 0.98 of pre-dam values, depending on reservoir size and operating rules [*Kondolf and Matthews*, 1993]. The progressive reduction in flood peaks resulting from construction of larger reservoirs is illustrated for the Mokelumne River in Figure 1.

Below large reservoirs that reduce all downstream floods, reduced flood peaks may result in such reduced sediment transport capacity that little gravel transport occurs and the principal downstream reservoir effects are encroachment of riparian vegetation and deposition in spawning gravel of fine sediment delivered to the main stem from tributaries. Downstream effects of reservoirs on bed sediment vary as a function of reservoir hydrologic effects. The coarsening observed on the Bear River occurred below Camp Far West Dam, where the Q_{10} is 0.98 of its pre-dam value. On the Merced River below New Exchequer Dam the Q_{10} is 0.12 of its pre-dam value, and spawning gravel are still abundant [*DWR*, 1994].

2.2.3 *Extent and distribution of reservoirs in California.* In California, only one major river (the Smith River) is undammed. There are 1212 dams that exceed 4.6 m in height and impound more than 37,000 m^3 of water, and, therefore, are large enough to be regulated by the state. There are another 144 dams operated by federal agencies. In the Sacramento and San Joaquin River basins, most tributaries have a large reservoir in the foothills (below 250 m elevation), transforming formerly lotic environments to

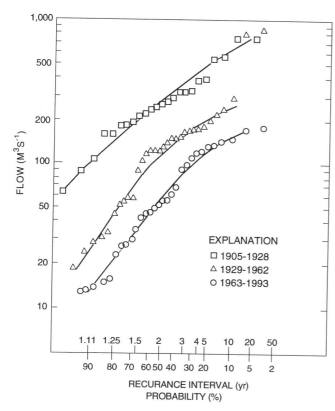

Fig. 1. Flood frequency curves for the Mokelumne River below Camanche Dam for three time periods: before Pardee Dam (1905-1928), after Pardee Dam but before Camanche Dam (1929-1962), and after Camanche Dam (1963-1993). (Adapted from the Federal Energy Regulatory Commission [1993]).

lentic environments upstream and causing profound changes in the hydrology and sediment load downstream. In addition to large reservoirs in the foothills, some rivers also contain numerous smaller dams upstream, most of which have little storage and serve only to divert water (in some cases to adjacent river basins) for hydroelectric projects. For example, there are at least 40 dams in the Stanislaus River basin, but the large foothills reservoir (New Melones Reservoir) accounts for 85% of the total reservoir capacity [*Kondolf and Matthews*, 1993].

2.3 Gravel Mining

In many areas, alluvial deposits of gravel and sand are mined for construction aggregate for use in concrete, asphalt, road base and sub-base, for fill, and for landscaping. Aggregate is mined from the active channel (*instream mining*) or from pits excavated in the floodplain or terraces (Figure 2). Active channel and unweathered floodplain deposits have been preferred sources of aggregate because river transport eliminates soft lithologies by abrasion and attrition, and the resulting deposits are of high quality: durable, rounded, well-sorted, and relatively free of interstitial clay. Also, transportation costs are a large part of the total production costs for aggregate, and many gravel-bed rivers are conveniently located near urbanizing areas and highway routes that are major markets for construction aggregate.

2.3.1 *Effects of gravel mining on stream channels.* Instream mining has commonly been viewed as sustainable, with the annual "harvest" of gravel "replenished" by gravel transported into the mined reach from upstream. However, the amounts annually extracted have been a function of local demand for aggregate, not an analysis of annual bed load transport. On rivers for which sediment budgets have been calculated, rates of instream extraction are commonly found to exceed rates of bedload transport by an order of magnitude or more [e.g., *Kondolf and Swanson*, 1993; *Collins and Dunne*, 1989, 1990].

If extraction rates were a small fraction of annual bedload, instream mining could probably be sustained without major impact on many rivers. However, extraction rates that are of the same order, or larger than, annual bedload discharge, will typically result in channel incision. This incision may propagate upstream for kilometers, and downstream of the instream pit, undermining bridges and other structures and causing channel instability such as bank undercutting and channel widening, and loss of spawning gravel [*Kondolf*, 1994a]. For example, on Stony Creek, a tributary to the Sacramento River, instream gravel mines are permitted to extract 1.3 million tonnes each year, an order of magnitude more than the estimated annual average of 0.16 million tonnes of bedload derived from the watershed prior to construction of Black Butte Dam in 1963, and 40 times greater than the estimated average annual post-dam gravel contribution from incision and bank erosion below the dam. The result has been channel incision of up to 5 m [*Kondolf and Swanson*, 1993].

2.3.2 *Extent of gravel mining from fluvial deposits.* The extent of aggregate extraction from alluvial deposits depends upon local geologic conditions. For example, in Washington State, alluvial deposits are very important in central parts of the state, such as the Yakima and Columbia River valleys, but less so in the Puget Sound area because excellent sand and gravel resources are available in extensive glacial outwash deposits [*Leighton*, 1919].

In California, which leads the nation in aggregate production (producing 30% of the national total), aggregate production is the largest mining industry in the state,

Fig. 2. Photograph of gravel mining along Clear Creek, California, showing excavation of floodplain pit to bedrock, January 1989.

exceeding even gold in value: over 1,000,000 tonnes are produced annually at an annual value of about $700 million in 1990 [*Carillo et al.* 1990; *Tepordei,* 1992]. Virtually all the aggregate is mined from alluvial deposits. The scale of this annual extraction can be grasped by a comparison with an estimate of annual bedload sediment yield from the state. Using a typical value of sediment yield of 200 m^3 km^{-2} y^{-1} (derived from reservoir sedimentation rates [*Kondolf and Matthews,* 1993]), a bulk density of 1.6 tonnes m^{-3} [*Vanoni,* 1975], and assuming that bedload constitutes 10% of this total, the annual bedload sediment yield from the area of California (400,000 km^2) is estimated at 13 million tonnes, which is roughly an order of magnitude smaller than current rates of bedload mining from river deposits. Much of this annual extraction is taken from storage in channel, floodplain, and terrace deposits.

2.4 Environmental consequences of gravel mining and reservoirs

Reservoir construction has affected formerly abundant anadromous salmonid populations in several ways. Many historically important spawning and rearing habitats upstream of the large downstream reservoirs are no longer accessible to fish. Overall, spawning habitat in the Sacramento-San Joaquin River system has been reduced to 4% of its historical total, largely because of reservoir construction and operation [*Peterson et al.,* 1982]. Spawning and rearing habitats downstream of large reservoirs have been compromised by alterations in seasonal flow regimes. For example, numerous salmon redds in the Mokelumne River below Camanche Dam were dewatered by an abrupt drop in flow after an unusually prolific spawning season in the fall of 1984.

In some cases spawning gravel have been lost by downstream transport without replacement from upstream because of trapping in reservoirs and/or instream gravel mining (e.g., the Upper Sacramento River [*Parfitt and Buer,* 1980]); while elsewhere fine sediment has infiltrated spawning gravel below large reservoirs such as on the Trinity River [*Kondolf and Wilcock,* 1993].

Incision resulting from instream gravel mining has undermined bridges, pipeline crossings, and other structures, necessitating expensive repairs. On Stony Creek (described above), the bed below the state Highway 32 bridge incised over 5 m since its construction in 1975, exposing the bridge piers. Emergency repairs have cost $2 million so far, and the bridge already requires replacement [*Kondolf and Swanson*, 1993].

Reduced delivery of bedload sediment to the coastal zone in southern California from upstream reservoirs and in-channel mining of sand and gravel has caused accelerated coastal erosion and resulted in millions of dollars in property loss and massive expenditures on erosion control structures [*Inman*, 1976]. The interrelated nature of upstream alterations to bedload sediment transport and downstream effects in the coastal zone has become more widely recognized, stimulating legal efforts to protect the sand supply to beaches [*Los Angeles Times*, 1991; *Gayman*, 1985].

3. MITIGATION STRATEGIES AND ISSUES

3.1 *Reservoirs*

3.1.1 *Sediment Pass-Through, Sluicing, and Gravel Replenishment.* The impacts of reservoir construction on downstream bedload sediment transport are increasingly recognized, but very few reservoirs are actively managed to minimize disruption of the continuity of sediment transport through the river. Below the the Barrage Iffezheim, the downstream-most dam on the Rhine River, 170,000 tonnes of gravel are artificially added to the river annually to prevent incision, an option that proved less expensive and environmentally damaging than construction of additional dams downstream [*Kuhl*, 1992]. On the Upper Sacramento River below Keswick Reservoir (a re-regulating reservoir downstream of Shasta Reservoir), gravel are added not to satisfy the river's bedload transport capacity but for use by spawning salmonids (Figure 3). From 1988-2000, the state is committed to spend $22 million on artificial spawning gravel enhancement in the Upper Sacramento River [*Denton*, 1991].

Some reservoirs in Germany and Austria [*Hack*, 1986; *Westrich et al.*, 1992] and Switzerland [*Zarn*, 1992], have been successfully designed to permit passage of sediment transport and the approach has been explored in reservoir planning for the humid tropics [*White and Bettess*, 1984]. In California, such a *sediment pass-through* approach has been proposed for two reservoirs on the North Fork Feather River, but none has been implemented as yet.

Small hydroelectric diversion dams could easily pass most sediment if the low level outlets were opened during high flows, but instead sediment is now allowed to accumulate. The California Department of Water Resources Division of Safety of Dams requires that low level outlets be maintained in operable condition. To remove accumulated sediment, the low level outlets have commonly been opened and flushed clean, releasing sediment downstream at baseflow, when flows are inadequate to disperse the sediment downstream, and deposition of gravel, sand, and finer sediment results. Elevated suspended sediment concentrations and reduced dissolved oxygen levels resulting from organic material in the released sediment can result in massive fish kills. This occurred on the Carmel River below Los Padres Dam in 1980 [*Buel*, 1980], where the dam operator has since been required to use a suction dredge to maintain the outlet (D. Dettman, Monterey Peninsula Water Management District, personal communication 1990). Because of the history of harmful effects of sediment sluicing from reservoirs, the California Department of Fish and Game is wary of sediment pass-through strategies (J. Mensch, California Department of Fish and Game, personal communication, 1991).

3.1.2 *Case study: sediment sluiced from Our House Reservoir.* Our House Reservoir (capacity 350,000 m³) on the Middle Yuba River (drainage area = 376 km²) illustrates the issues and attitudes surrounding sluicing. This reservoir was completely filled with sediment during high flows in February 1986. The following August, the Yuba County Water Agency began mechanically removing the accumulated sediment and transporting it by truck to a disposal site. During this process, a cofferdam directing the river flow around the work area failed. The sluice gate was open, and sediment began to flush from the reservoir through the gate, spreading out over the channel bed downstream. The gate was allowed to remain open for three days, until a plume of turbid water reached a popular bathing area 13 km downstream. The local fish and game warden was alerted, and he visited the dam and ordered the sluice gate closed immediately. By this time, an estimated 12,000 m³ of sand and gravel had deposited over a 1.2 km reach below the dam (Figure 4) (Ebasco Environmental, 1989). The Regional Water Quality Control Board cited the Yuba County Water Agency for release of a "deleterious material" (i.e., sand and gravel), threatened the Agency with a $1 million fine, and criminal charges were brought against the Agency's director under the state Fish and Game Code.

In subsequent negotiations with resource and regulatory agencies, Yuba County Water Agency and Pacific Gas and Electric Company (with whom the Agency has an operation and maintenance agreement) agreed to mechanically remove 6,000 m³ of sand and gravel. This operation required construction of a road into the channel, operation of heavy equipment in the channel bed, and rearrangement of boulders in the bed to permit excavation to proceed

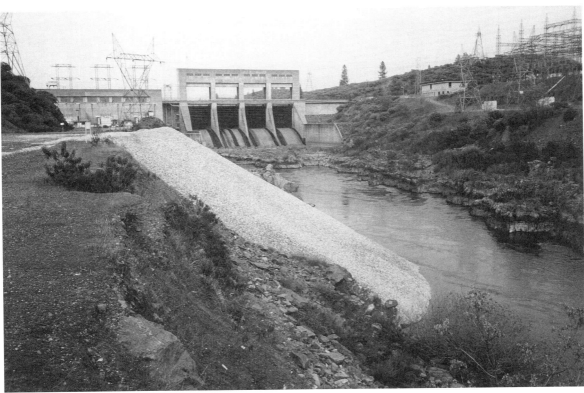

Fig. 3. Photograph of artificial gravel replenishment downstream of Keswick Dam (the re-regulating reservoir below Shasta Dam) on the Upper Sacramento River, California, January 1990.

(Figure 5). Total cost of this operation was $1.3 million, excluding costs of study design and subsequent monitoring. Much of the remaining sediment, estimated at 6,000 m^3, was transported downstream by uncontrolled spills in 1987 to 1989 (Figure 6). Spills during 1987 and 1988 (Figure 7) were relatively small and brief. Their net effect was to cause the tongue of sand to extend downstream in discrete steps of about 100 m, progressively filling pools. The larger spill of March 1989, however, effectively suspended sand, scoured pools, and dispersed sand over 5 km downstream. The passage of the tongue of sediment is illustrated by sequential surveys of a cross section across a pool 2 km downstream of the dam (Figure 8). This pool, unaffected by the initial sluice, was buried by the prograding tongue of sediment during the November 1988 spill, and then partially scoured by the March 1989 spill (Ebasco Environmental, 1989).

The progressive extension of a sediment tongue during the modest spills of 1987 and 1988 contrasts with the dispersion of sediment in the larger 1989 spill, and emphasizes that when sediment is introduced at low flows, little dispersion occurs. However, at high flows, sediment is readily dispersed, in a fashion similar to the natural movement of sediment through the river system. The failure of the Yuba County Water Agency to halt the sluicing for three days, the subsequent vigorous enforcement against the Agency and the fact that during this same period the Department of Fish and Game was purchasing gravel for deliberate release into the Upper Sacramento River, illustrate the degree to which management of gravel and sand transport in these regulated rivers is not based on a sound understanding of the continuity of sediment transport through river systems.

3.1.3 *Requirements for flushing flows*. Where tributary-derived fine sediment has accumulated in spawning gravel downstream of dams because transport capacity has decreased with decreasing flood flows, artificial *flushing flow* releases are often required of dam operators [*Reiser et al.,* 1988; *Milhous,* 1990]. These releases can be very expensive in lost hydropower and water revenues. Three experimental flushing flow releases from Lewiston Dam on the Trinity River in 1991-1993 cost about $4 million in lost hydropower revenues. Moreover, if flushing flows are too high, they may transport gravel from the channel below the

Fig. 4. Photograph of deposit of sediment sluiced from Our House Dam on the Middle Yuba River, California, October 1986.

dam, exacerbating loss of spawning gravel from sediment-starved releases. Despite the tremendous cost of flushing flow releases and the potential for accelerated spawning gravel loss, flushing flow releases are routinely set in negotiations between dam operators and resource agencies with little or no site-specific information on channel hydraulics and potential sediment transport, nor even a full understanding of the geomorphic processes underlying the problem.

3.1.4 *Reservoir sedimentation and dam safety.*
Reservoirs can be expected to fill with sediment eventually, at a rate controlled by basin sediment yield, reservoir capacity, and any sediment pass-through strategies employed. In California, small reservoirs in the erodible Coast Ranges have filled most rapidly, with sedimentation rates commonly over 300 $m^3 km^{-2} y^{-1}$. On the Carmel River, two small water supply reservoirs have lost 40% and 85% of original capacity since 1946 and 1921 respectively [*Matthews*, 1988]. In rivers draining the less erodible batholithic terrane of the Sierra Nevada, sediment yields are commonly under 200 $m^3 km^{-2} y^{-1}$, and the larger reservoirs are unlikely to fill completely for centuries. Once a reservoir has filled with sediment, bedload sediment can be expected to pass through outlet structures, potentially increasing the rate of abrasion.

Many dams were built prior to the requirement to have low-level outlets. Ironically, as these reservoirs fill with sediment, their storage capacity decreases below the minimum size required for state jurisdiction, and the Department of Water Resources ceases to make safety inspections. The Malibu Creek reservoir near Los Angeles is an example (D. Babbitt, California Department of Water Resources, personal communication 1991). It is unclear whether this loophole would apply only to older structures not equipped with low level outlets, or if it would apply to any reservoir that completely filled.

Discussions with many dam operators in California indicate that there has been essentially no planning for sediment management as reservoirs age. Options for managing a reservoir filled with sediment include leaving the reservoir as is and trusting that abrasion will not destabilize the dam outlet works, removing the sediment behind it, or removing the dam. Removal of sediment from reservoirs can cost as much as $3500 per m^3, depending upon access and distance to a disposal site (E. Stassevich, Pacific Gas & Electric Company, San Francisco, personal communication 1987). The potential for exploitation of this sediment as aggregate has been inadequately explored in

172 MANAGING BEDLOAD SEDIMENT IN REGULATED RIVERS

Fig. 5. Photograph of mechanical removal of sediment from the bed of the Middle Yuba River, California, November 1986. (Photograph by Ted Frink).

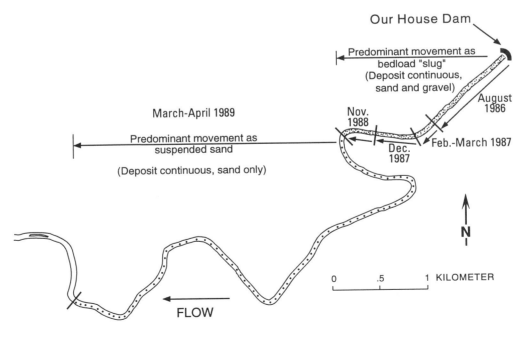

Fig. 6. Map showing progressive downstream transport and dispersion of sediment sluiced from Our House Reservoir on the Middle Yuba River. (Adapted from Ebasco Environmental [1989]).

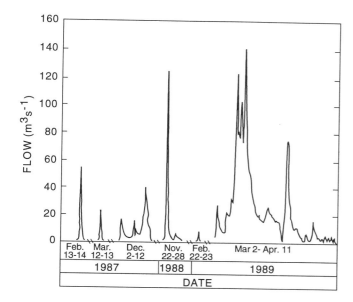

Fig. 7. Hydrograph of spills from Our House Dam on the Middle Yuba River, California, for water years 1987-1989. (Data from US Geological Survey gauge "Middle Yuba River below Our House Dam, near Camptonville, California". Adapted from Ebasco Environmental [1989]).

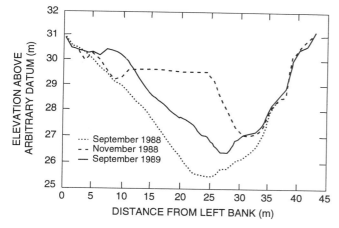

Fig. 8. Sequential surveys of a cross section across a pool 2 km downstream of Our House Dam on the Middle Yuba River, California, showing aggradation of the bed by downstream migration of bedload tongue in November 1988, and subsequent partial scour in March 1989. (Adapted from Ebasco Environmental [1989]).

most parts of the state, but, just as pollutants have been termed "resources out of place" [Miller, 1990], so the coarse sediment accumulated in reservoirs can be viewed as potential sources of aggregate.

Removal of dams completely filled with sediment can result in catastrophically increased delivery of sediment to downstream reaches. This could potentially overwhelm the channel, resulting in aggradation and destabilization. The consequences of these channel changes can be severe if there is extensive floodplain development below dams that was encouraged in part by the perception of flood protection provided by the reservoirs. The change in sediment load would not be fully apparent until the first high flow following dam removal. Thus, several years could pass before the newly available sediment migrated to downstream reaches.

In the case of concrete dams, it may be possible to remove them in stages, so that the sediment could be metered out to downstream reaches, reducing the risk of catastrophic channel change. In the case of earthen dams, however, no such option exists. Once the hardened superstructure is removed, the earthen material constituting most of the dam would be readily entrained by high flows and the dam would be rapidly destroyed, much as happened to the Auburn cofferdam on the American River near Sacramento in February 1986. Many of these issues are now being raised in connection with proposed dam removal on the Elwha River in the Olympic Peninsula of Washington [Stoker and Williams, 1991].

3.2 Gravel Mining

3.2.1 *Regulation in California.* Regulation and management of gravel mining vary considerably from state to state. In California, the principal statute is the Surface Mining and Reclamation Act of 1975 (SMARA), which is carried out primarily by designated *lead agencies*, usually county or city governments. Unfortunately, these local governments rarely possess the expertise or staff to analyze the cumulative geomorphic and environmental impacts of gravel mining. Nor would these agencies necessarily have the direct motivation to consider downstream effects. Moreover, to properly manage the gravel resource requires looking at bedload sources and transport on a basin scale, which commonly exceeds the jurisdictional boundaries of the lead agency [Kondolf, 1994b].

SMARA requires each mine to file a *reclamation plan* stating how the site will be restored to a productive use once the mine is closed. While the reclamation concept has some value for floodplain and terrace pits, the concept is meaningless when applied to the dynamic environment of the active channel. Operation of an instream mine can induce incision upstream and downstream of the actual extraction site, and there is no direct way that on-site reclamation can reverse off-site damage to bridges and other resources.

There are other regulatory requirements for gravel

mines, especially instream operations, which may require as many as thirteen permits or approvals from various agencies. These include permits from the US Army Corps of Engineers under Section 404 of the Clean Water Act, a Streambed Alteration Agreement from the California Department of Fish and Game, and review by the lead agency under the California Environmental Quality Act. Unfortunately, none of these agencies are charged with, or capable of conducting, the comprehensive analysis needed to assess the environmental impacts of gravel mining [*Kondolf*, 1994b].

The present practices and regulation of instream mining in California have evolved piecemeal over recent decades in reaction to problems resulting from increased extraction rates related to the state's rapid urban growth. Despite the fundamental economic importance of aggregate, and despite environmental impacts of its extraction from river channels, there has been little serious effort until recently to step back and view extraction in the larger context of aggregate resource quality and uses, and the environmental impacts of various alternatives. Unfortunately, in many areas, the data necessary for sound decision making are lacking or have not been compiled in a usable form.

There has also been little attempt in most of California to evaluate environmental costs of instream gravel extraction or to pass these costs on to consumers of the aggregate. As a result, instream sources may appear to be less expensive than alternatives such as dry terrace mines (which often require additional processing), quarries (from which rock must be crushed, washed, and sorted), or distant sources, such as reservoir deltas (which involve greater transportation costs). In many areas, competition for land from urbanization or agriculture, and provision of flood protection for these lands, has pushed aggregate mines into stream channels. However, with the loss of most of the riparian wetlands in the state over the past 150 years [*ESA/Madrone*, 1982] and increasing recognition of the ecological importance of these terrestrial and aquatic habitats, resource agencies are becoming reluctant to permit further instream mining operations without adequate environmental review.

4. COMPREHENSIVE ENVIRONMENTAL PLANNING IN BEDLOAD MANAGEMENT

While there has been increased recognition of sediment impacts from reservoir construction and operation, and increased awareness of environmental issues associated with gravel mining (SMARA has been amended ten times since its passage in 1975), coordinated management of bedload sediment resources has not been achieved. In large part this is because government agencies, the public, and industry do not view sediment as a dynamic resource, moving through the landscape just as the water that transports it.

Comprehensive planning and management of bedload sediment should include realistic assessment of the impacts of existing and proposed water development projects and gravel extraction operations, and the costs of these impacts should be incorporated into the financial evaluation of these projects. A sediment budget for present and historical conditions is needed as a fundamental basis for evaluation of these impacts. The notion of *sediment rights* (analogous to water rights) should be explored as a possible framework within which to assess reservoir operators and alluvial gravel miners for impacts such as bridge undermining, loss of spawning gravel, and loss of beach sand.

Comprehensive river basin management planning is needed to balance beneficial uses (as traditionally defined) with natural resource protection and restoration. This planning process would be unlike traditional river basin planning, in which hydroelectric generation, irrigation diversions, flood control, and gravel mining have been pursued at the expense of the integrity of the river ecosystem. Instead, the environmental impacts of existing uses should assessed, and the resulting insights applied to assessment of potential future projects, as now done in the Thames Region of the United Kingdom under the auspices of the National Rivers Authority [*Gardiner*, 1988, 1990]. The lack of a single agency with a mandate to plan comprehensively, as exists in the United Kingdom, is a major impediment to rational management of bedload sediment resources in California.

4.1 *Reaches downstream of reservoirs*

Any comprehensive bedload management plan should recognize the reduced gravel availability downstream of reservoirs. Potential gravel replenishment projects and flushing-flow releases require a coordinated evaluation. These reaches are inherently unsuitable for instream mining, with the possible exception of extraction based on sediment delivered from tributaries. Floodplain or terrace pit mining could be permitted, provided that the pits were located a sufficient distance from the active channel to minimize the risk of channel capture by the pits. In most cases, these pits intersect the water table, precluding reclamation to pre-existing agriculture, but making creation of wildlife habitat possible provided the pits are shaped to increase habitat values.

4.2 Aggregate extraction from reservoir deltas

The river delta at the upstream end of the reservoir pool provides a potential opportunity to obtain aggregate with relatively few environmental impacts, while reducing the loss of reservoir capacity. One approach would be over-excavation of a pit to trap bedload sediment. The upstream slope of the pit could be hardened to prevent upstream knickpoint migration, while downstream incision would not be an issue above the reservoir. Unfortunately, many reservoir deltas are distant from markets, resulting in long haul distances, and many reservoir deltas now fall within "recreation areas" in which industrial activity is prohibited.

Other alternative sources of aggregate should be aggressively pursued. If the true environmental costs of instream gravel mining are incorporated into the purchase price of instream-produced gravel, alternative sources will become more competitive.

Sediment pass-through strategies should be actively pursued wherever feasible to better mimic the natural transport of sediment through the river network, thereby reducing reservoir sedimentation and supplying bedload sediment to downstream reaches. The appropriate discharge and season for pass-through must be determined to disperse sediment adequately, protect spawning gravel, and protect juvenile cobble habitat.

Acknowledgements. The ideas presented here grew out of field work and many discussions with W.V.G. Matthews, Mitchell Swanson, Peter Wilcock, John Williams, and others. This paper has benefitted from critical comments from Karen Prestegaard and Peter Wilcock, and is based in large part upon research supported by the University of California Water Resources Center, as part of Water Resources Center project UCAL-WRC-W-748, administered by the Center for Environmental Design Research at the University of California, Berkeley.

REFERENCES

Brune, G. M., The trap efficiency of reservoirs, *Trans. Am. Geophys. Union,* 34, 407-418, 1953.

Buel, B., Effects of Los Padres Reservoir silt release, *Monterey Peninsula Water Management District,* Monterey, California, 1980.

Carillo, F. V., J. F. Davis, and J. L. Burnett, California: Annual Report, U.S. Dept. Int., Bur. Mines, Washington, D.C., 1990.

Collins, B. and T. Dunne, Gravel transport, gravel harvesting, and channel-bed degradation in rivers draining the Southern Olympic Mountains, Washington, USA. *Env. Geol. and Water Sci.,* 13, 213-224, 1989.

Collins, B. and T. Dunne, Fluvial geomorphology and river gravel mining: a guide for planners, case studies included, Special Publication 98, *Calif. Div. Mines and Geol.,* Sacramento, California, 1990.

Denton, D.N., Sacramento River gravel restoration project progress report, *Calif. Dept. Water Resour.,* Northern District, Red Bluff, California, 1991.

DWR (California Department of Water Resources), San Joaquin River tributaries spawning gravel assessment - Stanislaus, Tuolumne, and Merced rivers, DWR Northern District, Red Bluff, 1994.

Ebasco Environmental, Continued monitoring report for 1988/1989 for the cleanup and abatement of sediments sluiced from Our House Reservoir, Middle Yuba River, report to Yuba County Water Agency, Ebasco Environmental, Sacramento, Calif., 1989.

ESA/Madrone, Wetlands Policy Assessment: California Case Study, Report to U.S. Office Tech. Assesmnt. by *ESA/Madrone Consultants,* Novato, Calif., 1982.

Federal Energy Regulatory Commission, Final Environmental Impact Statement, Proposed Modifications to the Lower Mokelumne River Project, California, FERC project No. 2916-004, *Fed. Energy Reg. Comm.,* Office of Hydropower Licensing, Washington, D.C., 1993.

Gardiner, J. L., Environmentally sound river engineering: examples from the Thames Catchment, *Reg. Rivers,* 2, 445-469, 1988.

Gardiner, J. L., Influences on the development of river catchment planning in the Thames basin, Ph.D. thesis, University of Southampton, England, 1990.

Gayman, W., High-quality, unbiased data are urgently needed on rates of erosion, California's Battered Coast, Proc., *Conference on Coastal Erosion,* J. McGrath, ed., California Coastal Commission. pp. 26-43, 1985.

Hack, H. P., Design and calculation of reservoirs of run of river stations incorporating sedimentation, Transport of Suspended Solids in Open Channels, *Proc. Euromech 192,* W. Bechteler, ed., Munich, Germany, June 1985, 107-112, 1986.

Inman, D.L., Man's impact on the California coastal zone, summary report to Department of Navigation and Ocean Development, Sacramento, California, 1976.

Jobson, H. E. and W. P. Carey, Interaction of fine sediment with alluvial streambeds, *Wat. Resour. Res.,* 25, 135-140, 1989.

Kondolf, G. M., Geomorphic and environmental effects of instream gravel mining, *Landscape and Urban Planning,* 28, 225-243, 1994a.

Kondolf, G. M., Environmental planning in regulation and management of instream gravel mining in California, *Landscape and Urban Planning,* 29, 185-199, 1994b.

Kondolf, G.M., and W.V.G. Matthews, Coarse sediment management on regulated rivers, Report No. 80, *Calif. Wat. Resour. Center,* University of California, Davis, 1993.

Kondolf, G. M., and M. L. Swanson, Channel adjustments to reservoir construction and gravel extraction along Stony Creek, California, *Environ. Geol. Wat. Sci.,* 21, 256-269, 1993.

Kondolf, G. M., and P. V. Wilcock, The flushing flow problem on the Trinity River, CA, Hydraulic Engineering '93, Volume 1, Proc., 1993 Conference, H. W. Shen, S. T. Su, and F. Wen, eds., Am. Soc. Civil Eng., Hydr. Div., 1172-1177, 1993.

Kondolf, G. M., and M. G. Wolman, The sizes of salmonid spawning gravel, *Wat. Resour. Res.*, 29, 2275-2285, 1993.

Kuhl, D., 14 years of artificial grain feeding in the Rhine downstream the barrage Iffezheim, Proc., *5th Internat. Symp. on River Sedimentation*, Karlsruhe, Germany, 1121-1129, 1992.

Leighton, M.M., The road building sands and gravel of Washington, *Wash. Geol. Surv. Bull.*, 22, 1919.

Los Angeles Times, A new theory: a beach has a right to its sand, 29 Nov., 1991.

Mahmood, K., Reservoir sedimentation: impact, extent, and mitigation, *Tech. Pap. No. 71*, 118 pp., The World Bank, Washington D.C., 1987.

Matthews, W. V. G., Evaluation of reservoir sedimentation rates in the Upper Carmel River watershed, *Tech. Mem. 88-03*, Monterey Peninsula Wat. Management Dist., Monterey, Calif., 1988.

Milhous, R. T., The calculation of flushing flows for gravel and cobble bed rivers, *Hydraulic Engineering*, Volume 1, Proc. 1990 Conf., H. H. Chang and J. C. Hill, eds., 598-603, 1990.

Miller, G. T., *Living in the Environment,* 6th edition, Wadsworth Publishing Co., Belmont, California, 1990.

Parfitt, D., and K. Buer, Upper Sacramento River Spawning Gravel Study, *Calif. Dept. Wat. Resour.,* Northern Division, Red Bluff, California, 1980.

Peterson, D. F., D. F. Jacobs, B. J. Talley, R. N. Hinton, C. W. Pike, and A. F. Godwin, California's stream resources, Volume 1: Overview and Assessment, *Bull. 215, Calif. Dept. Wat. Resour.,* Sacramento, 1982.

Reiser, D. W., M. P. Ramey, and T. A. Wesche, Flushing flows, in *Alternatives in Regulated River Management,* pp. 91-135, CRC Press, Boca Raton, Florida, 1988.

Stoker, B. A. and D. T. Williams, Sediment modeling of dam removal alternatives, Elwha, Washington, in *Hydraulic Engineering,* Proc., 1991 Conf., Am. Soc. Civil Engr, pp. 674-679, 1991.

Tepordei, V. V., Construction sand and gravel: annual report, U.S. Dept. Int., Bureau of Mines, Washington, D.C., 1992.

Vanoni, V. (Ed.), *Sedimentation Engineering,* 745 pp., Am. Soc. Civil Engr., New York, 1975.

Westrich, B., S. Al-Zoubi, and J. Muller, Planning and designing a flushing channel for river reservoir sediment management, *Proc., Fifth Internat. Symp. on River Sedimentation,* Karlsruhe, Germany, 861-867, 1992.

White, W. R. and R. Bettess, The feasibility of flushing sediments through reservoirs, in *Challenges in African Hydrology and Water Resources,* edited by D. E. Walling, Proc., Harare Symp., July, 577-587, 1984.

Zarn, B., Numerical simulation of sediment management in reservoirs planned along the Rhine River upstream of Lake of Constance, *Proc. Fifth Internat. Symp. on River Sedimentation,* Karlsruhe, Germany, 853-860, 1992.

G. Mathias Kondolf, Department of Landscape Architecture, University of California, Berkeley, CA 94720

Regulated Streamflow, Fine-Grained Deposits, and Effective Discharge in Canyons with Abundant Debris Fans

John C. Schmidt

Department of Geography and Earth Resources, Utah State University, Logan, Utah

David M. Rubin

U. S. Geological Survey, Menlo Park, California

The fundamental channel unit of rivers that flow through canyons that have abundant debris fans is a channel complex composed of (1) a backwater upstream from the debris fan, (2) a debris fan and channel constriction, (3) an eddy or eddies and associated bars in the expansion downstream from the fan, and (4) a downstream gravel bar. These fan-eddy complexes exist at the mouths of nearly all debris-flow-generating tributaries. Such tributaries exist along many, but not all, of the narrow canyons of the Green and Colorado Rivers. Reaches affected by debris fans are steeper, have higher stream power per unit bed area, and have coarser beds than other narrow canyons of the same river system. A large proportion of fine-grained sediment in these canyons is deposited in eddies; this proportion is as large as 75 percent in Grand Canyon. Before construction of Glen Canyon Dam, many eddy bars along the Colorado River in Grand Canyon were more extensive than they are today, and separation and reattachment bars merged. Fine-grained deposits can be classified as (1) low-elevation eddy bars and channel-margin deposits formed by discharges less than or equal to the primary mode of the calculated product of streamflow frequency and sediment transport, and (2) high-elevation eddy bars and channel-margin deposits formed by floods that produce subsidiary modes of the streamflow-frequency-sediment-transport product.

1. INTRODUCTION

The relative channel-forming role of rare catastrophic floods and of frequent moderate-magnitude floods has been the subject of longstanding debate. Frequent events determine the shape of relatively unconstrained meandering streams whose floodplains are formed by lateral accretion [*Wolman and Leopold,* 1957; *Wolman and Miller,* 1960], but the role of catastrophic floods is more important in narrow canyons because flood flows are of higher stage, velocity, and turbulence [*Baker,* 1984]. Narrow canyons often have coarse bed material that can only be transported by high magnitude discharges [*Baker,* 1977]. Even where coarse bed material is absent, the tendency for streams in narrow canyons to build floodplains by vertical accretion increases the role of catastrophic floods in determining channel shape [*Nanson,* 1986]. The elevation of the active floodplain and of bankfull stage become more variable where there is a large range in flood magnitude and where recovery time is long [*Baker,* 1977; *Wolman and Gerson,* 1978; *Andrews,* 1980].

The role of catastrophic floods has been stressed in those narrow canyons of the Colorado Plateau where debris flows deliver large amounts of coarse sediment to the channel and valley floor. *Graf* [1979] and *Kieffer* [1985] showed that only rare floods significantly rework coarse bed material within rapids of the Green and Colorado Rivers. Numerous studies have described high-elevation slackwater deposits that contain the preserved evidence of rare high-magnitude discharges [*Baker et al.,* 1983; *O'Conner et al.,* 1994].

Natural and Anthropogenic Influences in Fluvial
Geomorphology
Geophysical Monograph 89
Copyright 1995 by the American Geophysical Union

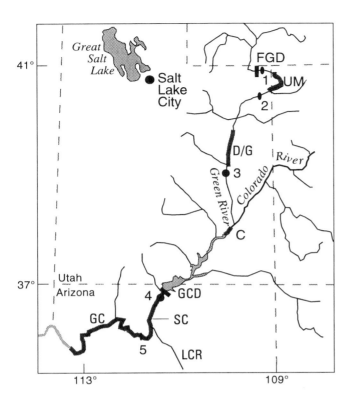

Fig. 1. Map showing major rivers of the Colorado Plateau. Canyons with abundant debris fans are shown as thick grey lines and are labelled as (GC) Grand Canyon, (C) Cataract Canyon, (D/G) Desolation/Gray Canyons, and (UM) Canyon of Lodore and Split Mountain Canyon in the Uinta Mountains. U.S. Geological Survey stream gaging stations used in flood-frequency analyses are located with black circles and labelled as: (1) Green River near Greendale, UT, (2) Green River near Jensen, UT, (3) Green River at Green River, UT, (4) Colorado River at Lees Ferry, AZ, and (5) Colorado River near Grand Canyon, AZ. FGD is the location of Flaming Gorge Dam and GCD is the location of Glen Canyon Dam. LCR is Little Colorado River and SC is Saddle Canyon.

Less attention has been given to fine-grained alluvial deposits that occur at lower elevations in these canyons, but environmental management considerations have recently redirected the attention of geomorphologists. These discontinuous deposits are important environmental resources along the Colorado Plateau's large rivers because recreational boating depends, in part, on the availability of these deposits for campsites [Kearsley et al., 1994]. These deposits are also substrate for riparian vegetation that may support an abundant and diverse ecosystem [Stevens et al., 1995]. These canyons are managed for their intrinsic environmental values because they are within the U.S. National Park Service system or are designated as Wild and Scenic Rivers.

Low-elevation alluvial deposits are affected by an extensive network of dams. These dams have greatly decreased the magnitude of floods and the volume of mainstem sediment transport of the Colorado River and its tributaries [Andrews, 1986, 1990]. The long-term fate of these deposits is the subject of substantial public concern, and management agencies wish to know if reservoir operations can be revised so as to provide more favorable downstream conditions (e.g. U.S. Bureau of Reclamation, 1993]. Thus, the attention of environmental managers has been drawn to the role of frequent moderate-magnitude hydrologic events in shaping attributes of the downstream channel because these are the hydrologic events that can be controlled by dam operations. Even though much of the river-corridor geomorphology may have been determined by rare catastrophic floods, low-elevation fine-grained alluvial deposits are an environmental resource that can be manipulated by normal dam operations. These management considerations led us to reevaluate the concepts of effective discharge and bankfull stage for those rivers where these concepts might otherwise be expected to have the least application.

The purposes of this paper are to (1) describe the large-scale geomorphic attributes of canyons that have abundant debris fans, (2) describe the reach-scale geomorphic organization of these rivers, (3) describe the processes and patterns of fine-grained sediment deposition, and (4) demonstrate that the concept of geomorphic effectiveness, defined as the product of streamflow frequency and sediment transport, provides useful geomorphic and management insights concerning the depositional regimes responsible for these fine-grained deposits.

2. METHODS

Reach-scale attributes of the Green River (Figure 1) were determined from 1:24000 scale topographic maps (Flaming Gorge Dam to the Colorado River confluence, 650 km) and by photogeologic interpretation of large scale (approx. scale 1:5000) air photography taken at low discharge in 1963 (16 reaches comprising 40 percent of the river between the downstream end of the Uinta Mountains and the Colorado River confluence, 520 km). Width of the alluvial valley floor and channel gradient were determined from topographic maps and published geologic maps. Reaches with abundant debris fans were identified, and meandering reaches were classified by channel pattern: restricted meanders occur where the outer limits of the meander belt impinge on confining valley walls, and incised meanders occur where the wavelength of the channel and the valley are similar. Active sand bars, gravel bars, vegetated terraces, and debris fans of the Green River were mapped. The different components of fan-eddy complexes, as described below, were also mapped. Average channel width for each mapped reach was determined by dividing the total area of low-flow channel and unvegetated bars by the length of each reach.

In Grand Canyon, mapping at a scale of 1:2400 was conducted in 2 reaches, each about 15 km long. These reaches were located between the stream gaging stations at Lees Ferry, Arizona, and near Grand Canyon, Arizona (Figure 1). The purpose of this mapping was to evaluate the adequacy of previously proposed classifications of eddy bars and to determine the discharges that form most of the alluvial deposits of the river corridor. Mapping in these reaches has included extensive field work as well as photogeologic interpretation; methods have been described by *Schmidt et al.* [1994a, b]. The methods used in sedimentologic analyses of many of these alluvial deposits in Grand Canyon are summarized by *Rubin et al.* [1990, 1994]. All mapping data for the Green and Colorado Rivers have been entered into a geographic information system, and area measurements for mapping units have been determined from these data.

Flood recurrence for 5 gaging stations on the Green and Colorado Rivers was determined for the period 1923 to 1962 using *U.S. Water Resources Council* [1981] methods. This period was prior to widespread completion of dams. The Green River stream gage near Jensen, Utah, was installed in 1947, and the flood record was extended by correlation with records of stations with longer periods of measurement [*Schmidt*, 1994].

The product of streamflow frequency and sediment transport for the Colorado River in Grand Canyon was determined from (1) flow duration data for hourly releases from Glen Canyon Dam [*U. S. Bureau of Reclamation*, 1990, written commun.], and (2) sand-transport relations for the Colorado River near Grand Canyon, Arizona [*Pemberton*, 1987]. The duration of hourly flow for each discharge increment of 5 m^{3}s^{-1} was determined and multiplied by the corresponding suspended sand-transport rate and summed by 25 m^{3}s^{-1} increments. Sand-transport data were determined from sampling conducted in 1983 and between 1985 and 1986 [*Garrett et al.*, 1993]. Transport data for discharges greater than 890 m^{3}s^{-1} were only collected in 1983. It is not known whether the same transport rates occurred during the high discharges that occurred between 1984 and 1986. As described later in this paper, deposits formed by high discharges that occurred between 1984 and 1986 are thin, suggesting that the 1983 transport rates may over-estimate transport conditions of 1984 to 1986. Daily flood waves caused by hydroelectric peak power production attenuate downstream (J. D. Smith and S. M. Wiele, U.S. Geological Survey, Boulder, written commun., 1994), but only hourly data at the dam were used in our analysis. Attenuation does not affect the general characteristics of the calculations described below, but it may affect the precision of the determination of the modal discharge increment.

3. REGIONAL SETTING

The earliest geomorphic investigations of the Colorado River system [*Powell*, 1875; *Dutton*, 1882; *Hunt*, 1969] recognized the disparity between present stream courses and trends of the dominant geologic structures. The Green and Colorado Rivers cross many geologic structures that expose formations of differing erosional resistance. The resulting width of the alluvial valley and the channel gradient of different segments of the Green and Colorado Rivers vary by an order of magnitude, and these differences partly control the characteristics of incised valley meanders [*Hardin*, 1990].

Many of the narrow canyons are affected by debris flows from tributaries. Although the ratio of width of the alluvial valley to width of the channel is similar to that of incised meander reaches, stream power per unit bed area in debris flow-affected reaches is much greater because the channel is narrower and because channel gradient is steeper (Figure 2). Stream power per unit bed area, ω, was calculated as

$$\omega = \rho g Q_{2yr} s w^{-1},$$

where ρ is the density of water, g is the acceleration of gravity, Q_{2yr} is the 2-yr recurrence flood at the nearest gaging station for the period 1923 to 1962, s is the channel slope determined from 1:24000 scale topographic maps, and w is the reach average channel width (Table 1).

In reaches with abundant debris fans, large parts of the river bed are composed of gravel and coarser material. On the Green River, the proportion of all alluvial deposits composed of gravel was calculated from surficial geologic maps (Table 1). Between 35 and 64 percent of all alluvial bars include gravel on Green River reaches with abundant debris fans. In Grand Canyon, *Wilson's* [1986] side-scan sonar surveys showed that the percentage of the bed of the Colorado River composed of bedrock or boulders varied between 30 and 81 percent during three surveys in 1984.

Although the stream bed includes significant amounts of coarse material in reaches with abundant debris fans, large loads of sand are transported as suspended load and as bed load in the form of ripples and dunes. Some of the

Table 1. -- Geomorphic Characteristics of the Green and Colorado Rivers

Location	Length of mapping, in meters	Pre-dam 2-yr flood, in cubic meters per second	Channel slope, in meters per meter	Channel width, in meters	Stream power per unit length, in watts per meter	Stream power per unit area, in watts per square meter	Average valley width, in meters	Ratio of valley to channel width, in meters per meter	Post-dam proportion of fine sediment deposited within eddies	Post-dam proportion of alluvial deposits that are gravel
Grand Canyon										
Little Colorado confluence	4020	2127	0.0016	105	33000	310	125	1.2	0.75	0.41
middle Marble Canyon	10500	2148	0.00067	115	14000	120	225	2.0	0.44	0.03
Green River										
Lower Stillwater Canyon	15540	789	0.00037	125	2900	23	600	4.8	0	0
Upper Stillwater Canyon	17490	789	0.00028	150	2200	15	1200	8.0	0	0
Lower Labyrinth Canyon	16990	789	0.00018	135	1400	10	1200	8.9	0	0
Middle Labyrinth Canyon	12730	789	0.00019	140	1500	11	700	5.0	0	0
Upper Labyrinth Canyon	16090	789	0.0002	135	1500	11	800	5.9	0	0
below Gunnison Valley	16490	789	0.00057	155	4400	28	1000	6.5	0	0.14
Green River Valley	14090	789	0.0008	180	6200	34	4700	26.1	0	0.15
Grey Canyon	14930	789	0.0013	110	10000	91	500	4.5	0.29	0.36
Lower Desolation Canyon	16090	789	0.002	110	15000	140	500	4.5	0.19	0.64
Middle Desolation Canyon	16120	789	0.0011	125	8500	68	700	5.6	0.12	0.35
Upper Desolation Canyon	16860	789	0.00025	165	1900	12	1400	8.5	0.01	0.03
southern Uinta Basin	16020	789	0.00019	215	1500	7	1800	8.4	0	0
central Uinta Basin (south)	16490	789	0.00019	205	1500	7	5000	24.4	0	0
central Uinta Basin (north)	15330	621	0.00024	140	1500	11	7000	50.0	0	0
northern Uinta Basin	13670	621	0.00024	185	1500	8	5000	27.0	0	0.02
downstream from Split Mountain	16090	621	0.00099	130	6000	46	800	6.2	0	0.87

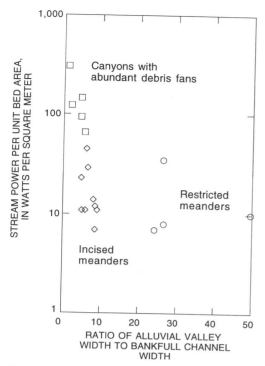

Fig. 2. Graph showing geomorphic characteristics of the Green and Colorado Rivers, and distinguishing restricted meanders, incised meanders, and canyons with abundant debris fans. Data are listed in Table 1.

suspended load, which also includes silt and clay, is deposited as bars and along the channel banks. The suspended load diffuses and is advected into eddies where it is deposited [*Andrews*, 1991; *Nelson et al.*, 1994]; thus, the size distribution of eddy bar sediments and measured sediment loads are similar [*Howard and Dolan*, 1981; *Schmidt*, 1990; *Schmidt et al.*, 1993]. *Andrews* [1986, 1990] has shown that the average annual sediment load has decreased by an order of magnitude since construction of Flaming Gorge and Glen Canyon Dams in 1962 and 1963, respectively.

4. THE FAN-EDDY COMPLEX

Although the meandering pattern of reaches with abundant debris fans may be similar to that of incised meanders, the characteristics of alluvial sedimentation are very different. Debris fans not only affect reach-scale channel attributes such as bed-material size and channel gradient, but fans also control the location and diversity of gravel and fine-grained deposits. Fine-grained deposits in narrow canyons unaffected by debris flows are less diverse and form long benches on alternating banks of the channel.

Alluvium is comprised of large proportions of vertical accretion deposits, and levees are common [*Nanson*, 1986].

The frequency of tributary junctions determines the number of debris fans that affect the channel [*Dolan et al.*, 1978]. In Grand Canyon, reaches may have many impinging debris fans (Figure 3). Upstream from each debris fan, a backwater of low-velocity flow may extend several kilometers [*Leopold*, 1969; *Kieffer*, 1985; *Miller*, 1995], and fine-grained alluvium may line these banks. Eddies exist in the lee of most constricting debris fans, and these eddies vary greatly in length. At high discharge, the downstream termination of these eddies (1) is caused by acceleration due to flow over or around a cobble/gravel bar, (2) is caused by narrowing of the bedrock or talus banks, or (3) occurs where the main channel flow impinges on curving channel banks. At low flow, many eddies terminate at exposed reattachment bars formed at higher discharges (Figure 4a and 4c). These channel irregularities cause eddies to be shorter than those predicted from laboratory experiments with similarly scaled constriction geometries [*Schmidt et al.*, 1993].

Eddy bars have distinctive topography and locations relative to the geometry of recirculating flow. *Schmidt* [1990] classified eddy bars based on observations of the Colorado River in Grand Canyon. Separation bars form near the flow-separation point and mantle the downstream parts of debris fans. Reattachment bars form under the primary eddy cell. Deposits not formed in eddies occur as channel-margin deposits that discontinuously line the banks.

Gravel bars are common (1) upstream from constrictions within backwaters of debris fans, and (2) downstream from large eddies. These bars either exist as mid-channel bars, or they may be attached to one bank. Attachment typically occurs on the bank opposite from the side where the debris fan enters the canyon. We refer to the geomorphic assemblage of backwater, constricting debris fan, eddy and eddy bars, and gravel bar as a fan-eddy complex (Figure 5). This assemblage is the fundamental geomorphic channel unit of canyons with abundant debris fans, and occurs at nearly every tributary mouth where debris fans constrict the river. The size of each channel element varies from site to site and is probably related to the size and characteristics of the associated debris fan, the frequency and magnitude of debris flows that replenish the fan, and the frequency and magnitude of main channel floods.

4.1. *Sedimentology of Fine-grained Deposits*

Separation and reattachment bars often have multiple topographic levels (Figure 4b and 4d). Typically, separation bars are of higher elevation and record evidence of

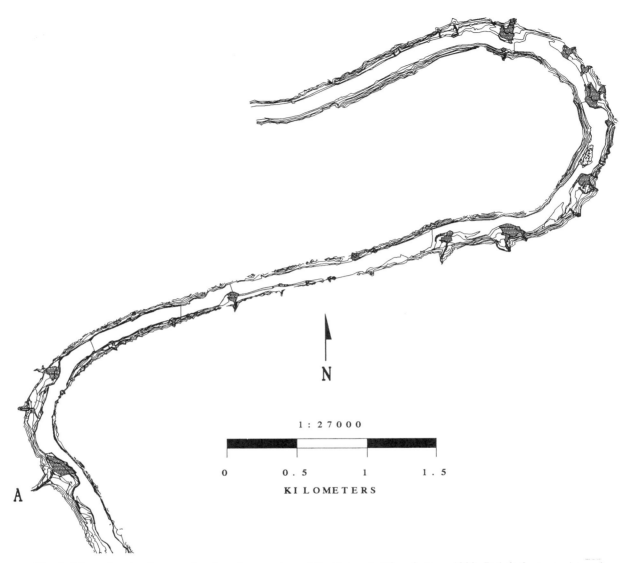

Fig. 3. Maps showing topography along two reaches of the Colorado River in June 1990. Debris fans are shown in dark shading. Contour interval is 2.5 m. (a) 9-km reach near Point Hansbrough and Saddle Canyon that begins 70 km downstream from Lees Ferry. Location A is shown in detail in Figures 4a and 4c. (b) 7-km reach near Little Colorado River confluence that begins 100 km downstream from Lees Ferry. Location B is shown in detail in Figures 4b and 4d.

higher formative discharges than do reattachment bars. Excavations of these deposits have been made at more than 20 sites throughout Grand Canyon [*Schmidt and Graf*, 1990; *Rubin et al.*, 1990, 1994]. In all cases, reattachment bars are composed of sedimentary structures indicative of rotary flow, similar to the pattern described by *Rubin et al.* [1990], or are composed of wave structures formed by processes described by *Bauer and Schmidt* [1993]. Separation bars are composed of a mixture of (1) fluvial structures consistent with secondary eddy cells and deposition in stagnating flow and (2) wave structures such as beach swash, wave ripples, and berms. Wave structures are more common in separation bars because these sites are closer to the wave source in the rapids.

Where debris fans are small or of low relief, alluvial deposits occur as continuous banks that extend downstream for several channel widths. These deposits may have ridges parallel or divergent to the orientation of main channel flow. Channel-parallel ridges are interpreted as levees formed by the same processes as on alluvial streams (Figure 6). Excavations indicate that these levees are composed of foresets indicating transport onshore and

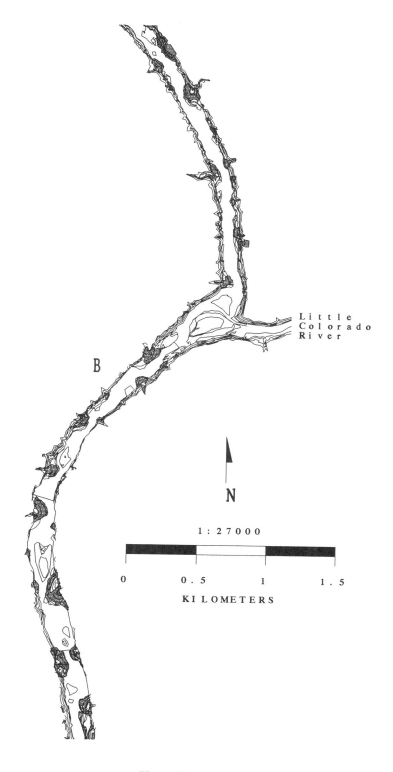

Figure 3 (continued)

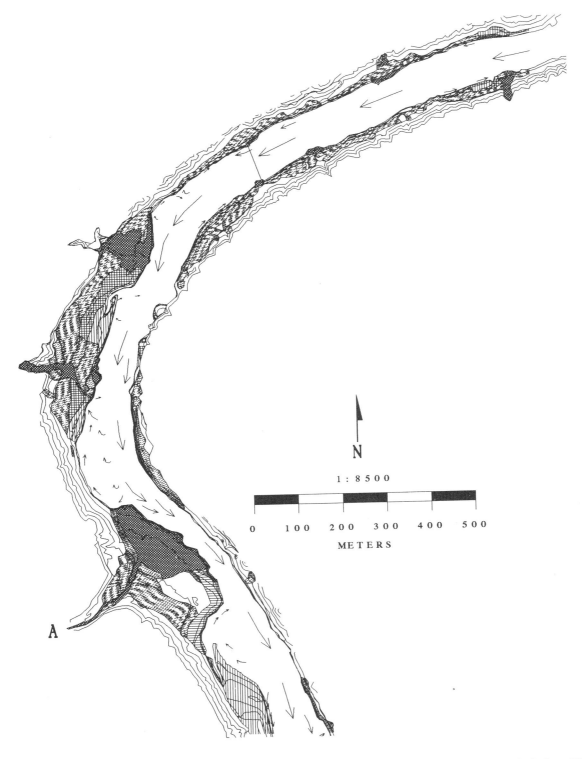

Fig. 4. Maps showing surface flow patterns at about 425 m³s⁻¹, major classes of river corridor deposits in June 1990, and topographic levels of fine-grained deposits in two reaches of the Colorado River in Grand Canyon. Surface flow patterns and major classes of deposits are shown in (a) and (b) where dark-shaded areas are debris fans, horizontal hatchures are separation bars, vertical hatchures are reattachment bars, cross-hatchures are undifferentiated eddy deposits, areas with broad lines are channel-margin deposits, and areas with large dots are gravel bars. Topographic levels of fine-grained deposits are shown in (c) and (d) where stippled areas are fluctuating flow sands deposited by

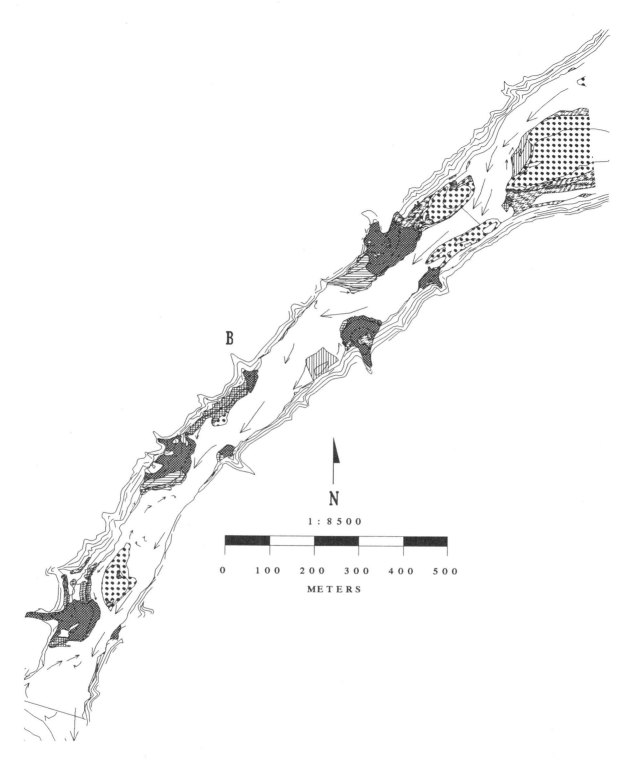

powerplant flows, arrowheads are high flow sands of 1984 to 1986, horizontal hatchures are flood sands of 1983, and areas with dashed lines are pre-dam deposits higher than those of 1983. (a) Major classes of river corridor deposits near Saddle Canyon. See Figure 3a for location. (b) Major classes of river corridor deposits downstream from the Little Colorado River. See Figure 3b for location. (c) Topographic levels of fine-grained alluvium near Saddle Canyon. See Figure 3b for location. (d) Topographic levels of fine-grained alluvium downstream from the Little Colorado River. See Figure 3b for location.

186 REGULATED STREAMFLOW, FINE DEPOSITS, AND EFFECTIVE DISCHARGE

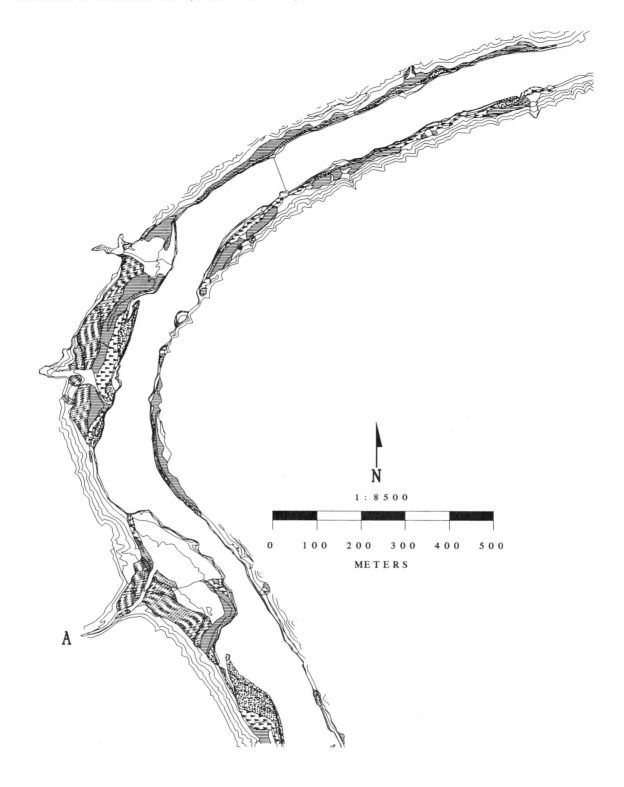

Figure 4 (continued)

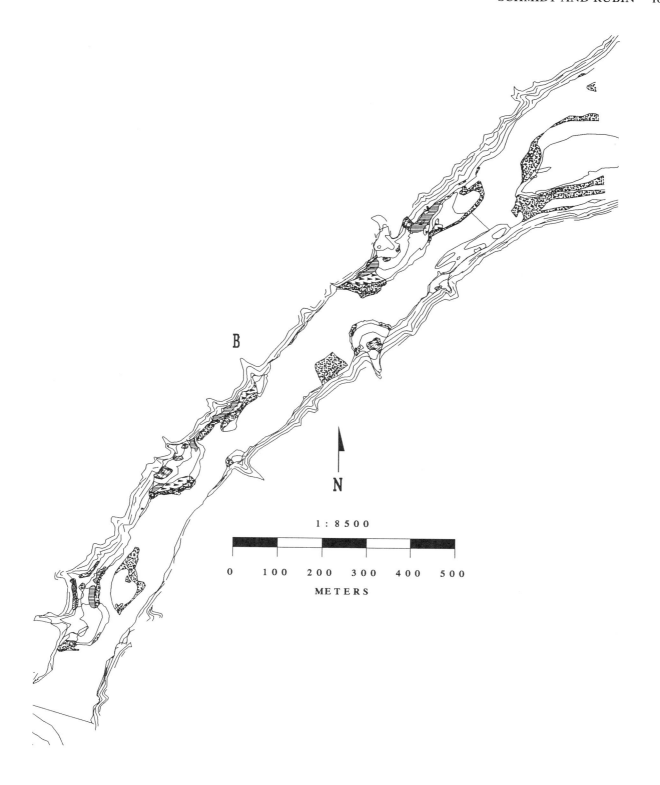

Figure 4 (continued)

Fig. 5. Photograph of a fan-eddy complex in Desolation Canyon, Green River, at low discharge, August 1992. Flow is towards the left side of photograph, and a 4.5 m long raft is shown in the lower part of the photo. Upstream from the debris fan is a central gravel bar, and fine-grained sediment lines the channel banks. Downstream from the debris fan is a reattachment bar, bounded on its upstream side by an eddy-return channel. The fine-grained sediment that mantles the downstream side of the fan is a separation bar. Further downstream is a gravel bar in the center of the channel.

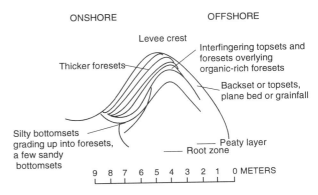

Fig. 6. Stratigraphy of a levee upstream from Little Nankoweap Creek, Grand Canyon. View is downstream, the river channel is to the right side of the levee, and vertical exaggeration is 3 times. This levee was formed in 1983.

downstream. Levees are typically composed of a single set of foresets that record the onshore migration and construction of the ridge. Divergent ridges occurring in series and that do not merge into higher downstream surfaces are also interpreted as levees. Channel-divergent ridges with sedimentary structures indicating rotary flow or where the crest of the ridge merges downstream with onshore alluvial surfaces are interpreted as narrow reattachment bars.

Alluvial deposits occur as distinct topographic surfaces throughout Grand Canyon, although the thickness of the associated deposits varies greatly (Figure 7). There are extensive topographic surfaces created by (1) the largest post-dam discharge, 2820 m^3s^{-1}, which occurred in June 1983, (2) high annual floods of about 1410 m^3s^{-1}, which occurred between 1984 and 1986, and (3) fluctuating flows within the capacity of the Glen Canyon Dam powerplant that are less than 890 m^3s^{-1} (Figure 8). The thickest deposits within eddy bars were formed by the 1983 flood, and contain large thicknesses of fluctuating-flow sands inset within their flanks. Deposits of the 1984 to 1986 floods are thin, despite the extensive area of the associated topographic surfaces.

High-elevation terraces composed of silty to very fine sand are common in some wider reaches of the Colorado River in Grand Canyon [*McKee*, 1938], and range in age from 50 yrs BP to at least 2000 yrs BP [*Hereford*, 1993; Hereford et al, 1993]. Our mapping did not focus on these deposits, and few excavations were made to establish sediment transport directions. Our estimates of the total proportion of fine-grained alluvium deposited within eddies is an underestimate because we classified all high terrace deposits as channel-margin deposits despite the fact that we observed rotary flow structures at some sites.

4.2. Distribution of Fine-Grained Deposits

Eddy processes are responsible for a large proportion of the fine-sediment deposition in canyons with abundant debris fans. In Grand Canyon, detailed mapping and sedimentologic analyses show that the proportion of fine-grained alluvium deposited within eddies is as large as 75 percent (Table 1). Along the Green River, reconnaissance photogeologic interpretation indicates that eddy deposits only occur in reaches with abundant debris fans, where they comprise between 1 and 29 percent of all fine sediment deposits (Table 1).

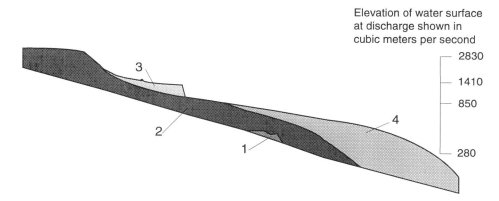

Fig. 7. Schematic diagram, not to scale, showing generalized internal structure and history of Grand Canyon reattachment bars. (1) Pre-dam deposits, eroded by high discharges of 1983. (2) Flood sands of 1983. This deposit truncates underlying pre-dam deposits and is truncated offshore by overlying deposits. Sedimentary structures are mainly fluvial dunes and climbing ripples. (3) Thin deposits of high flow (1984-1986) sands. These deposits truncate underlying 1983 flood sands and are of limited extent. They are typically bounded onshore by the 1983 deposits and are truncated offshore by younger deposits. Sedimentary structures are primarily climbing ripples, but are commonly trampled by humans or have been reworked by wind. (4) Deposits of recent (post-1986) discharges less than powerplant capacity. Sedimentary structures are primarily climbing ripples. Figure adapted from *Rubin et al.* [1994].

4.3 Depositional Patterns Prior to Reservoir Construction

The distinctions between separation and reattachment bars are not clear when large volumes of sediment are stored in eddies, such as occurred before dam construction. Aerial photographs of the Colorado River taken in 1935 show that the total amount of fine-grained sediments exposed at low discharge greatly exceeds the condition that has existed at any time since closure of Glen Canyon Dam. The channel bed within eddies typically was entirely covered with sand of sufficient thickness such that many eddy beds were entirely exposed at low discharge.

5. EFFECTIVE DISCHARGE

One of the goals of evaluating geomorphic effectiveness is to develop an understanding of the magnitude and frequency of discharges that determine the distribution and form of alluvial deposits. The comparison between effective discharge and modern alluvial deposits along Colorado Plateau rivers may also help define which dam-controlled discharges are of most importance in managing the downstream environment.

The effective discharge is defined as the modal value of the product of streamflow frequency and sediment transport. In the case of alluvial rivers, the suspended load or total load transport rate is used [*Andrews*, 1980, 1986; *Ashmore and Day*, 1988] because floodplain sediments are composed of the same sizes. In canyons with abundant debris fans, effective discharge calculations using suspended-sand transport rates apply to fine-grained alluvial deposits, and do not necessarily apply to coarse-grained alluvial deposits.

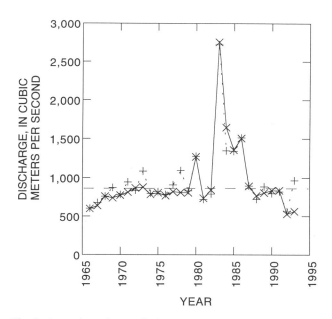

Fig. 8. Annual maximum discharge of the Colorado River at Lees Ferry, Arizona (stream gaging station 09380000), and near Grand Canyon, Arizona (09402500). Horizontal dashed line is maximum powerplant capacity of Glen Canyon Dam. Lees Ferry data are depicted with x's and Grand Canyon data with +'s. Until 1980, flood flows were entirely controlled because the upstream reservoir had not filled.

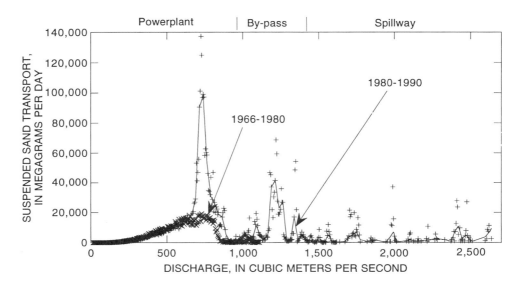

Fig. 9. Effective discharge curves for Colorado River near Grand Canyon, Arizona. Each symbol is calculated for a 25 m^{3}s^{-1} increment of discharge centered on the plotted point. X's are for 1965-1980 and +'s are for 1980-1990. Best fit smooth curves were calculated for each data set using the locally weighted least squared error method with a smoothing factor of 10 percent.

To accurately determine effective discharge, the evaluated time interval must be representative of the system in terms of the distribution of floods and the degree of sediment storage. On unregulated rivers, the time period over which these calculations are made must be sufficiently long so that rare floods are not given statistical importance beyond that associated with their expected recurrence. On regulated rivers where streamflow frequency is determined by basin hydrology as modified by reservoir operating rules, evaluations of geomorphic effectiveness should be consistent with the time period of a prevailing operating rule. If the operating rule changes, then the statistical distribution of downstream flow will change, and the effective discharge will change.

Between 1965 and 1980, the operating rule for Glen Canyon Dam was to completely control floods so as to fill its reservoir as quickly as possible; subsequently, the rule was to maintain a relatively full reservoir, and some floods were passed downstream [*U.S. Bureau of Reclamation*, 1993]. Between 1966 and 1979, annual maximum discharge of the Colorado River at Lees Ferry, Arizona, located 25 km downstream from Glen Canyon Dam, did not exceed powerplant capacity; annual maximum discharge near Grand Canyon, Arizona, exceeded powerplant capacity only in years when there was significant tributary flooding of the Little Colorado River (Figure 8). The cumulative duration of hourly releases from Glen Canyon Dam that exceeded powerplant capacity was 0.2 percent between 1966 and 1980. For this period prior to filling of the reservoir, the effective discharge curve of the Colorado River in Grand Canyon has a single mode (Figure 9). However, the curve is skewed with little transport at high discharges because of the operational restriction imposed by powerplant capacity. The effective discharge is about 700 to 750 m^{3}s^{-1}, but significant amounts of sand were transported by discharges as low as about 500 m^{3}s^{-1}. Because they are so infrequent, discharges greater than 850 m^{3}s^{-1} transported very little sand. The shape of this curve is similar to the post-dam effective discharge curve calculated by *Andrews* [1986] for the Green River.

Effective discharge for the period 1980 to 1990 was also about 725 m^{3}s^{-1}, but the dominant mode was more narrowly confined to discharges between about 650 and 825 m^{3}s^{-1}. The effective discharge curve for this period also shows that (1) high peak discharges between 1983 and 1986 transported large amounts of sand and (2) many increments of discharge transported little or no sand. Thus, the effective discharge curve has subsidiary modes. The existence of subsidiary modes is related to characteristics of dam operations at times when reservoir inflow was high and there was little available flood control capacity. In those circumstances, flows were maintained at maximum powerplant capacity. However, when overflow occurred, discharge was increased to the maximum capacity of the available overflow facilities. In the case of Glen Canyon Dam, when flows exceed powerplant capacity, discharge was increased to the full capacity of two by-pass tubes. Maximum discharge of the powerplant in combination with discharge from these tubes is about 1410 m^{3}s^{-1}, depending on reservoir elevation. Thus, the Colorado River rarely has had flows at discharges between 875 and 1150 m^{3}s^{-1}. In 1983, releases exceeded the capacity of the powerplant and by-pass tubes, and the emergency spillways were used. Dam releases in 1983 were held at approximately three dominant rates, and there are many increments of discharge greater than 1410 m^{3}s^{-1} that did not occur.

6. LONGITUDINAL CORRELATION OF GEOMORPHIC SURFACES AND RELATION TO EFFECTIVE DISCHARGE

The water surface elevation of the calculated effective discharge is typically compared with the elevation of the active floodplain of alluvial rivers in order to determine their formative flow [*Andrews*, 1980]. In the case of streams in canyons with abundant debris fans, the active floodplain and the bankfull channel are not obvious because channel migration does not provide an opportunity for development of lateral accretion deposits that are typical of many floodplains. Also, other parts of the channel, such as rapids, may be adjusted to extreme events [*Baker*, 1977; *Kieffer*, 1985]. Nevertheless, we can determine if any of the discontinuous fine-grained alluvial deposits are longitudinally correlative and have similar elevations to modes of the effective discharge calculation for sand transport. If such correlations exist, then the modal discharge increment may be considered to have produced these deposits even if other parts of the channel are shaped by other discharges.

In the case of Grand Canyon, discontinuous fine-grained deposits known to have formed by the same discharge can be identified, based on direct observation, repeated topographic surveys, and sedimentologic analysis [*Rubin et al.*, 1990; *Schmidt and Graf*, 1990]. Although these deposits have considerable relief, deposition approaches the water surface near stagnation points. Bars typically build to within about 0.3 m of the water surface [*Schmidt and Graf*, 1990; *Schmidt and Andrews*, unpubl. data]. Longitudinal correlation of the elevation of deposits formed near the reattachment point provides a consistent estimation of the water surface because the elevation of the water surface at the reattachment point is approximately the same as that of the adjacent main flow. Where internal stratification is visible, deposits formed near the reattachment point can be recognized by ripple structures produced by flow that reverses in an upstream-downstream direction [*Rubin et al.*, 1990]. Without visible internal stratification, the

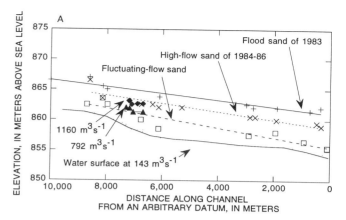

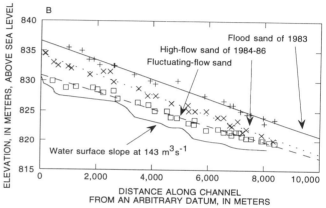

Fig. 10. Graphs showing longitudinal correlation of flood sand, high flow sands, and fluctuating-flow sands in two study reaches. (a) Reach near Saddle Canyon that is shown on Figure 3a. Dark circles and dark triangles are surveyed water surface elevations in 1985 at indicated discharges. (b) Reach near Little Colorado River that is shown on Figure 3b.

highest elevation part of reattachment bars can be used to approximate the water surface elevation.

The elevation of separation bars is not included in longitudinal correlations because the water surface of the upstream part of eddies, near the separation point, is considerably less than that of the adjacent main channel flow. Measurements of water surface in Grand Canyon at 1200 m^{3}s^{-1} show that the elevation of the water surface within an eddy near the separation point may be as much as 0.2 m lower that the elevation of the adjacent downstream-flowing water surface.

The elevation of reattachment point deposits known to have formed in 1983 and between 1984 and 1986 correlate well over long distances, and the average longitudinal slope of these deposits parallels that of the average low flow slope (Figure 10). The correlation of these deposits thus can be an useful tool for estimating formative discharges.

In the Colorado River in Grand Canyon, elevations of three longitudinally correlative surfaces are associated with different modes of the effective discharge calculation for the period 1980 to 1990. When averaged over a 10-yr period, discharges greater than powerplant capacity yield subsidiary modes and geomorphically recognizable deposits, and surfaces associated with these flows are widely exposed. In many reaches, deposits associated with the primary mode are less extensive than are these higher flood deposits. Deposits formed by floods in 1983 and between 1984 and 1986 do not constitute floodplains in the sense that they were not incrementally constructed, are not laterally continuous, and are not formed by channel migration. The 1983 deposits were constructed by one geomorphically effective event that has several subsidiary modes of the streamflow-frequency-sediment-transport product, and the 1984 to 1986 deposits were shaped by three successive years of similar peak discharges that have one mode. Each of these suites of high-flow deposits is now being modified by subaerial erosion, which subdues, but does not completely destroy, the fluvial attributes of the landforms.

Despite the correlative nature of these distinct high-elevation topographic surfaces, the thickness of the underlying deposits differs greatly. Bars and levees formed in 1983 commonly approach or exceed 1 m in thickness, whereas deposits formed between 1984 and 1986 are rarely more than 0.3 m thick. Thus, there is a significant disparity between the calculated effectiveness of these discharges, and the effectiveness as evaluated by the characteristics of the deposits themselves. The likely explanations for this disparity are that (1) main channel sediment transport was relatively low in 1984 to 1986 due to depletion of sediment by the 1983 flood, or (2) deposition rates in eddies were lower in 1984 to 1986 than in 1983 because the eddies were already partly filled with sediment. Thus, in fluvial systems where sediment supply and deposition rates may vary, the modes of the product of streamflow frequency and sediment transport may not successfully predict the thickness of associated deposits, despite the fact that extensive topographic surfaces are created.

The lower-elevation depositional surfaces along the Colorado River in Grand Canyon that formed by discharges within the range of powerplant capacity correlate with the modal discharge increment of about 725 m^{3}s^{-1}. These deposits are incrementally constructed by the river by discharges that occur nearly every year (Figure 8). The mechanisms of deposition are (1) the same eddy processes that occur at higher discharges, and (2) colonization by riparian plants and vertical accretion of silt and clay on top

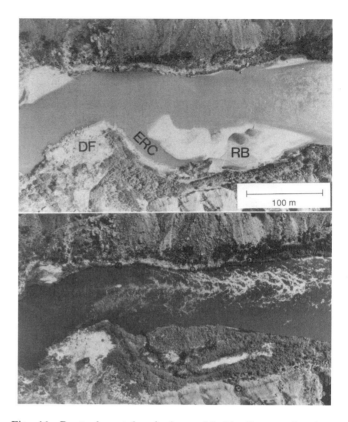

Fig. 11. Reattachment bar in lower Marble Canyon showing unvegetated and vegetated condition of the same site. DF is location of debris fan, RB is reattachment bar, and ERC is eddy return channel formed by upstream-directed eddy currents. (a) Unvegetated condition in October 1984. (b) Vegetated condition in May 1980.

of flood-formed sand bars. The former process was documented by *Rubin et al.* [1990] who demonstrated that a wedge of fluctuating-flow fine sand was deposited as an inset fill against the flood-formed main platform of the reattachment bar. The latter process is illustrated in Figure 11. This reattachment bar has alternated between an exposed sand condition, as existed in 1984 (Figure 11A) during the period of high discharges, and nearly complete vegetative overgrowth, as existed during the period of reservoir filling (Figure 11b). *Stevens et al.* [1995] described the succession of riparian plants on reattachment bars and show that such plants preferentially colonize silts and clays.

7. DISCUSSION

Alluvial deposits in canyons with abundant debris fans have numerous surfaces constructed by several discharges.

In the case of Grand Canyon, deposits formed by rare post-dam floods, and lower-elevation deposits constructed by powerplant discharges, both occur extensively. During periods of significant flood control, such as the period of initial filling of large reservoirs, reattachment bars may become vegetated, but subsequent floods that occur after reservoir filling can reactivate these surfaces in a process similar to the disequilibrium floodplain model of *Nanson* [1986].

Eddy bars persist in specific zones of recirculation because the coarse-grained debris fans that obstruct the river channel give rise to flow separation, and these fans are rarely modified. Although bars change shape with discharge, they remain within specific lateral separation eddies and do not migrate from eddy to eddy. Measurements and observations of the Colorado River in Grand Canyon based on aerial photography (dating to 1935) and oblique photography (dating to the 1880's) [*Webb*, 1995] show that the locations of eddy sand bars have been stable for long periods. Observations of relations between flow geometry and sand-bar location suggest that bars should be persistent over periods consistent with the frequency of events that reshape flow-separation-inducing debris fans. In Grand Canyon, that time scale is on the order of 10 to 100 yrs [*Melis et al.*, 1994].

Large floods that overtop low-relief debris fans may cause recirculation zones to diminish in size or disappear [e.g. *Kieffer and others*, 1989, fig. 3.5]. *Melis et al.* [1994] have shown that most debris fans in Grand Canyon are overtopped by discharges at or greater than the pre-dam mean annual flood, and photographs of the river at discharges greater than 2830 m^3s^{-1} show that many eddies are thin or non-existent at such discharges. Because eddy deposits contain sedimentary structures indicative of recirculation, eddy bars therefore must form in flows less than those that completely inundate the controlling constriction. Thus, most fine-grained alluvial deposits form at discharges less than those that overtop debris fans or during the descending limb of fan-overtopping floods after eddies have been reestablished.

Eddy bars are subject to scour and fill over various time scales. Interpretation of sedimentary structures shows that eddy bars are dynamic features, subject to erosion and deposition during floods and erosion after flood recession. The topographic form and internal stratigraphy of bars results from the range of eddy geometries that occur at each site, which are dependent on site-specific channel geometry-discharge relations. Eddy bars associated with low debris fans that are overtopped frequently by mainstem flooding are likely to have different scour-and-fill histories than eddy bars formed in the lee of high-elevation debris fans that are overtopped less frequently.

8. CONCLUSIONS

In canyons with abundant debris fans, the fundamental geomorphic unit is a complex of fan-related features: (1) a low-velocity backwater upstream from the fan, (2) a debris fan that constricts the channel, (3) eddies and eddy bars in the expansion downstream from the fan, and (4) a downstream gravel bar. Mapping of alluvial deposits in Grand Canyon demonstrates that eddies are the dominant environment for deposition of fine-grained sediment.

Previous work on alluvial rivers has shown that channel morphology is controlled by the effective discharge, which is calculated to be the modal value of the product of streamflow frequency and sediment transport rate. The present study indicates that the concept of effective discharge can also be applied to fine-grained deposits in narrow canyons with abundant debris fans.

Pre-dam effective discharge for the Green River had a single dominant mode [*Andrews*, 1986]; the Colorado River was probably similar in this regard. Post-dam effective discharge, however, is very different. *Hirsch et al.* [1990] have shown that the Colorado River basin has the highest proportion of reservoir volume to mean annual flow of any large drainage basin in the United States or Canada. Because the discharge in these rivers follows technological rules that are related to dam operations, streamflow frequency and the calculated effective discharge curve have multiple modes. Calculations of post-dam effective discharge that only use streamflow data for the period prior to filling of large reservoirs do not show subsidiary modes and do not anticipate the resulting geomorphic adjustment once large reservoirs fill.

Mapping of alluvial suspended-load deposits in Grand Canyon demonstrates that each of the longitudinally extensive post-dam geomorphic surfaces corresponds with one of the multiple modes in calculated effective discharge. Volume of these deposits, however, is not proportional to the area of the corresponding mode in the calculated effective discharge curve.

Acknowledgments. This work was partly supported by the U.S. Bureau of Reclamation Glen Canyon Environmental Studies program and the U. S. Fish and Wildlife Service Recovery Program for the Endangered Fishes of the Upper Colorado. Helpful comments on an earlier draft were provided by E. D. Andrews, A. D. Howard, T. S. Melis, and P. R. Wilcock. M. G. Wolman observed and discussed these processes with us in the field during a Grand Canyon river expedition in 1985.

REFERENCES

Andrews, E. D., Effective and bankfull discharge of streams in the Yampa River basin, Colorado and Wyoming, *J. Hydrol., 46,* 311-330, 1980.

Andrews, E. D., Downstream effects of Flaming Gorge Reservoir on the Green River, Colorado and Utah, *Geol. Soc. Amer. Bull., 97,* 1012-1023, 1986.

Andrews, E. D., The Colorado River: a perspective from Lees Ferry, Arizona, in *Surface Water Hydrology,* edited by M. G. Wolman and H. C. Riggs, pp. 304-310, The Geological Society of America, The Geology of North America vol. O-1, 1990.

Andrews, E. D., Deposition rate of sand in lateral separation zones, Colorado River, abst., Eos, 72, 219, 1991.

Ashmore, P. E. and T. J. Day, Effective discharge for suspended sediment transport in streams of the Saskatchewan River basin, *Water Resour. Res., 24,* 864-870, 1988.

Baker, V. R., Stream-channel response to floods, with examples from central Texas, *Geol. Soc. Amer. Bull., 88,* 1057-1071, 1977.

Baker, V. R., Flood sedimentation in bedrock fluvial systems, in *Sedimentology of Gravels and Conglomerates,* edited by E. H. Koster and R. J. Steel, pp. 87-98, The Canadian Society of Petroleum Geologists, Calgary, 1984.

Baker, V. R., R. C. Kochel, P. C. Patton, and G. Pickup, Paleohydrologic analysis of Holocene flood slack-water sediments, *Spec. Publs Int. Assoc. Sediment., 6,* 229-239, 1983.

Bauer, B. O. and J. C. Schmidt, Waves and sandbar erosion in the Grand Canyon: Applying coastal theory to a fluvial system, *Annals Assoc. Amer. Geogr., 83,* 475-497, 1993.

Dolan, R., A. D. Howard, and D. Trimble, Structural control of the rapids and pools of the Colorado River in the Grand Canyon, *Science, 202,* 629-631, 1978.

Dutton, C. E., The Tertiary history of the Grand Canyon district, 264 p., *Monograph 2,* U.S. Geol. Survey, 1882.

Garrett, W. B., E. K. VanDeVanter, and J. B. Graf, Streamflow and sediment-transport data, Colorado River and three tributaries in Grand Canyon, Arizona, 1983 and 1985-86, 624 p., *Open-File Rep. 93-174,* U.S. Geol. Surv., 1993.

Graf, W. L., Rapids in canyon rivers, *J. Geol., 87,* 533-551, 1979.

Hardin, D. R., Controlling factors in the distribution and development of incised meanders in the central Colorado Plateau, *Geol. Soc. Amer. Bull., 102,* 233-242, 1990.

Hereford, R., Map showing surficial geology and geomorphology of the Palisades Creek archaeologic area, Grand Canyon, Arizona, *Open-File Rep. 93-553,* U.S. Geol. Surv., 1993.

Hereford, R., H. C. Fairley, K. S. Thompson, and J. R. Balsom, Surficial geology, geomorphology, and erosion of archaeologic sites along the Colorado River, eastern Grand Canyon, Grand Canyon National Park, Arizona, 46 p., *Open-File Rep. 93-517,* U. S. Geol. Surv., 1993.

Hirsch, R. M., J. F. Walker, J. C. Day, and R. Kallio, The influence of man on hydrologic systems, in *Surface Water Hydrology,* edited by M. G. Wolman and H. C. Riggs, pp. 329-359, Geol. Soc. Am., The Geology of North America vol. O-1, 1990.

Howard, A. D. and R. Dolan, Geomorphology of the Colorado River in the Grand Canyon, *J. Geol., 89,* 269-298, 1981.

Hunt, C. B., Geologic history of the Colorado River, p. 59-130, Prof. Paper 669-C, U.S. Geol. Surv., 1969.

Kearsley, L. H., J. C. Schmidt, and K. D. Warren, Effects of Glen Canyon Dam on Colorado River sand deposits used as campsites in Grand Canyon National Park, USA, *Regulated Rivers, 9,* 137-149, 1994.

Kieffer, S. W., The 1983 hydraulic jump in Crystal Rapid: Implications for river-running and geomorphic evolution in the Grand Canyon, *J. Geol., 93,* 385-406, 1985.

Leopold, L. B., The rapids and the pools -- Grand Canyon, p. 131-145, *Prof. Paper 669-D,* U.S. Geol. Surv., 1969.

McKee, E. D., Original structures in Colorado River flood deposits of Grand Canyon, *J. Sed. Pet., 8,* 77-83, 1938.

Melis, T. S., R. H. Webb, P. G. Griffiths, and T. J. Wise, Magnitude and frequency data for historic debris flows in Grand Canyon National Park and vicinity, Arizona, *Water Res. Inves. Rept 94-4214,* U. S. Geol. Surv., 1994.

Miller, A. J., Debris-fan constrictions and flood hydraulics in river canyons: some implications from two-dimensional flow modelling, *Earth Surf. Proc. Ldfms., 19,* 681-697, 1995.

Nanson, G. C., Episodes of vertical accretion and catastrophic stripping: A model of disequilibrium flood-plain development, *Geol. Soc. Amer. Bull., 97,* 1467-1475, 1986.

Nanson, G. C. and J. C. Croke, A genetic classification of floodplains, *Geomorphology, 4,* 459-486, 1992.

Nelson, J. M., R. R. McDonald, and D. M. Rubin, Computational prediction of flow and sediment transport patterns in lateral separation eddies, abst., Eos, 75, 268, 1994.

O'Conner, J. E., L. L. Ely, E. E. Wohl, L. E. Stevens, T. S. Melis, V. S. Kale, and V. R. Baker, A 4500-year record of large floods on the Colorado River in the Grand Canyon, *J. Geol., 102,* 1-9, 1994.

Pemberton, E. L., Sediment data collection and analysis for five stations on the Colorado River from Lees Ferry to Diamond Creek, 159 p., U.S. Bur. Reclamation Glen Canyon Envir. Studies report, 1987. (NTIS No. PB88-183397/AS)

Powell, J. W., Exploration of the Colorado River of the West and its tributaries, 291 p., U.S. Govt. Printing Office, Wash., D. C., 1875.

Rubin, D. M., J. C. Schmidt, R. A. Anima, K. M. Brown, H. Ikeda, B. E. Jaffe, R. McDonald, J. M. Nelson, T. E. Reiss, R. Sanders, and R. G. Stanley, Internal structure of bars in Grand Canyon, Arizona, and evaluation of proposed flow alternatives for Glen Canyon Dam, 16 p., *Open-file Report OF 94-594,* U. S. Geol. Surv., 1994.

Rubin, D. M., J. C. Schmidt, and J. N. Moore, Origin, structure, and evolution of a reattachment bar, Colorado River, Grand Canyon, Arizona, *J. Sed. Pet., 60,* 982-991, 1990.

Schmidt, J. C., Compilation of historical hydrologic and geomorphic data for the upper Colorado River basin, Annual Report, Flaming Gorge Research Program Study No. 37, 5 p. w. append., 1994.

Schmidt, J. C., Recirculating flow and sedimentation in the Colorado River in Grand Canyon, Arizona, *J. Geol., 98,* 709-724, 1990.

Schmidt, J. C. and J. B. Graf, Aggradation and degradation of alluvial sand deposits, 1965 to 1986, Colorado River, Grand Canyon National Park, Arizona, 74 p., *Prof. Paper 1493*, U.S. Geol. Surv., 1990.

Schmidt, J. C., J. L. Mayers, and M. F. Leschin, Geomorphology of post-Glen Canyon Dam fine-grained alluvial deposits of the Colorado River near Point Hansbrough and Saddle Canyon in Grand Canyon National Park, 16 p., U.S. Bureau of Reclamation Glen Canyon Environmental Studies administrative report, Flagstaff, 1994a.

Schmidt, J. C., J. L. Mayers, M. F. Leschin, and D. M. Rubin, Geomorphic mapping of post-Glen Canyon Dam fine-grained alluvial deposits of the Colorado River in Grand Canyon, Arizona: Abstracts, Geol. Soc. Am. Annual Meeting, 1994b.

Schmidt, J. C., D. M. Rubin, and H. Ikeda, Flume simulation of recirculating flow and sedimentation, *Water Resour. Res., 29,* 2925-2939, 1993.

Stevens, L. E., J. C. Schmidt, T. J. Ayers, and B. T. Brown, Geomorphic influences on fluvial marsh development along the dam-regulated Colorado River in the Grand Canyon, Arizona, *Ecol. Applications, 5,* in press, 1995.

U.S. Bureau of Reclamation, Operation of Glen Canyon Dam, draft environmental impact statement, 324 p., Salt Lake City, 1993.

U.S. Water Resources Council, Guidelines for determining flood flow frequency, 183 p., Bulletin 17B, 1981.

Webb, R. H., *A century of environmental change in Grand Canyon,* in press, Tucson, Univ. Ariz. Press, 1995.

Wilson, R. P., Sonar patterns of Colorado River bed, Grand Canyon, in Proc. Fourth Fed. Interagency Sed. Conf., Las Vegas, Subcommittee on Sedimentation, Interagency Advisory Committee on Water Data, pp. 5-133 to 5-142, 1986.

Wolman, M. G. and R. Gerson, Relative scales of time and effectiveness of climate in watershed geomorphology, *Earth Surf. Proc. Lndf., 3,* 189-208, 1978.

Wolman, M. G. and L. B. Leopold, River flood plains -- some observations on their formation, p. 87-109, *Prof. Paper 282-C,* U.S. Geol. Surv., 1957.

Wolman, M. G. and J. C. Miller, Magnitude and frequency of forces in geomorphic processes, *J. Geol., 68,* 54-74, 1960.

J. C. Schmidt, Department of Geography and Earth Resources, Utah State University, Logan, UT 84322-5240.

D. M. Rubin, U.S. Geological Survey, MS 999, 345 Middlefield Road, Menlo Park, CA 94025.

Hydrologic Factors in Regeneration of Fremont Cottonwood Along the Fremont River, Utah

Benjamin L. Everitt

Utah Division of Water Resources, Salt Lake City, Utah

A regeneration of thickets of Fremont cottonwood along the middle portion of the Fremont River, Utah, followed the spring snowmelt flood of 1973. The 1973 spring flood was the largest May discharge in 25 years of record at Caineville Gage. Although smaller than late season floods, the 1973 spring flood was sufficiently large to cause channel scour and overbank deposition, and occurred during the floodplain construction phase of the geomorphic cycle on the Fremont River. The flood was followed by a season of high base flow, and several hydrographically quiet years. Cottonwood has not regenerated downstream from the confluence of Muddy Creek at Hanksville, where the spring flood becomes insignificant compared to late season floods. The ten storage reservoirs in the Fremont basin are high in the watershed, and are operated for irrigation storage only. Their passive regulation of the flow of the lower river most likely aided the 1973 cottonwood regeneration.

1. INTRODUCTION

Vegetation change along riparian corridors of the western United States is of increasing interest as land and resource managers wrestle with issues of environmental policy. The relationship between hydrology and riparian vegetation dynamics is sufficiently complex that general principles have been slow to emerge from the few case histories examined. A consensus is developing, however, that the life cycle of cottonwood flows from the seasonal pattern of flooding [*Fenner and others*, 1985; *Rood and Mahoney*, 1993a and 1993b; *Johnson*, 1994; *Everitt*, 1968].

In the late 1960's, Fremont cottonwood (*Populus fremontii*) appeared to be a relict species on the lower Fremont River (Figure 1) below Capitol Reef National Park [*Everitt*, 1971]. In 1967, the Fremont valley bottom contained two well-defined surfaces: the active floodplain, and a terrace three to four meters above the floodplain (Figure 2). Old cottonwoods and greasewood (*Sarcobatus vermiculatus*) occupied the terrace, while the introduced tamarisk (*Tamarix ramosissima*) occupied the active floodplain [*Everitt*, 1980b, Figure 4] with no contribution from cottonwood.

Within the last two decades a cohort of cottonwood has become established along the Fremont River, resulting in many acres of sapling thickets [*Graf*, 1982, Figs. 9.7 & 9.8]. The maturing trees are now shading out and replacing tamarisk (Figure 2), showing that cottonwood is still the dominant riparian species where the two are in competition for sunlight and moisture. Recent research has shown that, although both cottonwood and tamarisk depend on river flooding to complete their life cycles, their site preferences are different [*Irvine & West*, 1979], they bloom and seed in different seasons [*Warren and Turner*, 1975], and they respond differently to flooding [*Everitt*, 1980a]. The spread of tamarisk along southwestern rivers is at least partly due to its occupation of the vacuum left by the prior collapse of native riparian communities. Study of cottonwood, the dominant riparian woody species, is therefore of primary importance in understanding riparian ecology in general, and tamarisk ecology in particular.

2. LOCATION

The Fremont River rises at 3,300 m in the high plateaus of central Utah and flows eastward to the town of Hanksville, where it joins Muddy Creek (Figure 1) at an elevation of 1,300 m. As the Dirty Devil River, it then continues southeastward in 80 km of slick-rock canyon (Figure 3) to the Colorado River at the head of Lake Powell.

Natural and Anthropogenic Influences in Fluvial Geomorphology
Geophysical Monograph 89
Copyright 1995 by the American Geophysical Union

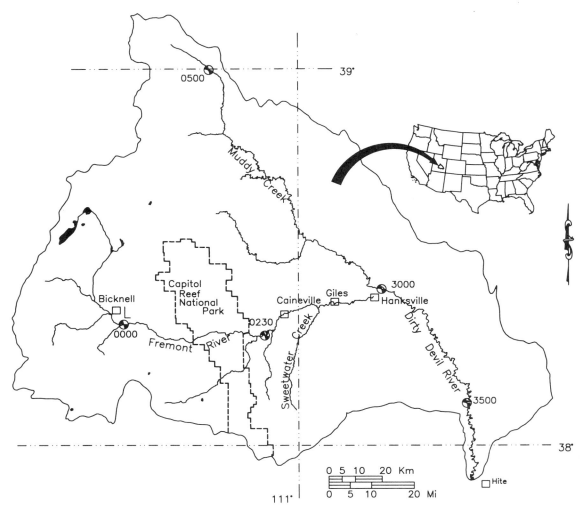

1. Location map. USGS Gaging stations are: 09330000 Fremont River near Bicknell, Utah, 09330230 Fremont River near Caineville, Utah, 09330500 Muddy Creek near Emery, Utah, 09333000 Dirty Devil River near Hanksville, Utah, and 09333500 Dirty Devil River above Poison Spring Wash, near Hanksville, Utah

In its descent from the plateau, it passes from spruce-fir forest watered by up to 1,000 mm of annual precipitation, mostly winter snow, to desert scrub-land of bi-seasonal precipitation averaging (at Hanksville) 130 mm per year. The river rises in highlands underlain by volcanic rock, and flows through lowlands eroded in sandstone, siltstone, and shale of Mesozoic age, which provide an abundant supply of fine-grained sediment to the lower reaches.

3. METHODS AND SOURCE OF DATA

This study describes the woody vegetation of the Fremont River and Dirty Devil Valleys downstream from Caineville, reviews the age structure of the cottonwood population, and discusses its relationship to hydrographic events. Data come from observations of the middle reach made between 1966 and 1993, and a reconnaissance of the Dirty Devil reach in May of 1984. Profiles were surveyed by pacing or chaining, and hand-level. Cottonwood ages were sampled by ring count as close to the present ground level as practical. At each locality, the samples included the largest tree and several of the smallest, plus several of intermediate size. Because most individuals have been beaver-cut, most ages were derived by adding the rings of beaver-cut stumps to the rings of stump-sprouts.

The primary hydrographic information for this study comes from USGS gaging stations "Fremont River near

2. Fremont River looking upstream, above Hanksville near State Highway mile post 113, May 27, 1993. Discharge is 1.5 m³/s. The 1973 cottonwoods occupy the riverbank in the foreground; older trees are on the 1896 terrace in the middle distance.

3. The Dirty Devil River, looking downstream about 32 km below Hanksville, May 28, 1984. The flow of 15 m³/s is barely to the edge of the cane thickets which border the channel, along with scattered tamarisk. Rabbitbrush and greasewood occupy the higher levels of the flood plain, out of view.

Caineville" and "Dirty Devil River above Poison Spring Wash" [*Christensen and others,* 1987], which bracket the study area. The Caineville gage was established in 1967 just upstream from the Caineville diversion. It is at an elevation of 1450 m and records drainage from 3.1×10^5 ha. The Poison Spring gage, 80 km below Hanksville and just above the mouth of the river in Lake Powell, was established in 1964, replacing an earlier gage at Hite, now beneath the lake. The Poison Spring gage is at an elevation of 1175 m, and records drainage from 1.1×10^6 ha. At Hanksville, a crest stage station above the mouth of Muddy Creek and a gaging station just below were operated briefly from 1945 to 1948. A gaging station has been operated intermittently at Bicknell beginning in 1909, but was out of service in 1973. Water budget data appear in *DWR* [1975].

To correlate streamflow characteristics with vegetation dynamics I have used mean daily discharge (MDD) data because riverbank vegetation is influenced by both peak flow and duration, and MDD provides a reasonable combination of peak flow and duration in a single parameter. Moreover, the MDD is an efficient way to compare floods of different seasons and on different reaches of river. *Whitaker* [1964, Figure 7] found that for all sizes of drainage basins in his region "C" the instantaneous peak discharges for rainfall floods exceed those for snowmelt floods. However, the summer rainfall floods tend to be local and of short duration (less than 24 hours). Therefore, although the instantaneous peak discharges can be very large, the duration and total volume of summer floods are modest compared to snowmelt floods or autumn floods. The amplitude (instantaneous peak gage height) and duration of flood waves is partly a function of channel geometry and sinuosity [*Burkham,* 1976; *Smith,* 1978], which have continued to change throughout the historic period.

The hydrographic data are grouped by water year, which runs from October through the following September. Thus in Table 2, for example, the flood at the Poison Spring gage listed for October 20, water year 1973, occurred in calendar year 1972.

4. ALLUVIAL FILL

Downstream from Capitol Reef the Fremont-Dirty Devil River flows on sandy alluvium. The maximum thickness of valley fill disclosed in water well logs is 15 m at Caineville and 27 m at Hanksville. The valley width is five to ten times the present width of the channel above Hanksville, but less below, leaving moderate room for meandering. Artificial grade control is provided only at the Hanksville diversion dam, one mile upstream from Hanksville.

5. PHYSIOGRAPHY OF THE VALLEY FLOOR

The physiographic description of valley surfaces is after *Osterkamp and Hupp* [1984, Figure 1] and is illustrated in Figure 4. Frequency of flooding, sediment size, depth to dry season water table, and other environmental factors vary with physiographic position, and influence vegetation survival.

On the Fremont, the *channel bed* (CB in Figure 4)

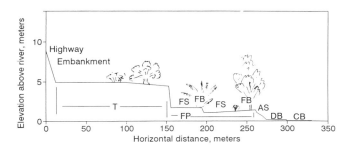

4. Pace and hand level traverse of the Fremont River valley, running S 16° W from mile post 111, Utah State Highway 24. The physiographic divisions are labeled with binomial initials:

CB	Channel bed
DB	Depositional bars
AS	Active channel shelf
FP	Active flood plain
FB	Flood bars
FS	Floodplain swales
T	Terrace (abandoned floodplain)

carries the low flow, when there is any, and is wet most of the time. *Depositional bars* (DB) are ephemeral features in or adjacent to the channel bed formed of newly deposited sediment, and unvegetated except by sparse annuals and seedlings (foreground, Fig. 3). The Fremont has an *active channel shelf* (AS) which is a composite of many flood-deposited treads separated by steps and rising 0.9 - 1.2 meters above low water. The break between the channel shelf and the more horizontal surface of the *active flood plain* (FP) is a matter of judgment. The floodplain surface undulates between *flood bars* (FB) and *swales* (FS), with a relief of 0.3 - 0.6 meters. Behind the flood plain are remnants of *terrace* (T), an abandoned floodplain surface. The terrace is usually about 3 meters above the active floodplain and separated from it by an erosional scarp. The substrate is friable fine sand with minor silt, with some clay locally in the swales.

The geomorphology of the valley is formed by the river and its floods, and is therefore ephemeral. Features closer to the river and lower in elevation are inevitably younger and more rapidly evolving than those higher and more distant. The floodplain vegetation follows this pattern, being younger and more rapidly evolving closer to the accreting riverbank, as described on the Little Missouri River [*Everitt*, 1968].

6. GEOMORPHIC HISTORY

During the last century the Fremont River passed through a geomorphic cycle analogous to the "arroyo-cutting" cycle of smaller drainages. In the 1890's a series of floods gutted the Fremont Valley, flushing large volumes of sediment downstream, and transformed the formerly narrow sinuous channel into a broad braided one [*Graf*, 1983; *Hunt and others*, 1953]. The first damaging flood occurred in September of 1896 [*Webb*, 1985; *Graf*, 1983]. A flood of September 1909 precipitated the abandonment of the towns of Giles and Caineville [*Webb*, 1985; *Godby*, 1981]. *Wooley* [1946] lists a large flood of November 6, 1916 "laying waste to farms and orchards" in Caineville, Giles, and Hanksville. Flooding, downcutting, and channel widening continued for at least a decade, and probably for several decades. *Webb* [1985] believes that these autumn floods were significantly larger than any recorded at the Caineville gage since 1967.

When *Hunt et al.* [1953] conducted field work in the 1930's the river channel was still wide and braided. Tributary arroyos had carried the pulse of erosion far upstream into the watershed, feeding large quantities of sediment into the river [*Graf*, 1983, Figure 5].

Since about 1940 the main channel has been gradually shrinking and constructing a floodplain. Observed changes are similar to those described for the adjacent Paria River [*Hereford*, 1986], although there is no topographic or sediment budget information for the Fremont. In 1993 the Fremont flowed mostly in a single, moderately sinuous channel about 30 m wide. Channel width increases gradually downstream to and beyond Hanksville. For 32 km below the mouth of Muddy Creek the channel of the Dirty Devil remains braided and is up to 150 m wide. Floodplain construction continues slowly with sediment derived from the waning pulse of erosion migrating headward in tributaries [*Graf*, 1982]. It is this "flood plain construction" as defined by *Schumm and Lichty* [1963] which has set the stage for the development of the floodplain vegetation to its present state.

The age of the valley surface is well constrained by historical documentation. The terrace (T in Figure 4) with its old cottonwoods, is the pre-1896 floodplain. The active floodplain inset within it post-dates the 1890's, but was apparently stabilized only about 1940. The oldest individual plant (tamarisk) found on the floodplain dates from the 1950's.

7. VEGETATION

The woody shrubs and trees that occupy the floodplain of the Fremont River between Caineville and Hanksville can be classified as either pioneer species, those which first occupy new ground, or seral (successionary) species, those more cosmopolitan species which succeed the pioneers as

TABLE 1. Woody Shrubs and Trees of the Fremont River Valley

Common Name	Binomial	Occurrence Fremont	Dirty Devil
Pioneer species:			
Big seeded plants:			
Russian Olive	*Eleagnus angustifolius*	common	rare
Small-seeded plants, spring seeding:			
Cottonwood	*Populus fremontii*	abundant	absent
Small-seeded plants, summer seeding:			
Tamarisk	*Tamarix ramosissima*	abundant	absent
Coyote willow	*Salix exigua*		
Seral species:			
Rabbitbrush	*Chrysothamnus nauseosus*	abundant	abundant
Greasewood	*Sarcobatus vermiculatus*	abundant	abundant
Squawbush	*Rhus trilobata*	rare	rare
Four-wing Saltbush	*Atriplex canescens*	common	common
Silver buffalo-berry	*Shepherdia argentea*	rare	absent
Torrey seepweed	*Suaeda torreyana*	common	common

the vegetation matures (Table 1). Nomenclature follows Welsh et al. [1987]. Blooming and seeding times are mostly from Vines [1960] and the author's observation. In addition to the woody perennials, two herbaceous perennials, river cane (*Phragmites australis*) and reed (*scirpus sp.*) form thickets and mats on accreting riverbanks, and probably play a role in the accretionary process.

Fremont cottonwood is the tallest and fastest growing of the floodplain species, and is the only plant large enough to be called a "tree." It blooms in early spring and sets seed in time with the spring snowmelt flood. Seeding time varies with geographic location, but on the middle Fremont it is early to mid-May.

The vegetation of the middle Fremont as recorded in 1982 at the mile-post 111 traverse (Fig. 4) is typical of the river between Caineville and Hanksville. Colonization by reed mats and seedlings of willow and tamarisk begins on the active shelf (AS in Fig. 4). Cottonwood probably appears here as well during favorable years. A row of cottonwood saplings to five meters in height occupies the proximal flood bar (FB) with some Russian olive, tamarisk, coyote willow, and rabbitbrush, with ground cover ranging from 30 to 50%. The distal flood bar is occupied by tamarisk and rabbitbrush, with ground cover estimated at 60%. In the intervening swales (FS) are rabbitbrush, tamarisk, four-wing saltbush, and Torrey seepweed to a lower density (10 to 20%), except for a dense fringe of tamarisk at the toe of the terrace. The terrace (T) is occupied by four-wing saltbush and greasewood, with some old cottonwoods, many dead or dying. A resurvey in 1993 showed very little change in either topography or vegetation.

The diversity of the floodplain vegetation is lower downstream from the mouth of Muddy Creek. Fremont cottonwood is absent from the floodplain of the Dirty Devil, and the introduced Russian olive and tamarisk are less common. The floodplain of the Dirty Devil is occupied by tamarisk, coyote willow, and greasewood, with an occasional Russian olive and cane thicket. The terrace is occupied by greasewood and rabbitbrush, with some four-wing saltbush and seepweed.

The age diversity of Fremont cottonwood decreases downstream as well. Many generations are present in Capitol Reef, but below Caineville cottonwood is represented almost entirely by the 1973 cohort on the floodplain, and the old trees on the 1896 terrace (Figure 2). A few kilometers downstream from the mouth of Muddy Creek there are no cottonwoods of any age in the main river valley (Figure 3), although cottonwoods in many tributaries

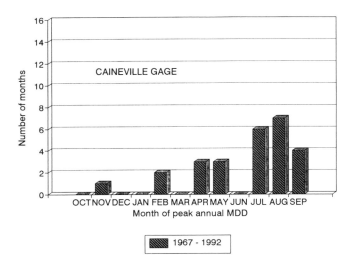

5. Monthly frequency of peak annual mean daily discharge, Fremont River near Caineville, Utah.

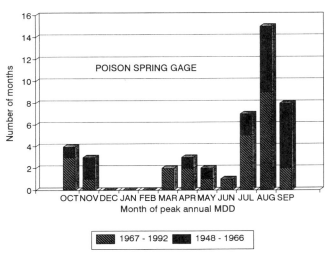

6. Monthly frequency of peak annual mean daily discharge, Dirty Devil River above Poison Spring Wash.

provide a seed source.

8. HYDROLOGY

8.1 Characteristics of floods

Floods on the Fremont-Dirty Devil River appear in three seasons: spring snowmelt, derived from high elevation headwaters; flashy runoff from summer convective storms on the low and mid-elevation tributaries; and the occasional big storms of autumn which generate runoff from the entire drainage basin. Each of these kinds of events can be seen in the annual distribution of peak mean daily discharge at both Caineville and Poison Spring Wash (Figures 5 and 6). Although late season (summer and autumn) floods dominate the record at both stations, at the Poison Spring gage, 100% of the annual peaks are late season floods. Table 2 provides a ranking of the 13 greatest peak annual Mean Daily Discharges (MDD's) for the last 25 years (1967-1992) at both stations.

At Caineville five annual peak MDD's equal to or greater than the median have occurred in April or May, more than one-third the total. At the Poison Spring gage the spring flood is insignificant. There are no spring

TABLE 2. Rank of the Thirteen Greatest Annual Peak Mean Daily Discharges at Caineville and Poison Spring Wash, 1967-1992 Water Years

	Caineville		Poison Spring	
Rank	MDD (m^3/s)	Date	MDD (m^3/s)	Date
1	34.0	July 19, 1985	280.0	September 10, 1980
2	24.4	April 29, 1973	81.6	October 20, 1973
3	21.2	August 21, 1986	76.5	September 6, 1981
4	19.1	September 9, 1980	71.9	November 12, 1979
5	19.0	April 12, 1987	45.6	July 25, 1977
6	15.7	July 18, 1974	40.5	July 19, 1985
7	11.0	May 13, 1975	34.0	August 8, 1982
8	9.9	May 11, 1983	32.0	July 28, 1968
9	9.6	September 9, 1981	28.3	June 24, 1969
10	8.8	August 7, 1991	26.8	July 3, 1983
11	8.4	August 18, 1982	22.3	July 26, 1984
12	8.0	November 2, 1979	22.1	August 4, 1970
13 (median)	7.3	April 8, 1988	21.4	July 13, 1992

TABLE 3. Reservoirs With Storage Greater Than 0.1×10^6 m^3, Fremont Drainage

Name	Date Constructed	Storage Capacity (10^6 m^3)	Elevation (m above msl)	Drainage Area (ha)
Johnson Valley	1910	12.8	2690	9320
Oak Creek	1918	1.1	3075	780
Forsyth	1922	4.5	2435	19,200
Donkey	1923	0.2	3100	780
Fish Creek	1933	0.4	3050	620
Fish Lake Diversion	1935	4.8	2710	---
Left Hand	1953	0.2	3025	1040
Mill Meadow	1954	6.5	2340	46,100
Lower Bowns	1960	4.3	2136	0
Teasdale B	1984	0.1	2255	3100
Total Capacity		34.5		
Net Drainage Area				51,800

MDD's greater than the median annual MDD, reflecting the greater proportion of low-elevation semi-arid drainage basin. This is true for the 1948-92 time series as well as the 1967-92 time series. The peak spring MDD of record at Caineville was 24.4 m^3/s, 70% of the peak MDD of 34 m^3/s. The peak spring MDD at the Poison Spring gage (1948-1992 record) was 30.9 m^3/s on April 2, 1949, 8% of the peak MDD of 396 m^3/s, November 4, 1958. In terms of the MDD, therefore, the spring flood, while of small but significant size at Caineville, is insignificant at Poison Spring gage.

The four largest annual peak MDD's at Poison Spring gage are from autumn storms. These storms were apparently basin-wide general rains because each can be identified at the Caineville gage as well. The October 1973 (water year) event produced an MDD of 11.6 m^3/s at Caineville, less than the spring flood, and therefore does not appear in the Caineville list in Table 2. The remaining nine peak MDD's at Poison Spring gage are summer events. The summer storms are more local; only the July 1985 flood is common to both gages.

Longer climatological and stream-gaging records for the adjacent Escalante drainage and other basins on the Colorado Plateau suggest that both the intensity of precipitation and the size of floods have undergone long-term changes, in the time domain of decades to centuries [*Webb and Baker*, 1987; *Hereford and Webb*, 1992]. It is likely that such climatic changes have played a role in the evolution of both landscape and vegetation in the Fremont River valley.

8.2 *Low flow*

Low-flow below Caineville averages 1.5 to 3 m^3/s in the winter when irrigators are not diverting. The Hanksville irrigation system has the senior water right on the river. During dry periods upstream users bypass flow to satisfy Hanksville's right. Therefore, the river above the Hanksville diversion has rarely been dry. The Dirty Devil below Hanksville is often dry in the summer months.

8.3 *Reservoirs*

There are ten reservoirs of significant size on the Fremont River upstream from the Caineville gage (Table 3). Total active storage in these reservoirs is 34.9×10^6 m^3, 50% of the average annual Caineville discharge of 69.1×10^6 m^3. The reservoirs control a net 51,800 ha of drainage basin, all above 2130 m elevation, and 16% of the drainage basin above the Caineville gage. All but one, Teasdale B, predate establishment of the Caineville gage. Small off-stream reservoirs through which flood flows are not routed are indicated in Table 3 with a zero area.

There are five significant reservoirs on the Muddy Creek drainage, all small reservoirs at high elevation (Table 4). Together, the 15 reservoirs of the Dirty Devil basin can store 38.0×10^6 m^3, 44% of the mean annual discharge of 86.9×10^6 m^3 at the Poison Spring gage. The drainage area controlled by the reservoirs is 56,000 ha, 5% of the drainage above Poison Spring gage.

All fifteen reservoirs are located at high elevation in

TABLE 4. Reservoirs With Storage Greater Than $0.1 \times 10^6 m^3$, Muddy Creek Drainage

Name	Date Constructed	Storage Capacity (10^6 m^3)	Elevation (m above msl)	Drainage Area (ha)
Emery	1934	0.2	2880	260
Sheep Valley	1948	0.6	2810	2850
Henningson	1951	0.6	3050	260
Julius Flat	1953	0.9	2700	260
Spinners	1978	0.8	2930	260
Total Capacity		3.1		
Net Drainage Area				3890

headwater tributaries and primarily regulate the snowmelt runoff. They are privately owned, are operated for irrigation storage only, and have no flood control or hydropower function. This means that the outlets remain closed until irrigation water is called for, usually in early to mid-May. If reservoirs fill before the irrigation season, the excess goes over ungated spillways as uncontrolled discharge. Early spring runoff can occur from release of a large low elevation snowpack downstream from the reservoirs. Late spring floods like that of 1973 occur only when the principal reservoirs fill before the irrigation season begins and excess run-off beyond reservoir capacity passes downstream. The entire reservoir system rarely fills; the 1973 event may be the only significant spill since Mill Meadow, the lowest large reservoir, was constructed in 1954.

9. AGE STRUCTURE OF THE FLOODPLAIN COTTONWOODS

9.1 Summary

Sampling of tree ages at several localities shows that the cottonwoods on the active floodplain between Caineville and Hanksville are of uniform age, and represent the survival of a single cohort, apparently the only regeneration of cottonwood by seed in the last century. The variety of tree heights and diameters is due to secondary growth from beaver-cut stumps and flood-trained saplings.

Ring counts among the Fremont River cottonwoods vary by two to three years. *Clark* [1987] believes that cottonwoods may have as many as 5% missing rings; however, I would not expect missing rings in young vigorous saplings. The observed variation is most likely due to limitations in sampling and accuracy of counting. Most tree ages were obtained by adding the rings of beaver-cut stumps to the rings of secondary shoots. Center dates cluster around 1974-1975. The spring flood of 1973 is most likely the sponsor of this successful cohort of cottonwoods.

9.2 Study site at the mouth of Sweetwater Creek

Just downstream from the mouth of Sweetwater Creek (Figure 1), two parallel rows of cottonwoods occupy flood bars near the accreting riverbank inside a bend. All trees sampled in 1993 showed ages of 18 to 19 years. Tamarisk, coyote willow, and some rabbitbrush occupy the flood bars with the cottonwood, as well as the intervening swales where cottonwood is absent. The few trees surviving intact are six meters tall; but most remaining individuals have been reduced to shrubby stump sprouts due to the combination of beaver activity and cattle browsing.

Two aerial photographic surveys bracket the creation of the flood bars and their cottonwoods; one in 1980 (Figure 7) and one in 1966 (Figure 8). In Figure 7 I have shown the boundary between the flood plain (FP) and the 1896 terrace (T). Passing toward the riverbank from the edge of the flood plain are many flood bars outlined by rows of vegetation. The dark-leaved shrubs are tamarisk (Tx). Present with the tamarisk and between the rows of tamarisk are rabbitbrush, saltbush, and greasewood, but no coyote willow. Only the last two rows next to the river include cottonwood (Pf). These flood bars and their cottonwoods were not present in 1966. The first landward row of tamarisk without cottonwood approximately follows the vegetated riverbank of 1966.

10. ANALYSIS OF SPRING FLOODS

10.1 Caineville gage

If the new cohort of Fremont River cottonwoods was sponsored by the 1973 spring flood, it is of value to examine the characteristics of that event. During the 25 years of record there have been six spring floods at the

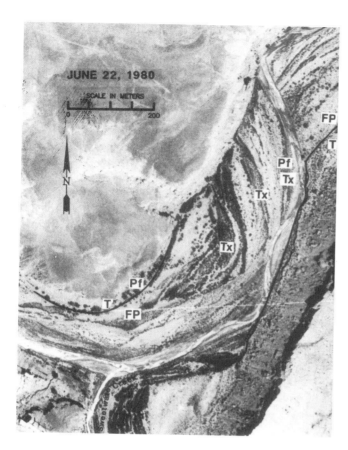

7. 1980 aerial photograph, Fremont River at mouth of Sweetwater Creek. Sweetwater Creek enters at the south edge of the photo.

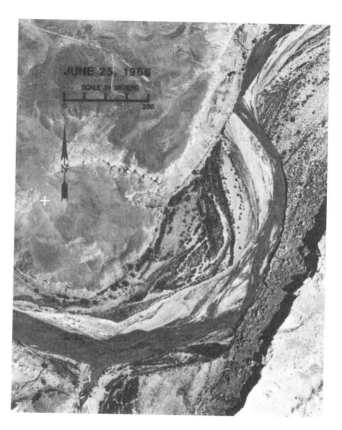

8. 1966 aerial photograph, Fremont River at mouth of Sweetwater Creek.

Caineville gage. Spring flood means a hydrographic event with a mean daily discharge greater than 5.7 m³/s occurring in April or May. Of these, the 1973, 1985, and 1987 events are significant, having an MDD greater than 8.5 m³/s of more than a week's duration. These three events were about the same size in volume of April plus May discharge. Their hydrographs are presented in Figures 9, 10, and 11. Both the floods of 1985 and 1987 receded in April, before cottonwood seed dispersal, whereas the 1973 flood did not recede below the median peak MDD until the 10th of May. Of the three largest floods, it alone overlapped with cottonwood seeding time in May.

Ranked by May discharge alone (Table 5), the 1973 flood is the outstanding event. The floods ranked second and third by total May discharge, those of 1984 and 1985, were followed closely by the flood of July 19, 1985, the largest mean daily discharge of record at Caineville (Table 2). The MDD of the 1973 flood was not exceeded until 1985, a span of 12 years. During the summer of 1973 the

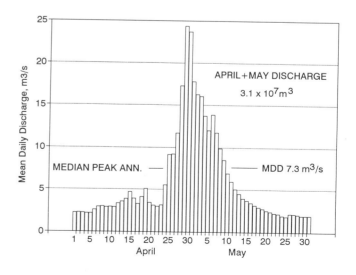

9. Fremont River near Caineville, Utah. Hydrograph of spring flood of 1973, mean daily discharge.

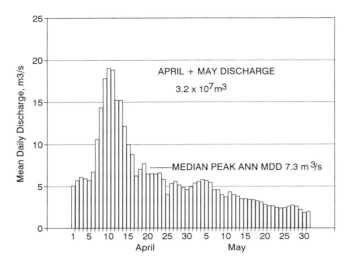

10. Fremont River near Caineville, Utah. Hydrograph of spring flood of 1985, mean daily discharge.

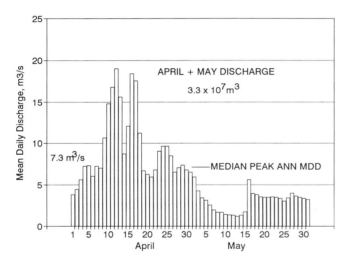

11. Fremont River near Caineville, Utah. Hydrograph of spring flood of 1987, mean daily discharge.

TABLE 5. Ranking of 13 Greatest May Discharges at Caineville 1967-1992

Rank	Year	Total May Discharge (10^6 m^3)
1	1973	16.2
2	1984	11.8
3	1985	9.6
4	1975	8.5
5	1987	8.4
6	1983	6.5
7	1988	6.1
8	1986	5.2
9	1992	4.6
10	1971	4.5
11	1982	4.4
12	1969	4.3
13 (median)	1979	4.0

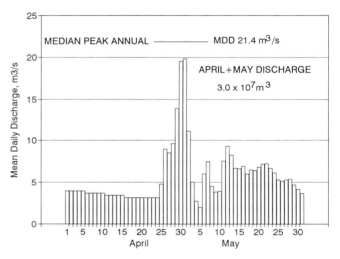

12. Dirty Devil River above Poison Spring Wash, Utah. Hydrograph of spring flood of 1973, mean daily discharge.

flow remained at 0.7 m^3/s or higher at Caineville, indicating good flow downriver to Hanksville.

10.2 Poison Spring gage

Figure 12 shows the hydrograph of the 1973 spring snowmelt at the gaging station above Poison Spring Wash. Its peak MDD is less than the median annual MDD, showing that even the 1973 flood is insignificant compared to late season runoff events. The differences between the Poison Spring hydrograph and the Caineville hydrograph (Figure 9) reflect the diversion at Hanksville beginning about May 1, and the late arrival of discharge from Muddy Creek, which enters below Hanksville.

11. REGENERATION AND THE SPRING FLOOD

11.1 General conclusions

The frequency of cottonwood regeneration has ranged from "rare" between Caineville and Hanksville to "none" downstream from Hanksville. The absence of cottonwood on the Dirty Devil River correlates with the absence of a spring flood larger than the median annual peak MDD. Other factors as well may be involved, such as sediment size, base flow, cyclic fluctuations in climate, and the enormous physical changes in the river channel which have affected the hydrographic characteristics of floods.

In the reach from Caineville to Hanksville, the conditions necessary for cottonwood regeneration can be deduced from the 1973 spring flood and its place in the general geomorphology and hydrology of the river.

11.2 *Antecedent conditions*

11.2.1. *Seed source.* Old cottonwoods remain on the high terrace of the Fremont River as they do along many streams of the Colorado Plateau, and can live more than 200 years [*Clark, 1987*, p. 18].

11.2.2. *Accretion.* Accretion of new ground for colonization of pioneer species requires an actively changing channel. *Everitt* [1968] attributed cottonwood colonization to lateral migration. However, reduction in flood size and consequent channel shrinking was the primary factor in the resurgence of cottonwood thickets along the Platte River [*Johnson*, 1994]. The flood bars colonized by both cottonwood and tamarisk on the Fremont below the mouth of Sweetwater Creek appeared during a period when both channel shrinking and lateral migration were active processes.

11.3 *Seed dispersal and germination conditions*

A late spring flood with bankfull or greater flow continuing into cottonwood seeding time (May on the Fremont River), as described by *Rood and Mahoney* [1993b], is necessary for preparing the seed bed, and distributing and watering the seed.

11.4 *Post-germination conditions*

Seedling survival requires a few hydrographically quiet years without extremes of either low or high water. The Caineville 1973 peak MDD was not exceeded for a decade. The fact that late season floods often exceed the spring flood is not a controlling factor, so long as the recurrence of occasional spring floods meeting the above requirements is less than the cottonwood life span.

11.5 *The effect of reservoirs*

Reservoir storage on the upper Fremont River undoubtedly affects spring runoff on the lower river, but the long-term effect of reservoir operation on cottonwood regeneration is not immediately obvious. Because the reservoirs are at high elevation, they partially regulate the flow from the last melting snow which provides the tail end of the spring runoff. During normal years, reservoir operation causes the spring flood on the lower river to peak lower and earlier, precluding cottonwood regeneration. However, in extraordinary years such as 1973, the reservoirs probably delayed the flood peak rather than advanced it, and most certainly contributed to the long tail of moderate summer flow which aided seedling survival. Lastly, a century of reservoir construction on the Fremont River has probably contributed to the channel shrinking and floodplain construction which was a necessary precondition for the 1973 cottonwood regeneration.

11.6 *Prospects for predicting vegetation change due to hydrologic manipulation*

On the Fremont River, cottonwood regeneration has been sustained by one single event during the last century. This regeneration was spawned by the coincidence of a rare hydrologic event with the floodplain construction phase of the geomorphic cycle on the central Fremont River. Great care is needed in predicting vegetation response from a few cases. Reduction in the spring snowmelt flood by reservoir operation has been credited with both reducing cottonwood forests along the Gila River in Arizona [*Fenner and others*, 1985], and expanding cottonwood woodland along the Platte River in Nebraska [*Johnson*, 1994]. It may be, to paraphrase Ian Malcolm [*Crichton*, 1990], that each case rolls off the knuckles of nature in a different direction. A comprehensive model can only be developed as more case histories are added to the data set.

Acknowledgements. I wish to thank R.H. Webb, U.S. Geological Survey, Tucson; A.E. Godfrey, U.S. Forest Service, Ogden; and L.H. Wullstein, University of Utah, Salt Lake City, for critical reviews of the manuscript, and C.A. Everitt, Hughes Company, Tucson, for her thoughtful attention to grammar, logic, and style. I appreciate the Utah Division of Water Resources for the opportunity to think about some of the broader aspects of water resource development. However, the ideas expressed herein do not necessarily reflect the policy of the State of Utah or its agencies. The 1984 reconnaissance of the Dirty Devil River would not have been possible without the assistance of Russell Paterson, who paddled the bow of "Old Ironsides"; Alan Hembree, who kept the journal [*Hembree*, 1984]; and Bruce Beck, who made sure we had enough to eat. Ben Franklin, Utah Natural Heritage program, and Lois Arnow, University of Utah Herbarium, assisted with plant identification. Plant collections are in the University of Utah Herbarium.

REFERENCES

Burkham, D. E., Effects of changes in an alluvial channel on the timing, magnitude, and transformation of flood waves, southeastern Arizona, *U.S. Geol. Sur. Prof. Paper 655-K*, 25 p, 1976.

Christensen, R.C., E.B. Johnson, and G.G. Plantz, Streamflow characteristics of the Colorado River Basin in Utah, *U.S. Geol. Sur. Open-File Report 85-421*, 674 p, 1987.

Clark, Susanmarie, Potential for use of cottonwoods in dendrogeomorphology and paleohydrology, M.S. thesis, University of Arizona, 52 p., 1987.

Crichton, Michael, *Jurassic Park*, Knopf, New York, 399 p., 1990.

DWR, Fremont River Study, Utah Div. of Water Resources, 71 p., 1975.

Donovan, L.A., James Mausberg, and J.R. Ehleringer, Seedling size and survival for Chrysothamnus nauseosus: *Great Basin Naturalist, 53*, 237-245, 1993.

Everitt, B.L., Use of the cottonwood in an investigation of the recent history of a flood plain: *Am. J. Sci., 266*, 417-439, 1968.

Everitt, B.L., A survey of the desert vegetation of the northern Henry Mountains region, Utah, PhD dissertation, Johns Hopkins Univ., Baltimore, 1971.

Everitt, B.L., Ecology of salt cedar: a plea for research, *Env. Geol., 3*, 77-84, 1980a.

Everitt, B.L., Vegetation and sediment migration in the Henry Mountains region, Utah, in Picard, M. Dane, 1990, Henry Mountains Symposium, *Utah Geol. Assoc. Pub. 8*, 209-214, 1980b.

Fenner, Pattie, Ward Brady, and D.R. Patton, Effects of regulated water flows on regeneration of Fremont cottonwood, *J. Range Management, 38*, 135-138, 1985.

Godby, Evangeline, After the floods, Caineville would never be quite the same, *Desert News, Today Section*, July 16, 1981.

Graf, W.L., Spatial variation of fluvial processes in semi-arid lands, in Thorn, C.E. ed., *Space and Time in Geomorphology*, George Allen & Unwin, London, 1982.

Graf, W.L., Downstream changes in stream power in the Henry Mountains, Utah, *Annals, Assoc. Am. Geog., 73(3)*, 373-387, 1983.

Hembree, Alan, Down the Dirty Devil in Canoes, Muddy Creek to Poison Springs, *Times-Independent*, p. B1, Moab, July 5, 1984.

Hereford, Richard, Climate and ephemeral stream processes: Twentieth century geomorphology and alluvial stratigraphy of the Little Colorado River, Arizona, *Geol. Soc. Am. Bull., 95*, 674-668, 1984.

Hereford, Richard, Modern alluvial history of the Paria River drainage basin, southern Utah, *Quat. Res., 25*, 293-311, 1986.

Hereford, Richard, and Robert H. Webb, Historic variation of warm-season rainfall, southern Colorado Plateau, southwestern USA, *Climatic Change, 25*, 239-256, 1992.

Hunt, C.B., Paul Averitt, and R.L. Miller, Geology and geography of the Henry Mountains region, Utah, *U.S. Geol. Sur. Prof. Paper 228*, Washington, 1953.

Irvine, James R., and Neil E. West, Riparian tree species distribution and succession along the lower Escalante River, Utah, *Southwestern Naturalist 24 (2)*, 331-346, 1979.

Johnson, W.C., Woodland expansion in the Platte River, Nebraska; patterns and causes, *Ecolog. Mono. 64(1)*, 45-84, 1994.

Osterkamp, W.R., and C.R. Hupp, Geomorphic and vegetative characteristics along three northern Virginia streams, *Geol. Soc. Am. Bull., 95*, 1093-1101, 1984.

Rood, Stewart B. and John M. Mahoney, River damming and riparian cottonwoods: management opportunities and problems: in Riparian management: common threads and shared interests, *USDA Forest Serv. General Tech. Rep. RM-226*, 134-143, 1993a.

Rood, Stewart B., and John M. Mahoney, A model for assessing the effects of altered river flows on the recruitment of riparian cottonwoods: in Riparian management: common threads and shared interests, *USDA Forest Serv. Gen. Tech. Rep. RM-226*, 228-232, 1993b.

Schumm, S.A., and R.W. Lichty, Channel widening and flood-plain construction along Cimarron River in southwestern Kansas, *U.S. Geol. Surv. Prof. Paper 352-D*, 71-88, 1963.

Smith, C.D., Effect of channel meanders on flood stage in valley, *J. of the Hydraulics Div., ASCE, 104 (HY1)*, 49-58, 1978.

Vines, Robert A., Trees, shrubs, and woody vines of the Southwest, Univ. of Texas Press, Austin, 1104 p, 1960.

Warren, Douglas K., and Raymond M. Turner, Saltcedar seed production, seedling establishment, and response to inundation, *Ariz. Acad. Sci., 10 (3)*, 135-144, 1975.

Webb, R.H., late Holocene flooding on the Escalante River, South-central Utah, PhD dissertation, University of Arizona, Tucson, 1985.

Webb, R.H., and Baker, V.R., Changes in hydrologic conditions related to large floods on the Escalante River, South-central Utah: in Singh, V.P., ed., Regional flood frequency analysis, D. Reidel, New York, 309-323, 1987.

Welsh, S.L., N.D. Atwood, Sherel Goodrich, and L.C. Higgins, eds., A Utah Flora, *Great Basin Naturalist Mem. No. 9*, BYU Press, Provo. 894 p, 1987.

Whitaker, G.L., Summary of maximum discharges in Utah streams, *Utah Dept. of Nat. Resour. Tech. Pub. 21*, 42 p, 1969.

Wooley, Ralf R., Cloudburst floods in Utah, 1850-1938, *U.S. Geol. Surv. Water Supply Paper 994*, 128 p, 1946.

Benjamin L. Everitt, Utah Division of Water Resources, 1636 W. North Temple, Salt Lake City, UT 84116

Fluvial Processes on an Urbanizing Alluvial Fan: Eilat, Israel

A.P. Schick

Institute of Earth Sciences, The Hebrew University of Jerusalem

Despite hyperarid conditions and the very small area of its up-fan catchments, the flooding erosion, and sedimentation hazards affecting the resort town of Eilat, Israel, are substantial. The hazards have increased with the advent of urbanization upslope on the fan and will soon reach their maximum potential as construction abuts the mountain slopes. A gradual decrease in the proportion of unpaved areas in the older parts of the town contributes as well. Although the main streets of the town were designed to convey the bulk of the floodwaters downslope, the system is unable to cope due to the much larger, sediment-laden flows. Jumpouts from the street floodways result, leading to the development of an unplanned semi-natural drainage network that resurrects temporarily portions of the natural pre-urbanized alluvial fan.

1. INTRODUCTION

Development on alluvial fans, which typically rim arid depressions, has recently undergone rapid expansion in many parts of the world, drastically upsetting the natural hydro-geomorphic balance between flooding, sediment conveyance, and channel morphology. Within, and downstream of densely urbanized fan segments, even medium frequency rainstorms produce catastrophic effects in the terrain. Protection by engineering intervention has, in many cases, low efficiency, entails disproportionate costs, and often inadvertently increases the damage when high-magnitude events occur [*Cooke et al.*, 1982; *French*, 1987].

Special problems in arid alluvial fan environments are posed by the steep slopes and high erodibility of the terrain, promoting massive erosion and sedimentation and extreme channel instability in time and space. A major obstacle to a prudent treatment of flood hazard on urbanizing fans is the paucity of dependable hydrologic measurements and geomorphic observations of flood events [*Graf*, 1988, p. 283-289]. Such observations are essential for improving the management of the flooding problem in alluvial fan towns and for developing realistic urban runoff models applicable to those unique environments [*Dawdy*, 1979; *Rhoads*, 1986]. Especially poor is information on initial losses of small impervious surfaces in arid areas, which may have highly variable values [*Hollis*, 1988].

This paper evaluates some of the fluvial processes affecting the town of Eilat as an example of an urbanizing alluvial fan. The interest in this evaluation stems from four aspects: first, the rapid development of the town from its founding in 1949 to its present population of 34,000 [*Taggar*, 1993]; second, the economic importance of Eilat as a major international resort with over 2 million tourist nights a year; third, the special combination of a hyperarid climate and a relatively short catchment hinterland; and fourth, the availability of a detailed data set on rainfall, runoff, and sediment at the nearby Nahal Yael research watershed [*Schick*, 1988; *Schick and Lekach*, 1993].

2. GEOGRAPHIC SETTING

Eilat is located at the northwestern corner of the Gulf of 'Aqaba, an extension of the Red Sea (Figure 1). It occupies an 8 km coastline extending from the Israel-Jordan border in the midst of the 'Arava valley to the Israel-Egypt border post at Taba.

The town has developed from the head of the Gulf in a northwesterly direction [*Wittelson*, 1993]. It has gradually covered most of the available alluvial terrain composed of alluvial fans that coalesce into a continuous apron. At several points development abuts on the steep rocky

Natural and Anthropogenic Influences in Fluvial Geomorphology
Geophysical Monograph 89
Copyright 1995 by the American Geophysical Union

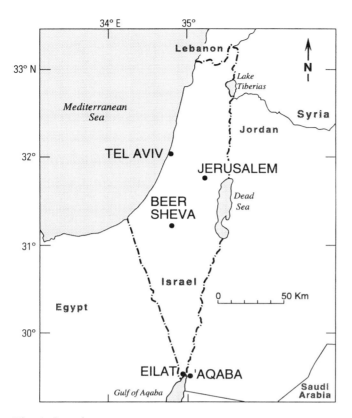

Fig. 1. Location map.

mountains to the west. Several small catchments drain these mountains towards the town (Figure 2).

Though mean annual rainfall in Eilat is only 31 mm, flooding, concomitant with urbanization, has increased substantially during the last decades. Rainfall that previously infiltrated completely into the alluvial fan now causes throughflow.

Three planning decisions made early in the town's development have a considerable impact on flooding problems. First, the airport, with its 2-km runway, was built parallel to the fan toe. Because the airport is the lifeline of the Eilat resorts, flooding of the airport cannot be tolerated, and drainage must be diverted south towards the Gulf or drained beneath the runway. Second, similarly to other desert towns (e.g. semiarid Tucson, AZ [*Resnick et al.*, 1983]), rather than invest in costly but infrequently needed storm sewers, the streets were designed as stormwater conveyors, directing the runoff to the narrow opening left between the southern end of the airport and the coast. And third, the main resort area of Eilat, with a very large investment in hotels and associated establishments, developed along the coast just beyond the southern end of the airport.

In general, measures to alleviate flooding damage were moderately efficient for small events, but larger storms resulted in massive relocation of sediment, gullying, flooding, and siltation of low-lying and closed areas. These effects grew in magnitude along with the expansion of the paved areas. In addition, areas previously only sparsely occupied become more and more congested. Encroachment into potential flood avenues resulted in jumpouts, flooding, and extensive siltation.

3. ENVIRONMENTAL DATA

3.1 *Nahal Yael*

The research watershed of Nahal Yael shares a divide with the Eilat watersheds (Figure 2). Hydrologic and sediment data collected there and elsewhere since 1965 [*Schick*, 1988; *Schick and Lekach*, 1993] are used below to evaluate the flooding and sedimentation characteristics of Eilat. A brief summary of the relevant data follows.

3.2 *Rainfall*

Despite the low annual average, the spatial and temporal distribution of rainfall in the Eilat region is such that flooding may be a serious problem even from very small catchments. The Ma'an flood of 1966 [*Schick*, 1971] is an example (Figure 3). Based on a very sparse raingauge network, the measured storm maxima of 54.2 and 71.6 mm were recorded 25 km north and 65 km northeast of Eilat, respectively. Inferential data suggest storm rainfall amounts in excess of 100 mm occurred in 1971 some 70 km southwest and, in 1993, some 95 km north of Eilat [*Schick and Lekach*, 1983; *Ginat and Yannay*, 1994].

Sixty percent of the Eilat region rainfall is associated with southerly weather systems, which tend to have high intensities for short periods [*Sharon and Morin*, unpublished manuscript]. The Nahal Yael data, based on high resolution raingauges suggest that 40% of the rain falls in intensities in excess of 20 mm/hr; 19% falls in intensities in excess of 56 mm/hr (Figure 4) [*Greenbaum*, 1986].

3.3 *Infiltration*

Sixty-eight tests on 0.25 m² plots were made to determine infiltration characteristics on diverse terrains for rainfall intensities ranging from 14 to 105 mm hr^{-1} [*Greenbaum*, 1986]. The lithologies tested included a variety of magmatic and metamorphic rocks found in the Eilat Mountains, colluvial surfaces, and surfaces of

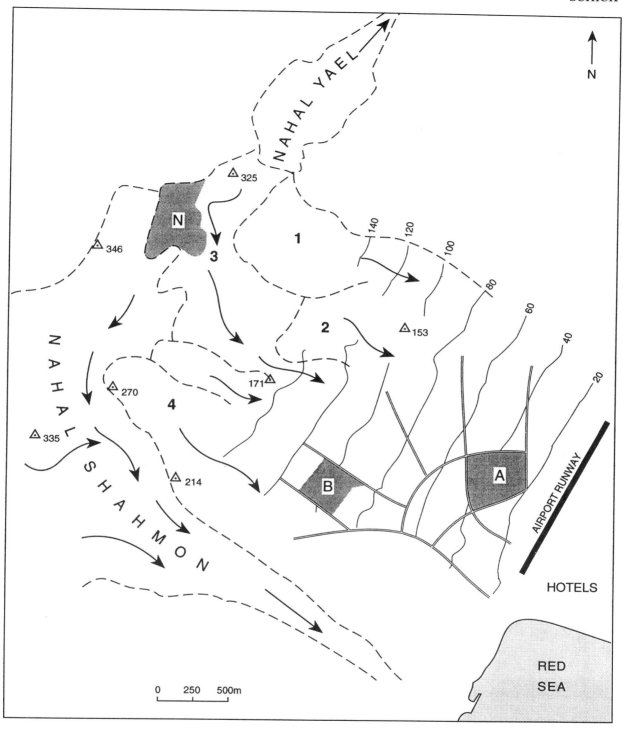

Fig. 2. Eilat in relation to Nahal Yael, Nahal Shahmon, and the small catchments (labeled 1-4) draining towards the town center. Divides are dash lines. Sub-catchment N was used for a comparative runoff evaluation based on the Nahal Yael data. Arrows show concentrated runoff trajectories. Contours on the alluvial apron and triangular elevation points in the mountains upslope are in meters. Within the urban area the figure shows the main streets as designed to convey stormwater runoff, blocks A and B (detailed in Figures 8 and 9, respectively), and the airport runway along the foot of the alluvial apron.

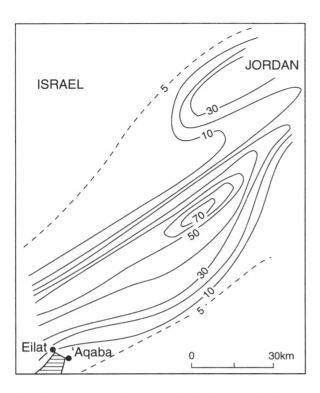

Fig. 3. The rainstorm which caused the Ma'an, Jordan flood of March 11, 1966. Isohyets in mm.

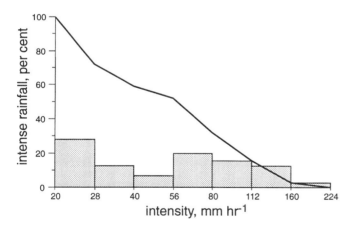

Fig. 4. Frequency distribution of intense rainfall, Nahal Yael near Eilat. Based on time intervals of 3 min. 'Intense' means intensity > 20 mm hr^{-1}. 40% of the total rainfall at Nahal Yael is 'intense'. Cumulative curve is also shown. [Data from *Greenbaum*, 1986].

Holocene and Pleistocene alluvium that are common on alluvial fans in the Eilat region. Effects of slope angle and of jointing were also tested. Within the urban area, tests were made on alluvial terrain compacted by vehicular use (unpaved parking lots).

The results of several infiltration tests are shown in Figure 5. Colluvial (over schist) and rocky (granite) surfaces were tested at simulated rain intensities of 30-35 mm hr^{-1}. The granite was found to have a similar response to gneiss and most other rock types in the Eilat catchments. Typically, final infiltration rates were reached 1-4 min after the onset of rainfall and the infiltration rate stabilized at 0.5-2 mm hr^{-1}. For the tests conducted on these rocks, including those on somewhat less impermeable limestones and sandstones, the runoff coefficient ranged from 52 to 86%. Continuing the simulation beyond stabilization would have increased this coefficient.

Colluvial surfaces, in contrast, require 5-10 minutes of rainfall to initiate runoff. Stabilization occurs typically only after 15-20 minutes at rates of 20-30 mm hr^{-1}. Yet, for the duration of testing and the range on intensities as simulated, the runoff coefficient ranged from 26 to 43%.

Somewhat surprisingly, old alluvial surfaces such as abound on most alluvial fans are closer in their hydrologic response to rocky surfaces than to active alluvial surfaces, with final infiltration rates as low as 10-20 mm hr^{-1} (Figure 5). Recent alluvial surfaces compacted by use as parking lots show similar characteristics. For both types of surfaces the final infiltration rates, typically achieved after 5-8 minutes of rain, were 5-15 mm hr^{-1}. The runoff coefficient of 40-60% typically found for the duration of the test would substantially increase for an actual intense rainstorm of 15-20 minutes (Figure 4).

As expected, active alluvial surfaces absorb all the direct rain over practically the entire range of intensities and reasonable storm durations. Only when an alluvial surface becomes detached from the floods on an interbraid bar and becomes a young Holocene surface (Figure 5) does a small amount of the rain become runoff; but this starts only after 6-8 minutes, with final infiltration rates of 60 mm hr^{-1}.

Results of the above tests agree, in general, with those made in the northern Negev [*Shanan and Schick*, 1980] as well as in the Southern Negev-Sinai regions [*Yair and Lavee*, 1974; *Salmon and Schick*, 1980].

3.4 Runoff

Prior to damming of the 0.5 km^2 rocky catchment of Nahal Yael at the apex of its alluvial fan, total runoff for a decade at the apex was 20% of the rainfall. For the same period, runoff at the toe of the fan was only 4% [*Schick*, 1977]. Transmission loss, controlled primarily by the depth of the coarse-grained alluvium as long as the channel is within the mountains (Schwartz, 1986), increases abruptly at the transition to the alluvial fan. The physiographic

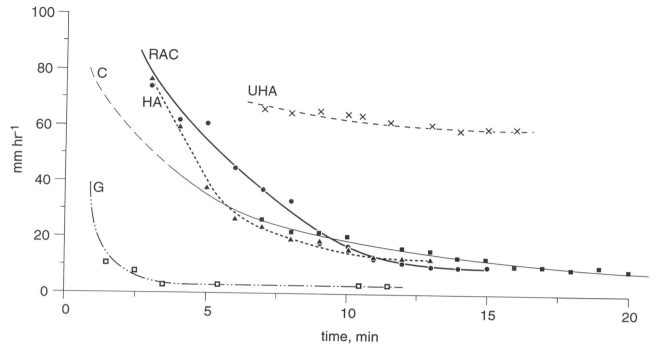

Fig. 5. Infiltration curves on miniplots in the Eilat - Nahal Yael area. The simulated rain was at a quasi-constant intensity of 30-100 mm hr^{-1}. Lithologies tested: C - colluvium; G - granite; HA - Holocene alluvium; RAC - Recent alluvium, compacted; UHA - Upper Holocene alluvium. [After *Greenbaum*, 1986.]

setting of the catchments draining towards Eilat is similar to that of Nahal Yael. The areas of these catchments vary between several hectares with essentially poorly concentrated slope flow to the fluvially well integrated 4 km^2 catchment of Nahal Shahmon (Figure 2).

An assessment of the flooding potential of tributary N, a 0.18 km^2 watershed in the mountains above Eilat, has been made by a lithohydrological comparison with catchment 05 of Nahal Yael (0.05 km^2) for storm 17 (March 22, 1985) [*Greenbaum*, 1986]. A 20 min low-intensity (1.5 mm hr^{-1}) pre-rain of 0.5 mm was followed by a 5 mm period of effective rain of 80 mm hr^{-1}, bringing the storm total to 7.2 mm. Excess rainfall amounted to 36% of the rainfall and runoff exported from the catchment to 23%, the balance being routing losses.

Based on these results and on the distribution of lithologies in catchment N (46% gneiss, 18% schists, 12% granite, 19% inactive alluvial surfaces, and 5% active alluvium), storm 17 yielded an excess rainfall of 57% and a runoff volume of 34% at the outlet.

Storm 17 was too short for all lithologies except gneiss to achieve the final infiltration rate. Frequency data on high resolution rain intensities are incomplete due to the short record and other factors. Nevertheless it is of interest to estimate the runoff with regard to an assumed low-frequency high-magnitude storm, with intensities of 60-100 mm hr^{-1} (Figure 4) lasting 20 min. Most infiltration tests were conducted in this intensity range. Catchments similar to watershed N are likely to yield, in such a storm, a runoff volume of 50-65% of the rainfall. For longer durations this rate may increase, but the point frequency of such long cloudbursts is not known.

Nahal Shahmon is a typical integrator of an assemblage of small rocky catchments to a channel that, soon after exiting the mountains into the Eilat fan apron, achieves widths of 20-40 m. Eight criteria were used to develop a peak discharge estimate for various frequencies [*Schick and Shanan*, unpublished report, 1983]. One likely interpretation of the data points to a 100 year flood with a peak discharge of 80 m^{3}s^{-1} for this 4 km^2 watershed (Figure 6). In the pre-urbanized stage losses to alluvium are much larger than in small catchments like N or Nahal Yael 05. Anticipated development in the upper parts of the Shahmon catchment is certain to affect this value.

3.5 *Sediment*

Nahal Yael provides elaborate data on sediment transported by flash floods [*Lekach and Schick*, 1982; *Schick and Lekach*, 1993]. At Nahal Yael 05, which

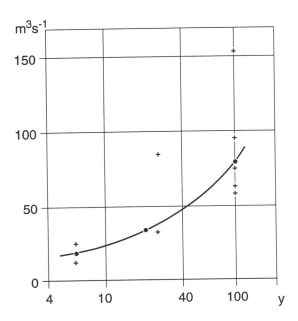

Fig. 6. Estimated peak discharges for Nahal Shahmon near Eilat, drainage area 4 km². Estimates are based on eight different sources or lines of reasoning, and show the large inconsistencies for the 100-year flood. The curve shown was adopted for planning purposes before the beginning of urbanization in the catchment. [After *Schick and Shanan*, unpublished technical report, 1983.]

represents mainly an assemblage of rocky, predominantly amphibolite slopes, typical concentrations of suspended sediment in the floodwaters are between 15,000 and 40,000 mg l^{-1}, and in a few instances over 100,000 mg l^{-1}.

The sediment yield of Nahal Yael has been monitored since 1977 at its outlet (0.52 km²) with the aid of a 100% trap efficiency dam constructed for this purpose. For the decade 1980-1990, 135 m³ of fines and 288 m³ of sand and gravel were deposited in the reservoir. The largest event during this decade (event 15, December 26, 1980), caused by a rainstorm of 45 mm (in part low intensity), deposited 57% of this decadal sediment yield [*Lekach*, 1993].

It is clear from these figures that high-magnitude low-frequency rainstorms, when centered on the mountains west of the town, yield not only substantial runoff, but also transport large amounts of sediment. Estimates of these follow.

4. ESTIMATE OF WATER AND SEDIMENT YIELD AT THE MOUNTAIN EDGE ABOVE EILAT

Runoff from catchments 1-4 (Figure 2) flows towards the town. Each of these catchments is comparable in size, lithology, and hydrogeomorphic attributes to Nahal Yael. They cover, together, an area of 1.8 km², of which 65% (catchments 2 and 3) are directed toward the town nucleus. Catchment 1 (20%) flows via the town cemetery towards the northern suburbs. Catchment 4 (15% of the rocky source area) flows towards the hospital and the southern suburbs.

Other parts of the town are also affected by flooding. Commercial enterprises sited northeast of the airport are exposed to flooding and sedimentation from Nahal Roded. Establishments along the coast south of the center, such as the Coral Nature Reserve, are subjected to floods from a number of smaller catchments.

One possible scenario for a high magnitude low frequency event, based on a 20 min 80 mm hr^{-1} rainstorm (above), would generate an outflow of 15,000-20,000 m³ at the upper boundary of the Eilat alluvial apron at a point about 2 km NW of the airport terminal. Irrespective of assumed lag times, the present urban setting is incapable of handling such amounts without substantial damage (below).

The sediment transported by the flow of such an event to the upper margins of the alluvial apron is also estimated by using the Nahal Yael data. The total volume of the flow of event 15 (3480 m³) required a mean total concentration (including bed load) of 104,000 mg l^{-1} to produce the sediment amounts deposited in the reservoir. Using a rough estimate of 10% sediment content by weight, the amount of sediment which may reach the upper margin of the Eilat alluvial apron (above) is 1500 to 2000 tons for such an event. Such an amount is capable of filling most reasonably-sized protective ditches, causing overflow and abandonment of the ditch. Deposition and associated damage in the economically most sensitive part of the town, at the toe of the alluvial fan, will result.

5. THE URBANIZATION OF EILAT

In 1962 the contiguous urban area of Eilat covered approximately one half of the area of the alluvial fan apron between Nahal Shahmon in the south and the northern divide of catchment 1 (Figure 2). At a few points the highest buildings upslope were within 300 m of the nearest granitic outliers. By 1982 the upper town limit was 200-600 m closer to the mountain front, and the alluvial strip between the granitic slopes and the edge of development was narrowed, at some points, to 100 m or less. Today development encroaches on the toes of the rocky slope along considerable lengths of the upper fan-apron boundary.

5.1 Effect on Runoff

The new configuration does not provide an opportunity for the run-on from the mountains to infiltrate substantially before reaching the built areas. Rather, these flows are reinforced by locally generated runoff from the paved areas (below). The main streets, originally designed to take care of stormwater runoff, are inadequate to cope with the amounts produced by high magnitude, low frequency events. The substantially shorter lag from urban areas [*Leopold,* 1991] promotes in-phase interference between peak flows of the run-on from the mountains and of the run-off from the town.

5.2 Effect on Sediment

The amount of sediment delivered by an event as described to the urban area has a marked effect on the floodwater conveyance capacity of the roads. The steep slope and the preponderance of bed material in the water result in a flow pattern that meanders from curb to curb, produces lateral superelevations, and deposits sediment in the margins (Figure 7). The result is a reduction of the curb to curb conveyance capacity to 20-30% of its theoretical curb to curb value. Thus, the probability of a jumpout of floodwater from such a street conveyor is substantial, and increases with distance down the urbanized fan. In case of a jumpout, the flow is likely to have a high erosive capacity, as it has already deposited some of its load in the town en route (below). Severe erosion in the middle portion of the urbanized fan has its counterpart in extensive sedimentation in the areas around its toe.

5.3 The urban terrain

Figure 8 shows the configuration of buildings of a low-cost housing development in the southeastern part of Eilat as it was in 1982. Paved areas covered 20% of the terrain. For a 40 mm hr^{-1} 30 min storm, estimated peak discharge at the drainage outlet of this low density development increased by a factor of 2 in comparison with the pre-urbanized natural state. Subsequent development since 1982 has at least doubled the paved areas, increased the hydraulic connectivity, and caused much of the remaining open space to become more compacted by vehicles, decreasing the infiltration capacity.

A situation more typical of Eilat as a whole is shown in Figure 9, which portrays an excerpt of an uptown development as in 1982. Estimates of peak discharge produced by a 40 mm hr^{-1} 30-min event for this 92,000 m^2 neighborhood, at an assumed common drainage outlet, range from 1.5 m^{3}s^{-1} (non-urbanized) to 6.0 m^{3}s^{-1} (1982 state, 63% paved) to 8.6 m^{3}s^{-1} (fully urbanized).

For the layout as in 1982, runoff volume generated on site by the event is estimated at 1000-1500 m^3. This amount is reduced by several water (and sediment) traps, mainly parking lots whose drains are mostly clogged by

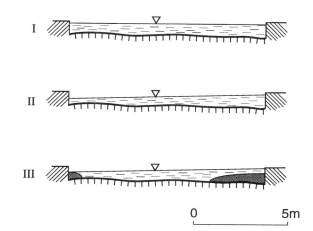

Fig. 7. An Eilat street as a floodwater conveyor. Sketch I shows the theoretical bankfull stage. Sketch II reflects the superelevation associated with in-street meandering, leading - for water emanating from the uptown mountains with a sizable sediment content - to lateral sediment deposition which further constrains the flow. The in-curb maximum discharge at situation III is roughly 25% of that in I.

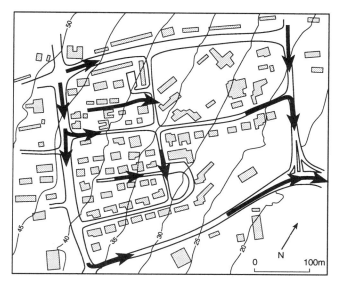

Fig. 8. Floodwater trajectories in relation to the streets, 1982. Block shown (A in Figure 2) is close to the lower end of the Eilat alluvial apron. Contours in meters.

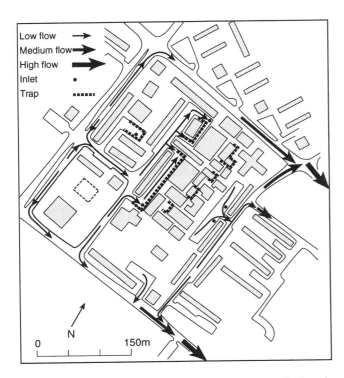

Fig. 9. Floodwater trajectories and relative magnitudes, in relation to streets, inlets, and water/sediment traps, 1982. Block shown (B in Figure 2) was in 1982 near the upper edge of the town. Today it is in the midst of the urbanized fan. The traps are caused by curbs of parking lots (assuming clogged inlets) and by fences. The natural contours have been substantially altered by construction; in general they run N-S.

sediment due to low-level maintenance during the long interflood periods. In the example cited, these traps account for 3600 m², i.e. 3.9% of the neighborhood area. The amount of water detained in these traps, for the above storm, could be 120-160 m³, i.e. 10-15% of the source runoff yield.

5.4 Streets as stormflow conveyors

The primary road network of Eilat, originally designed to serve as floodwater conveyors in case of a rainstorm, crosses the natural contours of the alluvial fan/apron at various angles (Figure 2). Streets orthogonal to the direction of maximum slope are efficient conveyors of stormwater; streets running diagonally to the contours offer opportunities for the stormflow to leave the road over the curb or at the intersection of a side street, to flow more directly downslope.

A sample of 69 points covering an aggregate length of 7.75 km of the main streets of Eilat resulted in the data

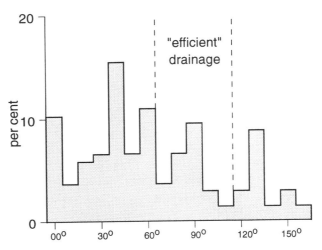

Fig. 10. Frequency distribution of the angle between street direction and slope direction for the main streets of Eilat designed to carry floodwaters. "Efficient" drainage in this figure is defined as having an angle less than 25° between maximum natural slope direction and street direction. Angles greater than 25° are regarded as promoting jumpouts.

shown in Figure 10. Only 21% of the town's main streets serve as "efficient" stormwater conveyors; 55% offer substantial opportunities for a "jumpout". The direction of 24% is roughly parallel to the contours.

The trajectories of the floodwaters through the town will be influenced by its buildings and roads. However, the larger the event, the greater the effect of the original fan topography whose constraints are difficult to remove. A new network, unique for every flood event, will develop. Similar effects were observed in urbanized areas in non-arid areas, where the urbanized, partly artificial drainage network tends to be more permanent than the one observed here [*Graf*, 1977].

6. A TENTATIVE SCENARIO FOR A FULLY URBANIZED EILAT FAN

Within a few years from now, the central part of Eilat (Figure 2) will coalesce hydrologically with the rocky catchments 1-4. The intensely urbanized area of about 5 km² will then generate, for the same 80-mm hr⁻¹ 20-min rainstorm, a runoff wave of 50,000 to 80,000 m³. This amount, assumed to have a very low sediment content at the source, will be reinforced by flows from the mountains upslope, the largest entering the town opposite its center, with an estimated total volume of 15,000 to 20,000 m³ and about 10 per cent sediment content by weight (above). The geomorphic work and urban damage caused by such an event depend on many factors. Factors to be considered include the temporal and spatial distribution of short interval rain intensities, size and resilience of protective engineering structures, and level of maintenance of local drainage devices.

7. CONCLUSIONS

The alluvial apron on which Eilat is situated is a unique urbanizing setting. On the one hand, the catchments feeding the alluvial fan are small and the climate is hyperarid, with flow-producing rainstorms occurring on average only once or twice a year. On the other hand, the intensity of these infrequent events, the economic value of the land on the fan and at its toe, and the functional layout of the town as established years ago, make it necessary to consider such an eventuality and prepare for it.

Applying hydrologic and geomorphic information gained over the years in nearby Nahal Yael provides the natural background to making these evaluations. The parameters of the urban hydrology of the town proper present more of a problem and are presented here only as a rough approximation. Future field studies and on-site, within-event observations must be made to bring these evaluations to an acceptable level of accuracy. Available models for flood hazard assessment [*Dawdy*, 1979; *French*, 1987, p. 183 ff.] are not applicable to the Eilat situation. The development of a conceptual geomorphic framework similar to one proposed a quarter of a century ago for erosion and sedimentation in urban channels [*Wolman*, 1967] may be an important step forward.

Acknowledgments. Of the numerous individuals who contributed over the years to the Nahal Yael Research Project, the author thanks especially H. Ginat, N. Greenbaum, T. Grodek, M. Hassan, X. Huang, N. Keshet, J. Lekach, D. Malmon, U. Schwartz, S. Taggar, S. Uhlenbrook, and several Earthwatch volunteers, whose Center for Field Research has funded some aspects of the work. The Eilat City Engineer helped in many stages. The review by Waite Osterkamp is gratefully acknowledged.

REFERENCES

Cooke, R.U., D. Brunsden, J.C. Doornkamp, and D.K.C. Jones, *Urban Geomorphology in Drylands,* 324 pp., Oxford Univ., Oxford, 1982.

Dawdy, D.R., Flood frequency estimates on alluvial fans, *J. Hydraulics Div.*, Am. Soc. Civil Engineers, 105, No. HY11, November, 1979, 1407-1413.

French, R.H., *Hydraulic Processes on Alluvial Fans,* 244 pp., Elsevier, Amsterdam, 1987.

Ginat, H., and S. Yannay, 22.12.93: A record event at Zihor junction -- precipitation, hail, floods (abs.), paper presented at Israel Geomorph. Res. Group, Univ. of Haifa, May 18, 1994.

Graf, W.L., Network characteristics in suburbanizing streams, *Wat. Resour. Res.*, 13, 459-463, 1977.

Graf, W.L., *Fluvial Processes in Dryland Rivers,* 346 pp., Springer, Berlin, Germany, 1988.

Greenbaum, N., Runoff in a hyperarid area - infiltration experiments on miniplots in the southern 'Arava and their hydrologic, pedologic, and paleogeomorphic implications (in Hebrew), M.Sc. thesis, 205 pp., The Hebrew Univ. of Jerusalem, 1986.

Hollis, G.E., Rain, roads, roofs and runoff: hydrology in cities, *Geography,* 73, 9-18, 1988.

Lekach, J., Bedload movement in a small mountain watershed in an extremely arid environment (in Hebrew), Ph.D. thesis, 131 + V pp., The Hebrew Univ. of Jerusalem, 1993.

Lekach, J., and A.P. Schick, Suspended sediment in desert floods in small catchments, *Israel J. Earth. Sci.*, 31, 144-156, 1982.

Leopold, L.B., Lag times for small drainage basins, *Catena,* 18, 157-171, 1991.

Resnick, S.D., K.J., DeCook, and R.A. Phillips, Hydrological and environmental controls on water management in semiarid urban areas - Phase II, *Rep. B-023-ARIZ, 14-31-001-3556*, 355 pp., Water Resources Res. Ctr., Univ. of Arizona, Tucson, Ariz., 1983.

Rhoads, B.L., Flood hazard assessment of land-use planning near desert mountains, *Envir. Management,* 10, 97-106, 1986.

Salmon, O., and A.P. Schick, Infiltration tests, in *Arid Zone Geosystems,* edited by A.P. Schick, pp. 55-115, Rep. DAJA-DAERO-78-G-111, 170 pp., Physical Geography, Inst. of Earth Sci., Hebrew Univ., Jerusalem, 1980.

Schick, A.P., A desert flood: physical characteristics, effects on man, geomorphic significance, human adaptation - a case study in the southern 'Arava watershed, *Jerusalem Studies in Geography,* 2, 91-155, 1971.

Schick, A.P., A tentative sediment budget for an extremely arid watershed in the Southern Negev, in *Geomorphology in Arid Regions,* edited by D.O. Doehring, pp. 139-163, State Univ. of N.Y., Binghamton, NY, 1977.

Schick, A.P., Hydrologic aspects of floods in extreme arid environments, in *Flood Geomorphology,* edited by V.R. Baker, R.C. Kochel, and P.C. Patton, pp. 189-204, J. Wiley, New York, NY, 1988.

Schick, A.P., and J. Lekach, An evaluation of two ten-year sediment budgets, Nahal Yael, Israel, *Physical Geog.*, 14, 225-238, 1993.

Schwartz, U., Water in the alluvial fill of ephemeral streams in the hyperarid zone (in Hebrew), M.Sc. thesis, 101 pp., The Hebrew Univ. of Jerusalem, 1986.

Shanan, L., and A.P. Schick, A hydrological model for the Negev Desert highlands; effects of infiltration, runoff, and ancient agriculture, *Hydr. Sci. Bull.*, 25, 269-282, 1980.

Taggar, S., Eilat, the jewel in the crown, 1949-1993 (in Hebrew), in *Eilat: Man, Sea, and Desert,* edited by M. Cohen and E. Schiller, Ariel, 93-94, 227-316.

Wittelson, Z., The urban development of Eilat, in *Eilat: Man, Sea, and Desert*, edited by M. Cohen and E. Schiller, Ariel, 93-94, 317-326.

Wolman, M.G., A cycle of sedimentation and erosion in urban river channels, *Geogr. Annaler,* 49A, 385-395, 1967.

Yair, A., and H. Lavee, Areal contribution to runoff on scree slopes in an extreme arid environment - a simulated rainstorm experiment, in *Geomorphology in Arid Regions,* edited by A.P. Schick, D.H. Yaalon and A. Yair, Zeitschr. Geomorph. Supplbd. 21, 106-121, 1974.

Asher P. Schick, Institute of Earth Sciences, The Hebrew University, Jerusalem, Israel

Spatial Controls on Patterns of Land-use Induced Stream Disturbance at the Drainage-basin Scale--An Example from Gravel-bed Streams of the Ozark Plateaus, Missouri

Robert B. Jacobson

U.S. Geological Survey, Rolla, Missouri

Seventy years of streambed elevation changes in the Ozark Plateaus of Missouri provide an historical record of spatial controls on patterns of land-use induced stream disturbance. Mean streambed elevation(MSBE) changes at stream gage stations indicate that drainage area is a primary control of the spatial distribution of MSBE response types. Gage stations at drainage areas of less than 1400 km^2 (square kilometers) are characterized by degraded, relatively immobile beds indicative of passage of a wave of sediment past the stations, or of channel incision. Gage stations at drainage areas of 1400 to 7000 km^2 are characterized by multiple waves of sediment that have moved past the stations during the last 70 years. In these mid-size drainage basins, local channel pattern and position in the drainage network also determine the magnitude of channel response to disturbance. Straight, stable reaches typically had smaller magnitude waves as compared to waves in sinuous, disturbance reaches. Regionally, gage stations that are located where the contributing drainage network is dendritic, with a nearly linear decrease of the logarithm of total stream length with increasing stream order, had larger waves of sediment than those gage stations where greater lengths of higher-order stream channels create a concave-upward relation. At the former sites, it is hypothesized that land-use induced sediment waves are likely to combine additively at stream junctions as they move downstream. At the latter sites, sediment waves are likely to reach stream junctions at different times, and greater lengths of higher- order channels allow more opportunity for storage in floodplain sediments.

1. INTRODUCTION

Land-use induced disturbances to stream channels can be a serious concern in stream and land management. Disturbance exists when channel conditions are outside of a normal or acceptable range of variation. Common examples of stream disturbance are channel widening, channel incision, bed aggradation, and changes in channel pattern. Increased sediment load, increased erosion of valuable riparian bottomland, and disruption of the aquatic ecosystem can be associated with channel disturbance. Disturbance can have a variety of causes, including changes in sediment or hydrologic budgets at the drainage-basin scale, or direct alterations such as channelization.

The conditions that control the evolution of channel disturbance through time and space are complex and depend at least on the type and size of the disturbance, sediment and hydrologic budgets, magnitude and frequency of successive hydrometeorological events, the location of disturbance in the drainage network, geomorphic thresholds, and biological processes that may have a role in healing or accelerating the disturbance. The large number of possible outcomes from single (or often multiple) disturbances have been referred to as complex response [*Schumm*, 1973]. In some alluvial systems, a fairly regular sequence of channel responses follows disturbance [*Simon*, 1989]. In many alluvial systems, however, the complexity is so great that direction of change may be predictable but the magnitude and timing are not [*Schumm*, 1969].

Of particular concern are cases in which stream disturbance is the result of land-use changes at the

Natural and Anthropogenic Influences in Fluvial Geomorphology
Geophysical Monograph 89
This paper is not subject to U. S. copyright. Published in 1995 by the American Geophysical Union

drainage- basin scale. In these cases, the source of disturbance is broadly disseminated over the landscape, rather than occurring at a specific location, as in the case of dam emplacements, gravel mining operations, or channelization. Many types of land-use changes at the drainage-basin scale create new sources of sediment with the potential for downstream aggradation. Common examples are agriculture and logging. Critical questions are the rate at which this sediment, and associated channel instability and water-quality degradation, translate downstream, and whether the disturbance grows or diminishes in the process.

In view of uncertainties in predicting disturbance response, long-term historical records are critical for providing information on how river systems have responded to past disturbances. Historical records provide basic observations of processes and rates from which predictive models can be developed. Also, the disturbance history of a river system places short-term observations of a river in a longer context. For many land- and stream-management decisions, it is often desirable to know if a river system has been disturbed, whether the disturbance is continuing or has been dampened to the level of natural variation, and whether the disturbance is moving through the drainage network. The question of whether a river channel system has reached a dynamic equilibrium about a stable mean condition (in the sense of *Hack,* [1960]) should be of concern to water-quality and biological monitoring studies because physical adjustments of the channel system may produce large changes in water quality and biological characteristics. Historical records serve to establish whether such changes are within the range of natural variation or if they are the effect of a transient response to an historical disturbance.

The purpose of this paper is to document and discuss an example of 70 years of stream-channel responses to land-use change. The example is from low-gradient, gravel-bed streams in the Ozark Plateaus of Missouri. These data provide historical evidence for spatial controls on the magnitude of channel response and lead to an improved understanding of sediment routing through basins of hundreds to thousands of square kilometers in area.

1.1 *Location and Physical Setting*

The Ozark Plateaus of Missouri (known locally as the Ozarks) is an area of humid, temperate climate. Uplands have gently rolling topography with slopes of typically 0-12° and local ridge-to-valley relief of tens of meters. Adjacent to major river valleys, vertical bluffs and steep slopes are common and local relief is 75-200 m (meters).

The Ozarks (Figure 1) are underlain by nearly flat-lying sedimentary rocks with small areas of meta-igneous rocks. Most of the rock is dolomite and cherty dolomite, with lesser contributions of limestone, sandstone, shale, and meta-volcanics [*Koenig,* 1961]. The chert-rich bedrock of the Ozarks has contributed abundant chert gravel to the bedload of streams. Residual soils range in thickness from 0 to as much as 10 m.

Much of the area is cavernous and supports a karst drainage system. The karst drainage system has resulted in some streams that are dry most of the time, whereas other streams with similar surface drainage areas have springs that provide substantial, relatively constant base flow. Because much of the residual soil of the region is thin, relatively impermeable, or both, intense rainstorms can produce runoff that bypasses the underlying karst drainage system, resulting in fast-rising floods.

Ozarks streams were classified by *Dury* [1964] as manifestly underfit because the modern streams meander at wavelengths much smaller than those of the valleys. The valley bottoms are wide, nearly flat, and filled with substantial thicknesses of alluvial sediment [*Jacobson and Pugh,* 1992]. The streams are typically gravel or cobble bed with either sand and gravel or mud (silt and clay) banks. They are characterized by long, straight reaches with nearly trapezoidal channel cross sections separated by short, steeper, sinuous reaches composed of alternating small pools and riffles. Because of the straight reaches, sinuosity measured over several kilometers of valley typically is quite low, about 1.1. However, in the sinuous reaches, sinuosity can be as much as 1.5 over several hundred meters. The sinuous reaches were termed sedimentation zones by *Saucier* [1983]. However, in this paper they are called disturbance reaches to emphasize that both erosion and deposition occur in these reaches, resulting in a high frequency of channel disturbance. The straight reaches are referred to herein as stable reaches. Disturbance reaches are characterized by rapid lateral channel migration of as much as 250 m in 50 years (Figure 2). Rapid channel migration in the disturbance reaches has produced extensive, unvegetated, rapidly aggrading gravel bars indicative of frequent bedload transport. Stable reaches can be practically static over the same time interval (Figure 2) and they lack extensive gravel bars. The pattern of alternating stable and disturbance reaches has no clear links to land-use history and seems to be the natural condition of Ozarks streams.

The drainage basins considered in this report were chosen because they are typical of much of the Ozarks, they cover a range of drainage areas, and the stream gage records are relatively long (20-70 years) (Figure 3, tables

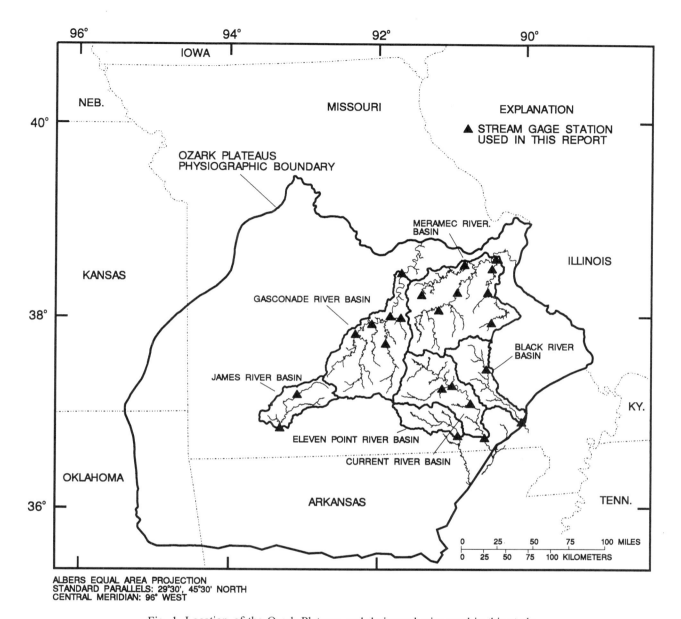

Fig. 1. Location of the Ozark Plateaus and drainage basins used in this study.

1, 2, and 3). Except for the two smallest gaged basins (Bourbeuse River near Highgate and Big River at Irondale) which are dominated by pasture, the basins are composed of a mosaic of nearly equal areas in forest and grazing land, with small total area in row crops. The row-crop land consists of small fields either in the valley bottoms or on relatively flat ridge crests. Small lakes and farm ponds are common throughout these basins. All of the streams in this study area have been unregulated with the exception of the Black River at Poplar Bluff. The dam for Clearwater Lake upstream of Poplar Bluff was closed in 1948 and intercepts 72 percent of the drainage area. Depths to bedrock at the gage stations are largely unknown, but bedrock is not an apparent channel control at any of these stations. Most of the stations are at bridges, with the exception of the Black River near Annapolis, the Current River near Eminence, and the Current River at Doniphan where the rivers are, or were, gaged at cableways. At bridge stations, lateral channel changes are constricted by bridge abutments, so channel responses are biased toward vertical changes. The Black River at Poplar Bluff is channelized and bordered by levees and floodwalls.

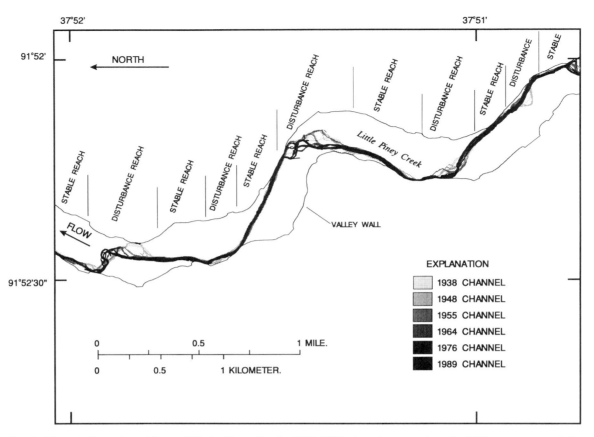

Fig. 2. Map of channel positions of Little Piney Creek, 1938-1989. Area is approximately 15 km upstream of the gage, Little Piney Creek at Newburg.

1.2 Methods

Discharge measurement notes of U.S. Geological Survey (USGS) stream gages provide one of the few sources of data for evaluating stream-channel changes over time intervals of decades. Discharge measurements are taken at stream gages to create stage-discharge rating curves. Typically in Missouri, discharge measurements are taken each month at most continuously recording gages; during 1920 to 1930 the frequency of measurement was less, sometimes only 4 or 5 times per year. Channel morphology data can be extracted from these data, but because the discharge data were collected for establishing discharge rating curves rather than for geomorphic monitoring, great care must be taken in compilation and interpretation. In this study, discharge measurement notes were compiled for 23 long-term stream gages in the Ozarks of Missouri. To evaluate dominant controls on the patterns of channel disturbance, these long-term records were compared with hydrologic characteristics (Table 1), basin-scale land use, geology and topographic characteristics (Table 2), and drainage network characteristics (Table 3).

Channel morphology data were summarized from discharge- measurement notes as mean streambed elevation (MSBE). The MSBE was calculated by subtracting the mean depth of flow from the water-surface elevation, where mean depth was calculated as the cross-sectional area divided by cross-section width, and water-surface elevation was calculated as local stage plus the gage datum, in meters above sea level (Figure 4). Only discharge measurements taken at replicated cross sections were used. This limited the number of data points available at many gages because the location of discharge measurements often varied with stage. Many low-flow measurements were discarded because they were waded at non-replicated cross sections. Cableway measurements are best for this analysis because there are no bridge abutments to constrain channel adjustment and the same location can be used over a wide range of discharges. However, only a small percentage of basins in the Ozarks are gaged by cableway. The most consistent cross sections were at bridges over large rivers

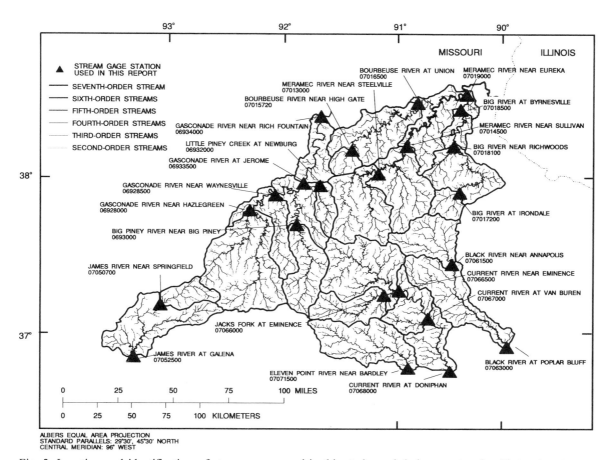

Fig. 3. Location and identification of stream gages used in this study, and drainage networks. First-order streams have been omitted for legibility.

where few, if any, wading measurements were made. Although discharge measurements at bridges ensure replicability in cross-section location, bridge abutments and pilings create unnatural hydraulic patterns and constrain channel changes. Fewer bridge measurements are made on small rivers because they can be waded more frequently, so datasets for smaller rivers tend to be more sparse. Measurements from overbank floods also were discarded because mean depths become uncharacteristically small when the valley bottom is flooded. The edited MSBE data and a five-point moving average of MSBE were plotted against time for the 23 gages along with the record of daily mean discharge (Figure 5).

Other data compiled for this report include hydrologic data, basin characteristic data, and drainage network data. Hydrologic parameters were calculated from USGS digital discharge records using standard techniques [Dempster, 1991]. Basin characteristic data were obtained from USGS digital basin characteristic files, supplemented with measurements from topographic and digital geologic maps.

Drainage network parameters were calculated using a Strahler stream ordering system [Strahler, 1957] and digital drainage network files that included all solid blue-line streams apparent on 1:100,000-scale topographic maps. For the Ozarks of Missouri, these digital blue-line streams are nearly identical to those on 1:24,000-scale topographic maps. No attempt was made at statistical hypothesis testing because of the small number of gage records available and the wide range of conditions represented. Rather, the data set was considered as the best available source for exploratory data analysis.

2. LAND USE IN THE OZARKS

Land-use changes with the potential to create landscape disturbance at the drainage-basin scale began in the Ozarks in the 1830's. Initial rural settlement in the valley bottoms was followed by a timber-boom period lasting from 1880 to 1920, during which highly valued yellow pine was harvested for the eastern timber markets and oak was

TABLE 1. Hydrologic Characteristics of Mean Streambed Elevation (MSBE) Stream Gage Stations.
[km², square kilometers; m m⁻¹, meters per meter; m³ s⁻¹ km⁻², cubic meters per second per square kilometer.]

Basin River	Station	Station Number	Period of Record	Drainage Area, km²	Channel Pattern Reach Type	Channel Slope Near Gage, m m⁻¹	Mean Annual Discharge Per Unit Area, m³ s⁻¹ km⁻²	Discharge Exceeded 10 Percent of the Time Per Unit Area, m³ s⁻¹ km⁻²	Maximum Daily Discharge of Record Per Unit Area, m³ s⁻¹ km⁻²	MSBE Response Type
Black River										
Black River	near Annapolis	07061500	April 1939 - Nov 1994	1239	Disturbance	0.00074	0.01	0.03	1.02	Depleted
	at Poplar Bluff	07063000	Oct 1936 - Nov 1994	3187	Stable	0.00026	0.01	0.03	0.39	Extremely wavy
Current River										
Jacks Fork	at Eminence	07066000	Oct 1921 - Nov 1994	1019	Disturbance	0.00147	0.01	0.02	0.67	Depleted
Current River	near Eminence	07066500	Aug 1921 - Sep 1975	3256	Stable	0.00054	0.01	0.02	0.49	Slightly wavy
	at Van Buren	07067000	July 1921 - Nov 1993	4268	Disturbance	0.00074	0.01	0.02	0.42	Extremely wavy
	at Doniphan	07068000	July 1921 - Nov 1994	5217	Stable	0.00049	0.02	0.03	0.49	Stable/degrading
Eleven Point River										
Eleven Point River	near Bardley	07071500	Oct 1921 - Nov 1994	2030	Stable	0.00047	0.01	0.02	0.37	Extremely wavy
Gasconade River										
Little Piney Creek	at Newburg	06932000	Oct 1928 - Nov 1994	512	Disturbance	0.00141	0.01	0.02	1.08	Depleted
Big Piney River	near Big Piney	06930000	Oct 1921 - Nov 1994[1]	1434	Stable	0.00092	0.01	0.02	0.45	Slightly wavy
Gasconade River	near Hazlegreen	06928000	Oct 1928 - Sep 1971	3200	Stable	0.00057	0.01	0.02	0.52	Slightly wavy
	near Waynesville	06928500	June 1921 - Sep 1971	4301	Stable	0.00041	0.01	0.02	0.53	Slightly wavy
	at Jerome	06933500	Jan 1923 - Nov 1994	7270	Stable	0.00064	0.01	0.02	0.47	Slightly wavy
	near Rich Fountain	06934000	Oct 1921 - Nov 1994[1]	8141	Stable	0.00056	0.01	0.02	0.32	Stable/degrading
James River										
James River	near Springfield	07050700	Oct 1955 - Nov 1994	630	Stable	0.00112	0.01	0.02	1.10	Slightly wavy
	at Galena	07052500	Oct 1921 - Nov 1994	2527	Stable	0.00086	0.01	0.02	0.64	Slightly wavy
Meramec River										
Bourbeuse River	near Highgate	07015720	July 1965 - Nov 1994	354	Disturbance	0.00116	0.01	0.02	1.68	Depleted
	at Union	07016500	June 1921 - Nov 1994	2063	Stable	0.00039	0.01	0.02	0.86	Slightly wavy
Big River	at Irondale	07017200	July 1965 - Nov 1994	464	Stable	0.00069	0.01	0.02	1.01	Extremely wavy
	near Richwoods	07018100	Oct 1942 - Nov 1994	1923	Disturbance	0.00067	0.01	0.02	0.57	Extremely wavy
	at Byrnesville	07018500	Oct 1921 - Nov 1994	2366	Stable	0.00032	0.01	0.02	0.69	Extremely wavy
Meramec River	near Steelville	07013000	Oct 1922 - Nov 1994	1991	Disturbance	0.00064	0.01	0.02	0.63	Extremely wavy
	near Sullivan	07014500	Oct 1921 - Nov 1994[1]	3831	Stable	0.00045	0.01	0.02	0.52	Extremely wavy
	near Eureka	07019000	Oct 1921 - Nov 1994	9841	Stable	0.00023	0.01	0.02	0.40	Stable/degrading

[1] Incomplete or partial record

TABLE 2. Drainage-basin Characteristics Upstream of Mean Streambed Elevation (MSBE) Gage Stations.
[km, kilometers; m m^{-1}, meters per meter.]

River	Station	Maximum Channel Length from Gage to Divide, km	Channel Slope at 15 and 85 Percent of Maximum Channel Length, m m^{-1}	Forest Cover	Percentage of Area[1]								
					Dolomite	Cherty Dolomite	Interbedded Dolomite & Shale	Limestone	Cherty Limestone	Interbedded Limestone & Shale	Sandstone	Shale	Meta-igneous
Black River	near Annapolis	79	0.00206	81	48	25	17	0	0	0	0	0	10
	at Poplar Bluff	169	0.00114	89	42	47	8	0	0	0	0	0	4
Jacks Fork	at Eminence	85	0.00180	77	36	64	0	0	0	0	0	0	0
Current River	near Eminence	115	0.00144	79	32	68	0	0	0	0	0	0	0
	at Van Buren	168	0.00112	76	31	68	0	0	0	0	0	0	1
	at Doniphan	228	0.00094	85	24	75	0	0	0	0	0	0	1
Eleven Point River	near Bardley	76	0.00259	69	40	60	0	0	0	0	0	0	0
Little Piney Creek	at Newburg	61	0.00265	55	8	92	0	0	0	0	0	0	0
Big Piney River	near Big Piney	121	0.00107	56	49	51	0	0	0	0	0	0	0
Gasconade River	near Hazlegreen	156	0.00075	35	68	29	0	0	1	0	0	2	0
	near Waynesville	229	0.00060	37	63	34	0	0	1	0	0	1	0
	at Jerome	277	0.00057	46	49	50	0	0	1	0	0	1	0
	near Rich Fountain	362	0.00051	49	47	51	0	0	1	1	0	1	0
James River	near Springfield	80	0.00123	21	16	0	0	0	60	0	0	23	0
	at Galena	162	0.00090	16	10	0	0	0	0	79	1	10	0
Bourbeuse River	near High Gate	31	0.00218	0	34	1	0	0	65	0	0	0	0
Big River	at Union	216	0.00052	35	20	10	53	0	0	0	5	0	13
	at Irondale	36	0.00366	0	42	12	38	0	0	0	5	0	4
	near Richwoods	132	0.00088	68	13	86	0	0	0	1	0	0	0
	at Byrnesville	203	0.00064	60	28	30	0	0	0	42	0	0	0
Meramec River	near Steelville	123	0.00119	50	47	14	31	0	0	0	5	0	3
	near Sullivan	170	0.00094	59	27	71	1	0	0	1	0	0	0
	near Eureka	296	0.00065	53	34	43	8	1	0	10	2	0	1

[1] Percentage area of lithologies may not sum to 100 percent because of rounding error.

TABLE 3. Drainage-network Characteristics Upstream of Mean Streambed Elevation (MSBE) Gage Stations.
[km^2, square kilometers; km km^{-2}, kilometers per square kilometer; km, kilometers.]

River	Station	Drainage Area, km^2	Drainage Density, km km^{-2}	Total Channel Length of Streams, km						
				Order 1	Order 2	Order 3	Order 4	Order 5	Order 6	Order 7
Black River	near Annapolis	1239	0.96	787	201	122	65	20	0	0
	at Poplar Bluff	3187	0.99	2062	533	281	150	53	76	0
Jacks Fork	at Eminence	1019	1.12	711	191	127	48	61	0	0
Current River	near Eminence	3256	1.10	2263	551	356	220	153	30	0
	at Van Buren	4268	1.08	2920	704	452	283	153	84	0
	at Doniphan	5217	1.06	3529	894	514	315	153	149	0
Eleven Point River	near Bardley	2030	0.98	1276	349	170	129	56	0	0
Little Piney Creek	at Newburg	512	1.12	389	81	63	12	30	0	0
Big Piney River	near Big Piney	1434	1.02	938	250	134	71	76	0	0
Gasconade River	near Hazlegreen	3200	1.07	2241	546	262	143	228	0	0
	near Waynesville	4301	1.06	2965	736	324	236	228	74	0
	at Jerome	7270	1.06	4994	1228	614	354	393	126	0
	near Rich Fountain	8141	1.06	5626	1391	671	372	393	213	0
James River	near Springfield	630	1.11	487	109	53	48	0	0	0
	at Galena	2527	0.98	1663	377	190	122	84	41	0
Bourbeuse River	near High Gate	354	1.08	276	57	42	6	0	0	0
	at Union	2063	0.98	1263	329	160	81	183	0	0
Big River	at Irondale	464	1.11	352	89	46	25	5	0	0
	near Richwoods	1923	1.05	1320	335	168	68	124	12	0
	at Byrnesville	2366	1.07	1650	412	207	70	124	76	0
Meramec River	near Steelville	1991	1.07	1389	308	165	147	81	32	0
	near Sullivan	3831	1.04	2598	619	333	226	115	60	27
	near Eureka	9841	1.04	6613	1606	823	415	448	159	163

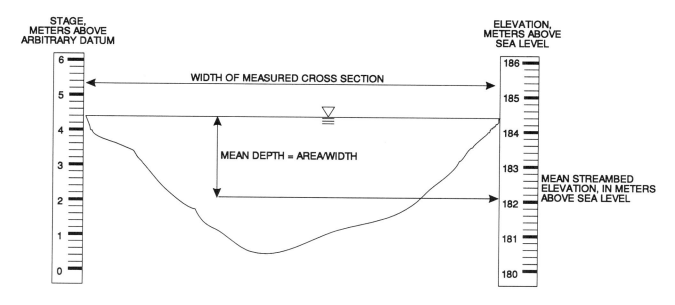

Fig. 4. Schematic drawing illustrating calculation of mean streambed elevation (MSBE).

harvested for railroad ties. After the timber-boom period, cut-over land was used for open-range grazing and small railroad tie and specialty timber operations. During the 1940's and early 1950's there was a peak in row crop agriculture in the uplands, but droughty, infertile soils limited the viability of cash crops. By the late 20th century most of the Ozarks had become a mosaic of about 60 percent forest and 40 percent grassland. Additional information about the land-use effects on stream disturbance history is described by *Jacobson and Primm* [1994].

Land-use disturbance has been much more intense in isolated parts of the Ozarks subjected to lead and barite mining. Mining in the "Old Lead Belt" region of the Ozarks in the Big River Basin (Figure 3) produced large quantities of erodible mine tailings until the lead ore was mined out in the late 1960's; some stream instability has been linked specifically to mine-tailings point sources [*Smith*, 1987].

Extraction of gravel from streams has been pervasive throughout the Ozarks. Most of the gravel-mining operations are small, consisting of borrow sites for farms and county roads. On a regional scale, they can be considered to be a disseminated, non-point sources of disturbance. Two possible exceptions considered in this paper are Black River and Meramec River where extensive gravel mines and gravel dredging operations have been common (Figure 3). Historically, gravel dredging has occurred about 10 km upstream of the gage at Poplar Bluff on the Black river and has extended as much as 38 km upstream from the gage at Eureka on the Meramec River [*Dake*, 1918].

Beginning in the 1930's, residents of the Ozarks began to note that streams were becoming choked by gravel. The most common anecdotal observation was that favorite fishing holes were filling in. Aggradation of stream channels was accompanied by perceived increases in lateral channel migration rates. These observations were made across the Ozarks region, and popular opinion assigned the responsibility to logging practices during the timber-boom period and to indiscriminate post-timber-boom land-use practices on marginal, logged land [*Hall*, 1958; *Rafferty*, 1980].

Jacobson and Pugh [1992] evaluated stratigraphic evidence for stream disturbance. They found that, although more coarse sediment (cobble, gravel, and sand) is present in post-settlement alluvial sediment in Ozarks valley bottoms, the most striking difference is the lack of fine sediment (silt and clay) in the recently deposited strata. These observations are consistent with a model of riparian land-use practices that destroyed or diminished riparian vegetation, thereby reducing flow resistance and deposition of fine sediment. That model is supported by an analysis of the land-use history of the Ozark of Missouri [*Jacobson and Primm*, 1994]. Jacobson and Primm integrated a variety of lines of evidence to determine that valley-bottom land-clearing and agriculture and open-range grazing in the riparian zones of Ozarks rivers had the greatest potential among other land-use changes to create regional stream

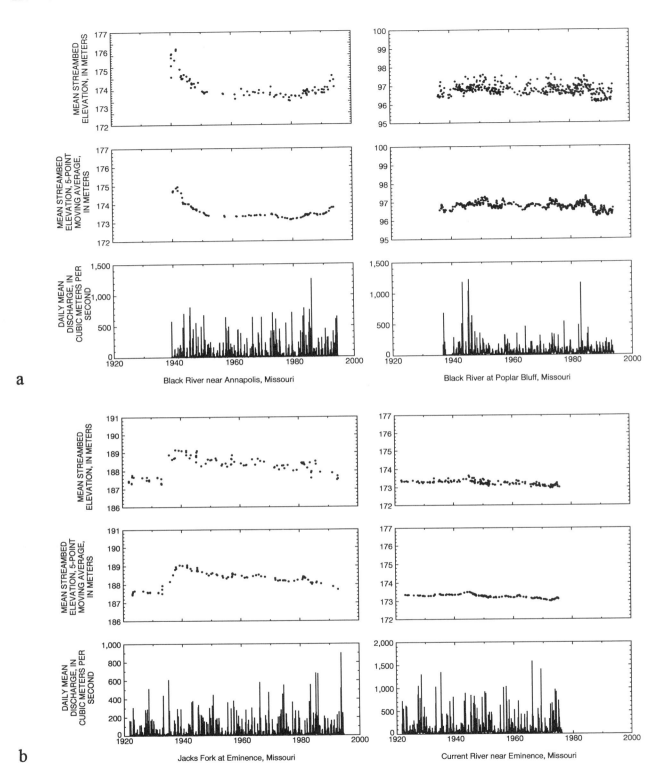

Fig. 5. Mean streambed elevation (MSBE) data, five-point moving averages, and daily mean discharge records plotted against time, for 23 stream gages in the Ozark Plateaus of Missouri. Datum is mean sea level.

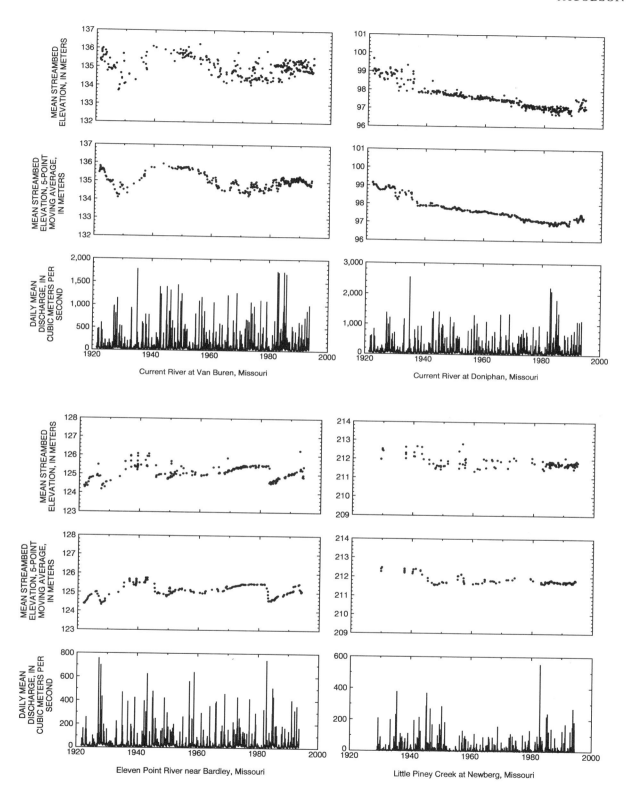

Figure 5 (continued)

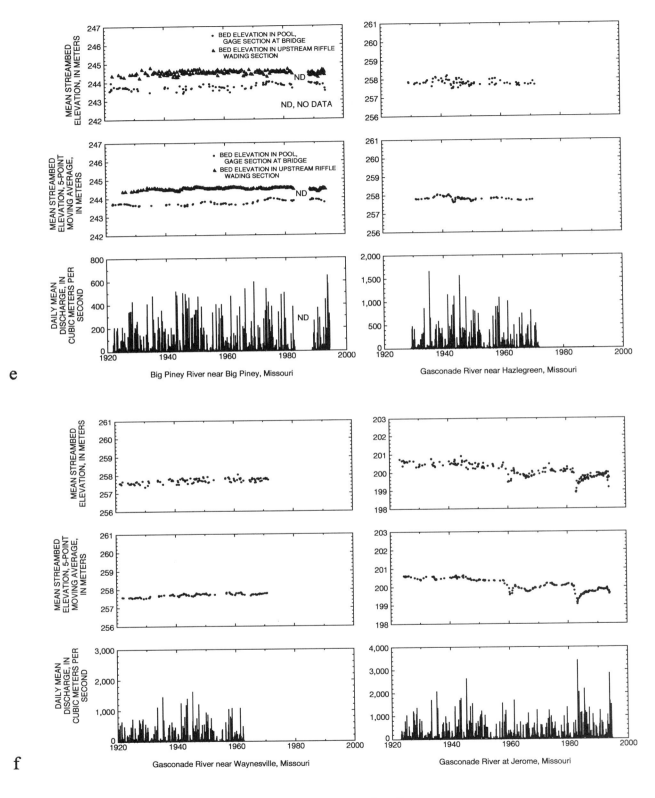

Figure 5 (continued)

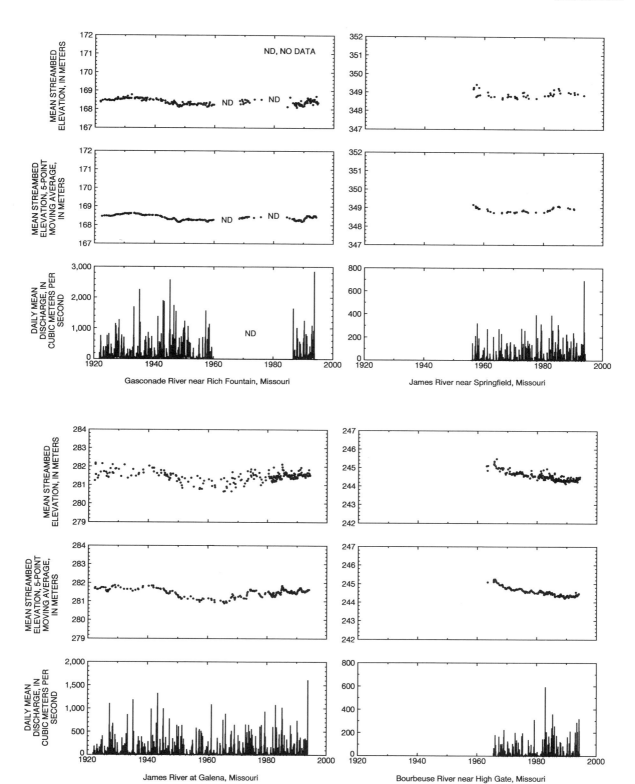

Figure 5 (continued)

232 SPATIAL CONTROLS ON STREAM DISTURBANCE

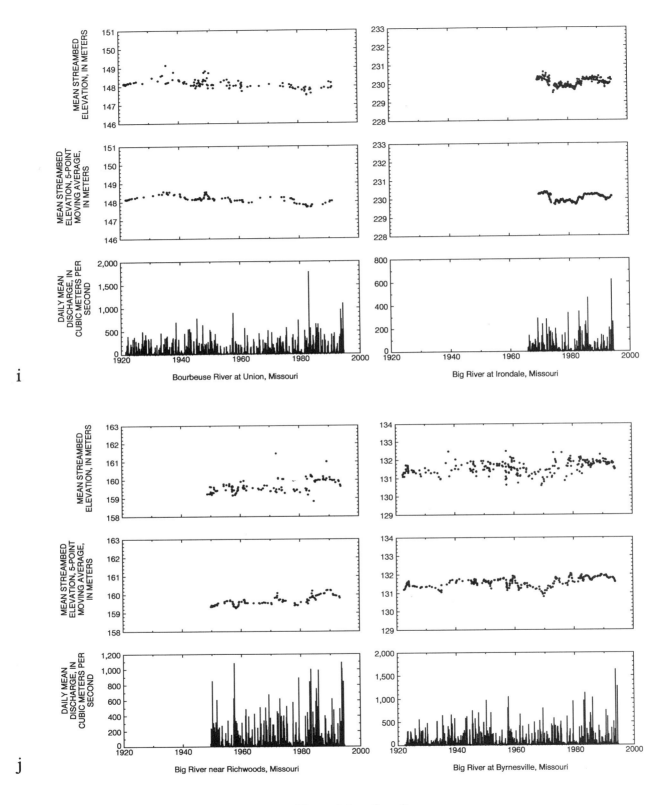

Figure 5 (continued)

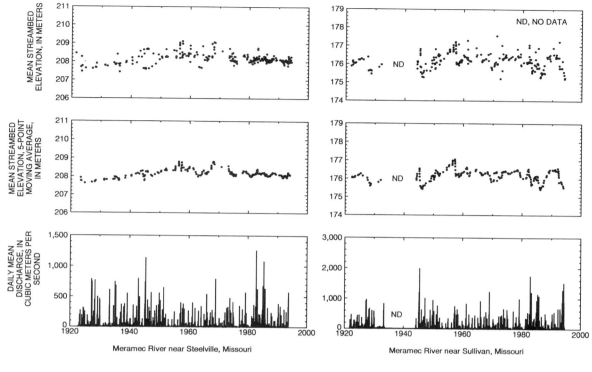

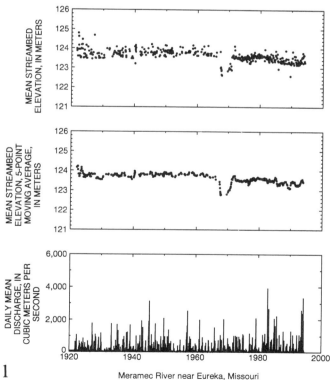

Figure 5 (continued)

disturbance. They hypothesized that clearing and browsing of riparian woody vegetation reduced bank strength and lowered flow resistance, thereby allowing fine sediment to be transported through the stream reaches. They further hypothesized that gravel additions to stream reaches were by headward extension of the stream network into previously unchanneled upland valleys, rather than from hillslope gullying accompanying upland land-use practices. A period of intense row-crop farming during 1940 to 1950 was considered a possible additional source of regional disturbance.

In the period since the turn of the century, the Ozarks have declined in rural population, substantial land has been incorporated into national forest, parkland, and conservation land, and the economy has shifted to small logging operations and grazing. As a result, the disturbance potential from riparian land use and row crops has diminished [*Jacobson and Primm*, 1994]. As in many areas affected by logging and agriculture, the landscape disturbance in the Ozarks varied in time and space. This is a typical situation that complicates interpretation of the record of disturbance and recovery of stream channels. Changes in populations, row crops, grazing, and logging have been fairly uniform across the Ozarks when viewed at the scale of counties [*Jacobson and Primm*, 1994,] which are comparable to the scale of major drainage basins. Therefore, for the purposes of this paper, land-use changes are considered to have been distributed uniformly over space, although variably over time.

3. MEAN STREAMBED ELEVATION RECORDS

To analyze the broad range of data presented in Figure 5, the MSBE records have been classified qualitatively into four responses types: depleted, slightly wavy, extremely wavy, and stable/degrading (Table 1). The classification is based primarily on the pattern of MSBE changes over time, and secondarily on the relative magnitude and rate of MSBE changes (waviness). As in any classification of natural phenomena, the response types are discrete categories that serve to organize and simplify a complex and continuous system; many actual MSBE records fall in the transitions between defined types.

Depleted MSBE records are characterized by bed elevations that degraded rapidly 1-2 m during the 1930's and 1940's, and then stabilized or degraded more gradually until 1994. These are mainly smaller drainage basins with less than 1400 km^2 drainage area. Examples are the Black River near Annapolis, Jacks Fork at Eminence, and Little Piney Creek at Newburg. The record of Jacks Fork is complicated by relocation of the gage and bridge construction in 1935. Although the streambed elevations have been corrected for the associated datum shift, the cross section had a narrower and deeper shape before 1935. Hence, the increase in MSBE in 1935 probably is the result of gage relocation.

The MSBE records at depleted stations can be interpreted as a response to a wave of land-use induced sediment passing through the gage reach by the late 1940's. Sediment waves would result from transient bed aggradation over time intervals of several years to a decade. Alternatively, the response may have resulted from increased rates of channel degradation 1920 to 1940. Although there can be no definitive cause-and-effect link between the history of land-use practices and these gage records, the timing of MSBE decline strongly supports a land-use connection. Lacking gage records for the period 1830 to 1920, it is not possible to demonstrate whether land-use changes produced a sediment wave or channel degradation.

In addition to the lowering of the bed elevation, the streambeds at depleted gage stations show lessened sensitivity of MSBE to floods during 1950 to the present (1994). Sensitivity is used here to describe the relative magnitude of MSBE change that results and persists for months to several years after a flood. Sensitivity does not include scour and fill of the streambed during an individual flood. Despite large floods that affected Ozarks streams in the 1980's, streambed elevation changes at depleted stations were minor. Floods of comparable magnitude during 1920 to 1940 had much more effect in either local aggradation or scour at the gage. Diminishing sensitivity of stream bed responses through time supports the idea that the stations had been previously aggraded and are now depleted of excess sediment. During the period of aggradation pre-1940, the abundance of mobile sediment in the wave would have enhanced MSBE variation by producing local bed aggradation and channel planform changes.

The second type of streambed response is classified as slightly wavy. This response is characterized by multiple, low-amplitude sediment waves moving through the gage reach, and the response lacks the rapid degradation 1920 to 1940 typical of depleted stations. Examples are the Big Piney River near Big Piney, the Current River near Eminence, and the James River at Galena (Figure 5). Slightly wavy types tend to occur in mid-size drainage basins of 1400-7000 km^2, and they are relatively insensitive to individual flood events. For example, the MSBE of Big Piney River near Big Piney has at most only 40 cm of degradation associated with a record flood in November, 1993 while multiple sediment waves of 20-50 cm amplitude occur in the record without association with individual floods.

The MSBE record of Big Piney River near Big Piney (Figure 5) illustrates that variability in sediment waves also depends on reach-scale channel features near the gage. At this station, streamgagers have consistently measured flow during wading conditions at one riffle approximately 150 m upstream of the gage. During non-wading conditions, flow was measured from a bridge directly at the gage in a pool. The elevations at the riffle are relative to the datum at the gage, so they are plotted consistently higher than their true elevations. The MSBE waves at the two sections are distinctly out of phase, indicating aggradation at the upstream riffle at times when the pool section at the gage was being deepened, and degradation at the riffle when the pool was being filled. Comparison with the discharge record indicates that the pool deepened and the riffle aggraded during multiple-year periods characterized by larger floods. Conversely, during periods of smaller floods, the pool filled and the riffle degraded.

The third type of streambed elevation response is classified as extremely wavy. Similar to slightly wavy response stations, extremely wavy stations are associated with mid-size drainage areas of 1400 to 7000 km^2 in this data set. These stations show multiple, high-amplitude waves of streambed elevation changes, and the waviness does not diminish over time. Differentiation between slightly wavy and extremely wavy MSBE records is arbitrarily based on amplitude and wavelength. Generally, records with waves of a meter or more amplitude and wavelengths of one to five years are considered extremely wavy. Typical examples are the Current River at Van Buren, the Eleven Point River near Bardley, the Meramec River near Steelville, and the Meramec River near Sullivan (Figure 5). All these stations exhibit greater sensitivity to the discharge record than slightly wavy stations exhibit, but in non-systematic ways. For example, the Eleven Point River near Bardley and the Meramec River near Sullivan eroded their beds as a result of large floods and slowly aggraded during intervening periods. The Current River at Van Buren aggraded its bed directly after a large flood in 1936, slowly degraded its bed during a period of smaller floods during 1950 to 1980, and aggraded again during a period of larger floods in the 1980's. In contrast, the Meramec River near Steelville aggraded its bed during 1950 to 1980. These different types of responses may be due to particular hydraulic conditions at the gage stations or to the timing of waves of sediment moving past the gages, or both.

The fourth type of MSBE response is classified as stable/degrading. This type applies to the two gages with the largest drainage area, the Gasconade River near Rich Fountain and the Meramec River near Eureka, and to the Current River at Doniphan. These records are characterized by minimal waviness and insensitivity to large floods. The distinct lowering of MSBE on the Meramec River in the late 1960's was caused by local channelization intended to improve conditions for a canoe race. That channelization was maintained for about a year, after which the streambed recovered to its initial mean elevation after about 2 years. The Meramec River reach also is the most likely to have been affected by in-stream gravel and sand dredging, but the record shows no clear effects. In contrast, the Current River at Doniphan has no large gravel removal operations near it, but shows steady degradation of the bed of nearly 2 m over 70 years.

4. COMPLEX CHANNEL RESPONSE IN TIME AND SPACE

The wide array of channel response types evident in the MSBE data exists because of the wide array of spatial controls acting on sediment moving downstream from diffuse, spatially uniform sources. The responses can be explained in part by how much drainage area is upstream of the gage station, the channel pattern at the gage station, and the location of the gage in the drainage network. Some of the responses apparently are not explainable by measured factors and may instead arise from unmeasured factors, or seemingly random interactions among factors. Data in Table 1 and Table 2 indicate that many possible factors do not satisfactorily explain differences in MSBE response. Except for relations that covary with drainage area, no clear relations exist between MSBE response history and hydrologic variables (Table 1) or basin-scale characteristics, such as percent forest, channel slope, maximum channel length, or bedrock lithology (Table 2).

The most evident relation among the MSBE data stations is downstream progression from depleted, to slightly wavy or extremely wavy, to stable/degraded response types (Figure 6). This downstream progression is consistent with a land-use history wherein coarse sediment has been eroded from floodplain or channel storage in upper reaches of the drainage basins and moved downstream through the system. At depleted stations of drainage areas of about 1400 km^2 or less, the earliest gage records from 1920 to 1940 record the last part of the passage of a discrete wave of sediment past the gages, or the last of a period of bed degradation. Since these stations have become depleted, the channels have less mobile sediment to form local aggradational features and channel planform changes. As a result, large floods like those of the early 1980's have been relatively ineffective in altering the streambed elevation.

Downstream, at drainage areas approximately 1400 to 7000 km^2, the MSBE responses are slightly wavy to extremely wavy. The differences between slightly wavy and extremely wavy stations can be substantial. Gages at the Big Piney River near Big Piney, the Current River near Eminence, and the Gasconade River near Hazlegreen and Gasconade River near Waynesville have notably low-amplitude waves of sediment passing by them, whereas the Eleven Point River near Bardley, the Current River at Van Buren, the Meramec River near Steelville, and the Meramec River near Sullivan have much greater amplitude waves. Amplitude of the waves does increase systematically with drainage area as would be expected if sediment waves always tended to interact additively downstream. Nor does the amplitude decrease systematically with increasing drainage area, as would be expected if waves of sediment produced in small streams diffused downstream from their initial points of delivery in the channel system. In the case of the Current River Basin, the depleted station at Jacks Fork at Eminence is upstream of the slightly wavy station of the Current River near Eminence, which, in turn, is upstream of the extremely wavy station at the Current River at Van Buren. Hence, in addition to downstream translation of waves of sediment, other factors are contributing to the magnitude of channel response.

One factor that clearly operates to determine magnitude of response is local channel pattern type. Ozarks streams

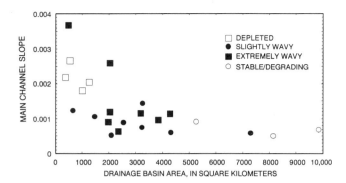

Fig. 6. Plot of local channel slope by drainage area with classification of MSBE response types.

are characterized by alternating stable and disturbance reaches (Figure 2). When the MSBE gage stations are classified into these two types of reaches, it is clear that the MSBE response type is related to channel pattern type of the station. Among all stations in mid-sized drainage basins, all stations with slightly wavy response types are in stable reaches. However, extremely wavy stations are nearly evenly divided between stable and disturbance reaches (Table 1). This effect of channel pattern can be seen in the downstream variation of channel response type in the Current River basin discussed earlier. The gage at the Current River near Eminence is located in a long, straight, stable reach whereas the gage at Jacks Fork at Eminence and the Current River at Van Buren are both in disturbance reaches. Apparently, stable reaches transport sediment through them without exhibiting substantial lateral or vertical channel changes. Conversely, channel response to sediment transport is enhanced in the steeper, more sinuous disturbance reaches.

The underlying reason for alternating stable and disturbance reaches is unknown. In contrast to the accepted model that continuously meandering streams are the most probable channel pattern because meandering minimizes the variance of energy dissipation [*Langbein and Leopold*, 1966], these rivers are characterized by two types of channel planforms with seemingly different potentials for energy dissipation. The dichotomy of stable and disturbance reaches is similar to the transport reaches and sedimentation zones described by *Church* [1983] in British Columbia. However, Church determined that these sedimentation reaches were caused mostly by outside factors such as increased sediment load at tributary junctions. In the Ozarks, disturbance reaches are independent of tributary junctions or inputs of sediment from hillslope erosion. Stable reaches typically are located adjacent to a straight valley wall (Figure 2). Disturbance reaches occur at almost all locations where the channel crosses from one valley wall to the next, but they also may occur away from crossings. Stratigraphic data indicate that both types of reaches existed before settlement of the Ozarks, and there is no evidence to support a connection between land-use changes and the origin of disturbance reaches.

Gages located in or near disturbance reaches would record the effects of greater channel migration rates and vertical bed changes associated with transient movements of riffles, bars, and the thalweg. These vertical changes would be expected to have greater amplitude than in stable reaches because of flow divergence and convergence around mid-channel bars and islands, and frequent channel avulsion. Flow complexity in time and space in disturbance reaches provides more opportunities for transient storage of bedload sediment. In contrast, the straight stable reaches lack hydraulic variation, commonly lacking well-defined thalwegs. Straight reaches therefore tend to lack the flow complexity that produces bars and other temporary bedload sediment storage sites. MSBE changes are accordingly minimized.

The variation of MSBE responses ascribed to disturbance and stable reaches may be sufficient to mask the effects of land-use induced sediment passing through the reaches. It is possible that sediment waves in disturbance and stable reaches are independent of land-use history in the drainage basins; presence of land-use derived sediment can only be inferred from the presence of depleted MSBE records at upstream gages. Nevertheless, the comparison of responses between the two types of reaches indicates that whatever their origin, waves of sediment in these rivers have much different effects on channel morphology in the two types of reaches.

Differences in channel form do not explain all of the differences in channel response between slightly and extremely wavy MSBE records, however. The differences between MSBE records from the Gasconade River Basin and those from the Meramec and Big River Basins are notable (Figure 5). Although the underlying bedrock of the Meramec and Big River Basins is substantially different from the other drainage basins (Table 2), other hydrologic variables and basin characteristics do not indicate a reason for the different channel responses. Bedrock in the Meramec and Big River Basins is not known to contribute sediment types that are substantially different from what is found in the other basins [*Koenig*, 1961]. Lead and barite mining in the Big River Basin was rejected as a causal factor for extreme waviness because the Meramec River Basin, which has substantially less mining, also showed extremely wavy MSBE responses.

One hypothesis is that channel response to sediment inputs is in part a function of location of the station in the drainage network. For sediment that enters the drainage network from disseminated sources (mostly low-order tributaries), the downstream effect will vary depending on how the drainage network routes the sediment. In a dendritic drainage network with consistent channel bifurcation ratios and link lengths, sediment waves might be expected to join synchronously at tributary junctions resulting in more noticeable channel responses downstream. As the network departs from a perfect dendritic pattern, lags in transit time of the sediment waves would decrease the likelihood that sediment waves would meet synchronously at channel junctions. Hence, trellis or contorted drainage networks would be expected to have less reinforcement of sediment waves, and consequently, smaller downstream channel responses, than dendritic drainage patterns.

Larger drainage basins in the Ozarks tend to be spoon shaped (Figures 1 and 3). Plots of the logarithm of total stream length in each order by stream order for the MSBE basins show the effect of the basin shape (Figure 7A). Typically, the plots of logarithm of total stream length by stream order are convex upward, indicating an increasing rate of addition of higher order stream length in the downstream direction. This is also indicated by the total length bifurcation ratio (TLBR), calculated as the ratio of total stream length in each stream order divided by the total stream length in the next higher order (Figure 7B). These calculations differ from those of *Strahler* [1951] in using total stream lengths in each order rather than average stream length. All basins show a consistent decrease in the TLBR from order 1 to order 2, indicating the relatively small contribution of order 2 to order 3 streams compared to order 1 to order 2. From the TLBR of order 3 to order 7, the ratios show a great deal of variability, but generally level out from 1 to 2. Some misleadingly large ratios are calculated where the gage is located on a short segment of a stream order close to where that order begins.

Three classes of stream-length distributions are evident (Figure 7 A, B). Class 1 distributions apply at MSBE stations located just downstream of the junction of two streams of the same order. The stations typically are at small drainage areas, with maximum stream orders of 4 or 5, and the total length of the stream order where the gage is located is small. Examples of class 1 stations are the Bourbeuse River near Highgate, the Black River near Annapolis, and the Big River near Irondale; these stations have relatively linear plots of logarithm of total stream length by stream order (Figure 7A)., and the TLBR plot increases rapidly at the downstream termination of the plots

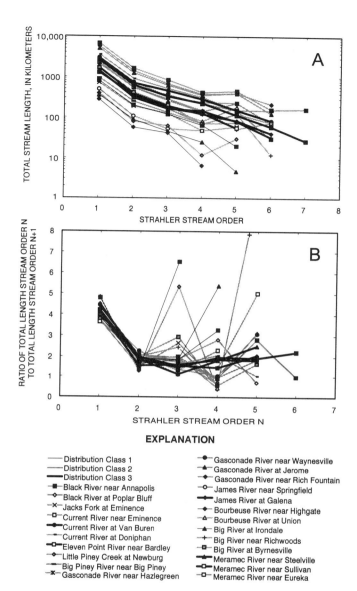

Fig. 7. Drainage network characteristics of MSBE stations. (A) Plot of total stream length (logarithmic scale) in each stream order. (B) Plot of ratio of total stream length for each stream order to total stream length of next higher stream order.

(Figure 7B). At these stations, the potential for synchronous additions of sediment from the upstream drainage network would be high; however, because most of these stations are at small drainage areas (Figure 6), they are usually depleted sites.

In class 2 distributions, the total stream length increases or levels off at the stream order where the gage is located. Gages in the Gasconade River Basin are examples. The Big Piney River near Big Piney, and the Gasconade River near Hazlegreen, near Waynesville, at Jerome, and near Rich

Fountain have stream-length distributions that level off in the range of order 4 to order 5 (Figure 7A). This is the result of increased lengths of high-order channels without comparably increased lengths of contributing low-order channels. The TLBR plots (Figure 7B) have low ratios of order 3 to order 4 or order 4 to order 5 stream lengths. The minima in the distributions are indicative of large contributions of order 4 and order 5 stream lengths. In these segments of the river basin, land-use induced waves of sediment can be expected to enter the higher order channels at different times and places, thus minimizing the cumulative channel response. Also, long lengths of stream channel at order 4 and 5, relative to contributing, lower stream orders, provide abundant sites for storage of sediment that would minimize downstream channel change. Some class 2 distributions have rapid decreases in stream length and consequent increases in TLBR at the highest order, similar to class 1 distributions. However, in these larger streams, the rapid decrease in total stream length occurs after total stream lengths have levelled off at order 4 and 5. This is true for the Big River near Richwoods, the Current River near Eminence, the Gasconade River near Rich Fountain, the Gasconade River at Jerome, and the Gasconade River near Waynesville.

The third class of length distribution plots nearly linearly on the log-linear plot through orders 4 to 7. Class three distributions have TLBR plots that reach minima at order 3 and rise gradually to between 1.5 and 2 at the highest order. This distribution results from dendritic drainage networks in which total stream length in each stream order decreases nearly logarithmically with increasing order, and, for orders 4 through 7, the ratio of the total length of stream in each order to the total length in the next higher order is nearly constant. Examples are the Black River at Poplar Bluff, the Current River at Van Buren, the Eleven Point River near Bardley, the James River at Galena, the Meramec river near Steelville, and the Meramec River near Sullivan. Five of these six stations have an extremely wavy MSBE response type. This result is consistent with the hypothesis that a more dendritic drainage network is likely to show greater effects of channel disturbance from sediment waves because of additive interactions in the downstream direction.

5. SUMMARY AND CONCLUSIONS

A predictive understanding of the processes and rates of sediment movement through drainage basins is complicated by the spatial variability of most drainage basins. The historical record of stream-channel responses to land-use changes presented in this report illustrates some of the factors that should be taken into account to fully understand routing of sediment and disturbance through drainage basins.

In small drainage areas (less than 1400 km^2) channel response histories are indicative of depletion of sediment produced by land use. From this, it can be inferred that a wave of sediment moved past gages on small watersheds in the period 1920 to 1940. Another possible interpretation is that the channels at these stations have degraded, making them sources of gravel to downstream reaches. Decreasing mobility of the streambed with time and decreasing MSBE sensitivity to floods support the interpretation that a wave of mobile sediment has moved past gages at small drainage areas.

Wavy MSBE response types in mid-sized basins (1400-7000 km^2) also are consistent with a model of downstream translation of sediment waves. The waviness presumably is the result of multiple pulses of land-use derived sediment passing these gages. The magnitude of waviness is poorly related to drainage area in this data set. The MSBE response types in the largest drainage areas studied (8000-10,000 km^2) are either extremely stable or slightly degrading. This response would be expected if the waves of sediment had not yet reached the downstream reaches of these drainage basins, or if the sediment waves diminished in magnitude as they moved downstream.

Differences in the amplitudes of sediment waves depicted in the MSBE records of mid-sized drainage basins (1400-7000 km^2) can be attributed to at least two spatial factors. The MSBE response history is related strongly to the local channel pattern. Most MSBE records in stable reaches are only slightly wavy, whereas extremely wavy records occur in both stable and disturbance reaches. Straight channels in stable reaches seem to transport sediment efficiently through them whereas sediment tends to accumulate and cause larger streambed elevation changes in disturbance reaches.

The other factor that seems to explain the amplitude of waviness is the location of the gage in the drainage network. Small drainage basins in the Ozarks tend to have a dendritic drainage network of low-order streams, whereas larger, spoon-shaped drainage basins have greater contributions of high-order streams to total stream length. In drainage basins where the dendritic network persists to higher drainage areas, there is greater chance for waves of land-use derived sediment to accumulate synchronously at stream junctions to produce larger waves downstream. In drainage basins where higher-order streams contribute substantially to the total stream length, there is less chance for additive interactions of sediment waves, and more sites are available for storage of sediment in floodplains.

Acknowledgments. This study was carried out in cooperation with the Missouri Department of Conservation and the National Biological Survey Global Change Research Program.

REFERENCES

Church, Michael, Pattern of instability in a wandering gravel bed channel, *Spec. Pub. Internat. Assoc. Sedimentologists, 6,* 169-180, 1983.

Dake, C. L, *The sand and gravel resources of Missouri,* 15, 2d series, 274 pp., Missouri Div. Geol. and Land Surv., 1918.

Dempster, George, R., *National water information system user's manual,* Vol. 2, Chap. 3. Automated data processing system, 248 pp., U.S. Geol. Surv., 1991.

Dury, G. H., Principles of underfit streams, *U.S. Geol. Surv. Prof. Paper 452-A,* 67 pp., 1964.

Hack, J. T., Interpretation of erosional topography in humid temperate regions, *Am. J. Sci., 258-A,* 80-97, 1960.

Hall, Leonard, *Stars upstream--Life along an Ozark River,* 262 pp., Univ. of Missouri Press, Columbia, Missouri, 1958.

Jacobson, R. B., and A. T. Primm, Historical land-use changes and potential effects on stream disturbance in the Ozark Plateaus, Missouri, U.S. Geological Survey Open-File Report 94-333, 95 pp., 1994.

Jacobson, R. B., and A. L. Pugh, Effects of land use and climate shifts on channel instability, Ozarks Plateaus, Missouri, U.S.A., *Proc. of the Workshop on the Effects of Global Climate Change on Hydrology and Water Resources at the Catchment Scale,* 423-444, Japan-U.S. Committee on Hydrology, Water Resources and Global Climate Change, 1992.

Koenig, J. W., The Stratigraphic succession in Missouri: 185 pp., *Missouri Div of Geol. and Land Surv.,* 1961.

Langbein, W. B. and L. B. Leopold, River meanders--Theory of minimum variance, *U.S. Geol Surv. Prof. Paper 422-H,* 15 pp., 1966.

Rafferty, M. D., *The Ozarks land and life,* Univ. of Oklahoma Press, Norman, 282 pp., 1980.

Saucier, R. T., Historic changes in Current River meander regime, Proc of the Conf., *Rivers '83,* Am. Soc. Civil Eng., 180-190, 1983.

Schumm, S. A., River metamorphosis, *J. Hydrau. Div., HY1*: 255-273, Am. Soc. Civil Engr., 1969.

Schumm, S. A., Geomorphic thresholds and complex response of drainage systems, in Fluvial Geomorphology-- *Proc. of the Fourth Annual Geomorphology Symposium Series,* edited by M. Morisawa, State Univ. N.Y., Binghamton, N.Y., 299-310, 1973.

Simon, Andrew, A model of channel response in disturbed alluvial channels, *Earth Surf. Proc. and Landforms,* 14, 11-26, 1989.

Smith, B. J., Assessment of water quality in non-coal mining areas of Missouri, U.S. Geol. Surv. *Water Res. Invest. Rep. 87-4286,* 50 pp., 1987.

Strahler, A. N., 1957, Quantitative analysis of watershed geomorphology, *Am. Geophys. Union Trans.,* 38, 913-920, 1957.

Robert B. Jacobson, U.S. Geological Survey, 1400 Independence Road, MS 200, Rolla, MO 65401